CYCLODEXTRIN - A VERSATILE INGREDIENT

Edited by **Poonam Arora**
and **Neelima Dhingra**

Cyclodextrin - A Versatile Ingredient
http://dx.doi.org/10.5772/intechopen.69187
Edited by Poonam Arora and Neelima Dhingra

Contributors

Laura Romero-Zerón, Xingzhi Jiang, Hidetoshi Arima, Agustín G. Asuero, Julia Martín, Enrique Jacobo Diaz-Montaña, Pinalysa Cosma, Paola Semeraro, Josè Antonio Gabaldon, José Antonio Pellicer, Paola Fini, Vito Rizzi, Estrella Nunez, Charlotte Bonisiwe Seshabela, Mandla Brian Chabalala, Bhekie Mamba, Sabelo Mhlanga, Edward Nxumalo, Artur Valente, Tânia Firmino Cova, Sandra M. A. Cruz, Paulo Abreu, Jorge Marques, Alberto Pais, Bruno Medronho, Anabela Romano, Sandra Gonçalves, Raquel Rodríguez-Solana, Jesus Simal-Gandara, Jaruporn Rakmai, Benjamas Cheirsilp, Antonio Cid, Ana Torrado-Agrasar, Juan C. Mejuto, Márcio Antônio Fiori, Francieli Dalcanton, Ana Paula Capelezzo, Laura Cassol Mohr, Josiane Maria Muneron De Mello, Mohsen M. Zareh, Juan Eduardo Megías-Vericat, María José Company-Albir, Ana Alejandra García-Robles, Jose Luis Poveda, Aida Maria Gonçalves Moreira Da Silva

First published in London, United Kingdom, 2018 by IntechOpen
IntechOpen is the global imprint of INTECHOPEN LIMITED, registered in England and Wales, registration number: 11086078, The Shard, 25th floor, 32 London Bridge Street
London, SE19SG – United Kingdom
Printed in Croatia

British Library Cataloguing-in-Publication Data
A catalogue record for this book is available from the British Library

Additional hard copies can be obtained from orders@intechopen.com

Cyclodextrin - A Versatile Ingredient, Edited by Poonam Arora and Neelima Dhingra
p. cm.
Print ISBN 978-1-78923-068-0
Online ISBN 978-1-78923-069-7

For EU product safety concerns:
IN TECH d.o.o., Prolaz Marije Krucifikse Kozulić 3, 51000 Rijeka, Croatia,
info@intechopen.com or visit our website at intechopen.com.

Meet the editors

Dr. Poonam Arora is presently working as an Assistant Professor at GHG Khalsa College of Pharmacy, Gurusar Sadhar, Ludhiana, Punjab, India. She obtained her Bachelor of Pharmacy and Master of Pharmacy (Pharmaceutical Chemistry) degrees from Panjab University, Chandigarh, India, and PhD (Pharmaceutical Chemistry) degree in 2011 from the same institute. She has more than 10 years of teaching and research experience. Her major areas of research are cyclodextrin complexation studies of pharmaceutical drugs, polymorph/solvatomorph studies, and design and synthesis of cocrystals and eutectic mixtures. Her research work has been credited with 2 national patents and more than 30 research and review articles in peer-reviewed national and international journals. Besides this, she has one book and two book chapters to her credit.

Dr. Neelima Dhingra has been associated with the University Institute of Pharmaceutical Sciences, Panjab University, Chandigarh, for the past 23 years. She received her PhD (2007), master's (2000), and bachelor's (1998) degrees in Pharmaceutical Sciences, and she then joined the same institute as an Assistant Professor (2007) in the Department of Pharmaceutical Chemistry. She is an expert Medicinal Chemist and synthesizes new chemical entities after relating molecular biology and pharmacology of hormone-dependent cancers (prostate/breast/lung). She has published over thirty scientific articles in peer-reviewed journals, one book, and one book chapter and holds two US and two European patents in new molecules as cytotoxic agents. She has been conferred with a DST Young Scientist Award and a UGC Basic Scientific Research Grant and eight national and one international awards in the field of cancer research. Dr. Neelima Dhingra is a lifetime member of various national scientific bodies like APTI, IPGA, PAS, PUPS, IABMS, and SPER; is an active reviewer and an editorial board member of various national/international journals.

Contents

Preface

The history of cyclodextrins (CDs) began in France in the late nineteenth century, and they were obtained from the enzymatic degradation of one of the most essential polysaccharides, starch. Their remarkable inclusion complexation properties led to a "host-guest"-type relationship, thus modifying or improving the physicochemical properties and/or bioavailability characteristics of the guest molecules. Negligible cytotoxic effects enabled scientists and researchers to explore their applications ranging from pharmaceutical to biotechnology industries. Further expansion of their application in the different field of cosmetics, food, agriculture, textile, separation process, analytical methods, catalysis, environment protection, and diagnostics was made possible by the production of the three CDs known as "native," that is, α-CD, β-CD, and γ-CD.

The book *Cyclodextrin - A Versatile Ingredient* is the result of collective work that addresses, in a clear and comprehensive way for readers, the fundamental aspects of CDs as well as their applications. Twelve chapters have been placed under two heads. The first section deals with the overview of the properties of cyclodextrins along with primary issues related to aggregation, whereas specific applications in various fields are covered in the second section, for further advancement in the physical sciences. Efforts have been made to provide an invaluable resource and pedagogical support for academician in general and researchers in specific as well as for industrial scientists with the introduction of the latest ongoing trends in research with cyclodextrins.

The Editors are thankful to all the contributors of this book for their cooperation. My special thanks go to Ms. Marina Dusevic, Publishing Process Manager, and InTechOpen Publishing for bringing out the book in the present form.

Dr. Poonam Arora (MPharm, PhD)
GHG Khalsa Colleges
Gurusar Sadhar, Ludhiana
India

Dr. Neelima Dhingra (MPharm, PhD)
University Institute of Pharmaceutical Sciences
University Institute of Pharmaceutical Sciences
Panjab University, Chandigarh
India

Section 1

Cyclodextrins: An Overview

Chapter 1

Cyclodextrins: Past and Present

Julia Martin, Enrique Jacobo Díaz-Montaña and
Agustín G. Asuero

Additional information is available at the end of the chapter

http://dx.doi.org/10.5772/intechopen.72736

Abstract

Cyclodextrins (CDs) are cyclic oligosaccharides produced by enzymatic degradation of starch. The most common CDs are the main natural ones, α, β and γ, which are constituted of 6, 7 and 8 glucopyranose units, respectively. The CD structure forms a torus or doughnut ring and the molecule actually exists as a truncated cone. The outer side of the toroid is hydrophilic in nature due to the hydroxyl groups of the glucopyranose units while the internal cavity is relatively apolar. Thus, CDs have a high potential to entrap entirely or partially a wide variety of compounds in a process known as complexation. This gives them new physico-chemical properties and characteristics. The main applications of CDs in drug formulation rely on CD complexation and include the protection of easily oxidizable molecules or the improvement of aqueous solubility. The use of CDs in analytical chemistry is based on his host-guest recognition property, known as supramolecular complex formation. Currently, CDs are successfully used in molecular recognition-based methods like chromatographic separations, spectroscopic and electroanalyses. Quiral analytical separations are a CD area of special relevance. In this work, attention is paid to more recent references, especially to selected reviews.

Keywords: cyclodextrins, applications, encapsulation, controlled release, nano, food, cosmetic

1. Introduction

Cyclodextrins (CDs) at times referred as Schardinger sugars or cycloamylose dextrins, were fortuitously discovered [1, 2] by Vielliers in 1891, who named these compounds as "cellulosing." Later on Schardinger, who is considered the founder of CD chemistry, gave a detailed description about preparation and separation of CD and, more recently, Kurkov and Loftsson [3] also made significant contributions to CD science.

Franz Schardinger, studying microorganisms which play a role in the deterioration of foods and by action of cyclodextrinase-Bacillus macerans amylase on the starch, obtained two distinct crystalline substances with similar properties to the already known partial degradation products of starch, the dextrins, so he named them α-, and β-dextrin. The separation of the cycloalkyls may be carried out by selective precipitation by means of organic compounds or by high temperature chromatography on a cellulose column. French et al. demonstrated that CDs are cyclic oligosaccharides composed of several D-(+)-glucopyranose units in the form of a saddle [4]. In the second half of the 1930s, Freudenberg and his co-workers elucidated the cyclic structure of α-, and β-dextrin [5]. They consist of (α-1,4)-linked glucose units. A Greek letter preceding the abbreviation CD—for cyclodextrin—indicates the number of glucose units (α for 6, β for 7, and γ for 8) entering the composition of the cycloamylose. CDs constituted of less than 6 glucopyranose units cannot be formed due to steric hindrances [6]. Approximately, 1500 CD derivatives have been reported [7] in the literature.

CDs have a truncated cone appearance [7–12], and a doughnut, toroidal- or cylinder-like shape, due to the spatial arrangement characteristic of the various functional groups of the glucose units. As a consequence of this conformation, all the secondary hydroxyl groups (corresponding to the C2 and C3 carbon atoms of the glucose units) are at one of the edges of the cavity, whereas the primary hydroxyls are in the other end of the cavity. Rotation of these –OH groups reduces the effective size of the cavity, making it have a more open conical truncated aspect [13] toward the side of the secondary hydroxyls (**Figures 1** and **2**).

This spatial arrangement gives an apolar character to the interior of the cavity, whereas the presence of the –OH groups at the edges of the cone trunk makes them very water soluble. For instance, hydrophobic hosts will be housed inside the cavity because of the hydrophobic van der Waals type interactions, whereas simultaneously polar interactions

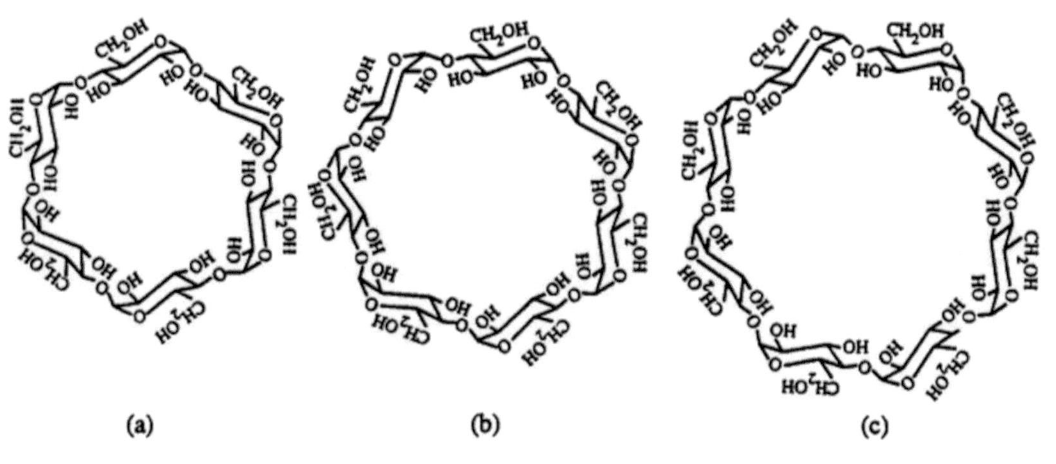

(a) (b) (c)

Figure 1. Molecular structure of (a) α, (b) β, and (c) γ-CDs.

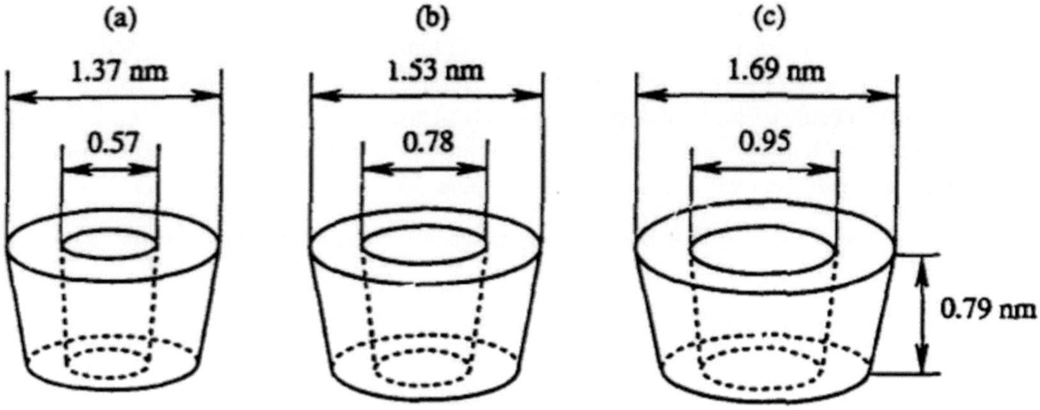

Figure 2. Steric structures of molecules from (a) α, (b) β, and (c) γ-CDs.

can be established by the formation of hydrogen bridges between polar hosts and –OH of the primary hydroxyls. An endless number of physical and chemical processes [10, 14] are usually facilitated, that is, volatile substances may be stabilized by transforming in crystalline substances; oxygen-sensitive materials could find protection against oxidation; solubility and bioavailability of drugs could be improved [15–20] by participating in controlled delivery processes.

CDs have been the subject of a large number of studies dealing with complexation and molecular catalysis [21–25], as well as studies about hydrophobic effects and fine-tune models of biological processes. In 1953, the first patent on CDs and complexes was registered, but until 1970, only small amounts of relatively pure CDs were produced for industrial use due to their high production cost. Although in the beginning it was erroneously thought that CDs were toxic, currently, it is difficult to imagine a world without CDs [3] due to their potential use [26]. The number of possible applications seems to be unlimited, i.e., computer-aided drug design, pharmaceutical, medical, biomedical and biotechnological, drug and gene delivery, foods, foods additives and ingredients, food processing, cosmetic, textiles, industrial and analytical. Currently, patents on CDs are counted by thousands.

2. Inclusion complexes

An inherent interest surrounds these compounds due to their physical and chemicals properties [26–38]. The common feature of CDs is their ability to form inclusion complexes with a variety of molecules and ions, both in the solid state (crystalline substances) and in solution. As results of the structure of CDs, they can establish apolar-apolar interactions encapsulating other apolar molecules which may undergo structural changes [33–38], acting as molecular capsules [27–32]. However, the idea that one molecule could envelop another one to form a new compound (adduct, inclusion complex) was not accepted until X-ray diffraction showed the formation of an inclusion complex between α-CD and iodine [37]. They constitute a significant example of relatively simple organic compounds showing complex formation with other organic molecules. They are excellent models of enzymes that lead to their use as catalysts [21, 24, 39], both in enzymatic and non-enzymatic reactions. Additionally, they are natural products and readily available to most researchers.

It is accepted [18, 38, 40–43] that the binding forces involved in complex formation are, in general:

i. van der Waals type interactions (or hydrophobic interactions) between the hydrophobic unit of the guest molecules and the CD cavity.

ii. Hydrogen bond between the polar functional groups of the guest molecules and the hydroxyl groups of the CD.

iii. Release of high energy water molecules from the cavity in the complex formation process.

iv. Release of strain energy into the ring structure system of the CD.

The role of the hydrogen bond is not universal since stable complexes are formed with hosts such as benzene, which do not form hydrogen bonds.

2.1. Factors affecting stability

Regardless of which type of stabilizing force is involved, the most important factors in determining the stability of the inclusion complex are [36, 40–45]:

the geometric capability

polarity of the guest molecules

the medium

temperature

Geometric, rather than chemical factors, are critical in determining the type of "guest" molecules that can penetrate into the cavity. If the guest is too small, it passes easily through the cavity and the bond will be weak or will not occur. The formation of complexes with molecules significantly larger than the cavity is also possible, but only some limited groups or side chains penetrate into the CD cavity.

The stability of an inclusion complex also depends on the polarity of the "guest" molecule. Only substrates that are less polar than water may form inclusion complexes with the CDs. The stability of a complex is proportional to the hydrophobic character of the "guest" molecule. Highly hydrophilic molecules form complex CDs very weakly or do not complex at all.

On the other hand, stability depends heavily on the nature of the medium used for complexation. In principle, the inclusion complexes may be formed either in solution [46–49] (generally carried out in the presence of water) or in the crystalline [40, 50–52] state. Although the formation of inclusion complexes also takes place [53] in an organic solvent, the guest molecules are weakly complexed. Additionally, although a 1:1 stoichiometry between the substrate and the CD molecule is typical [46, 54–56], with certain systems (**Figure 3**), 1:2 and 2:1 complex formations are possible. Experimentally determined formation constant can be the function (**Figure 4**) of the formation constants of the isomeric complexes [46]. In addition, substitution of one or more hydroxyls results in most cases in better water-soluble derivatives. For example, CDs can be polymerized [32, 36, 40, 42, 44, 45, 57] by suitable bio- or polyfunctional agents to oligomers, long-chain polymers or crosslinked or immobilized networks in various supports. Low molecular weight oligomeric CDs are readily soluble in water. Polymers (molecular mass over 10,000) are swollen gels which can be prepared in bead forms. The rigid structure of CDs "host" translates into well-defined and differentiated inclusion complex depending on the nature of the "guest" molecule.

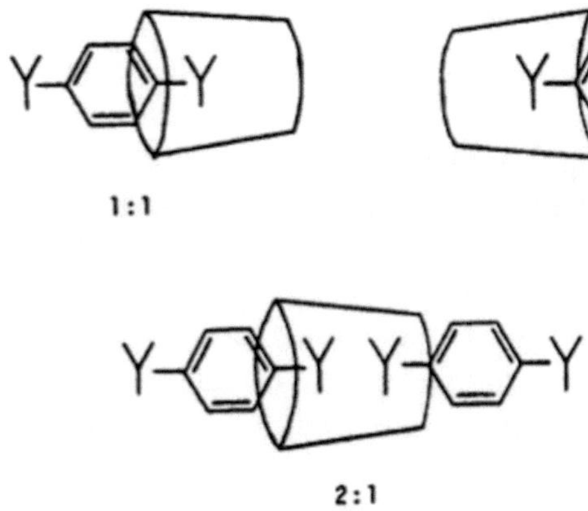

Figure 3. Complexes of α-CDs and 1,4-disubstituted benzene [13].

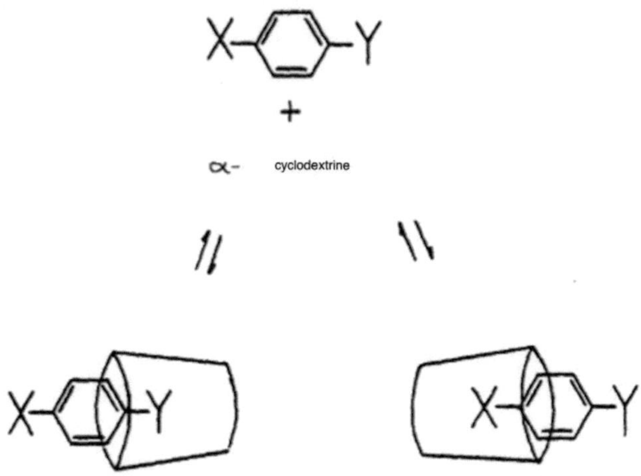

Figure 4. Isomeric complexes from substrate and free ligand [55].

Finally, the stability of the inclusion complex, in general, decreases when temperature increases [46]. Enthalpy and entropy changes can be obtained from the temperature dependence of the equilibrium constant. An important issue, often overlooked in the CD field, is that the magnitudes of the standard free energy and entropy changes are dependent on the standard state chosen by the experimentalist.

3. Analytical and physicochemical applications

In the last years, CDs and their derivatives have been used in a variety of fields of analytical chemistry, especially in analytical separations [45, 58–63]. Spectral properties of CD and guest molecules can be altered due to the changes of the electrons distribution in the CD hole. CDs are used as reagent in different analyses such as UV-visible spectrophotometry, fluorescence [64, 65], phosphorescence [66, 67], and nuclear magnetic resonance methods [45, 68, 69].

The complexation of the analyte and/or the colored reagent can effectively change its properties. Among the most notable uses of this effect are: (i) enhancing the solubility of polar or non-polar analyte; (ii) enhancing the stability in polar or non-polar solution of reagents and colored complexes; (iii) increasing UV-visible absorption which improves the sensitivity of the colored reactions; and (iv) enhancing colored reactions selectivity. Luminescence techniques, in terms of fluorimetry and phosphorimetry, have reached a rapid development in routine analysis. However, many compounds luminesce very weakly in aqueous solution and the addition of CDs protects the excited (singlet or triplet) states of the possible dampers present in the solution since the rotation of the molecules is impeded due to the formation of the complex of inclusion with the result of a decrease in vibrational relaxation processes. The formation of inclusion complexes also increases the quantum fluorescence yield and hence the fluorescence intensities of numerous compounds. Sensitivity to certain characteristic reactions also increases.

CDs also increase the emission intensity of the chemiluminescent reactions. This improvement can be attributed to a number of factors, including an increase in the reaction rate and a greater efficiency in the process of excitation and protection of species that emit quenching phenomena. One of the most relevant applications of CDs is to allow the observation of phosphorescence at room temperature [67]. This is because they protect the excited triplet state of the molecules of the shock absorbers present in the solution, and in the case of molecular quenching phosphorescence. They are used as chiral reagents in NMR. In many cases, the formation of inclusion compounds modifies the general characteristics and chemical shifts of two enantiomers. Differences in the chemical shifts of two diastereoisomers can be used for the determination of the isomeric purity of the samples. The formation of inclusion complexes can very significantly modify the redox characteristics [13, 70, 71] of the included molecules. Voltammetric sensors capable of responding to anionic compounds have been developed. The changes produced after the complexation (selective interaction) allow the voltammetry to be used in the study of the complexation between CDs and organic molecules.

CDs increase the selectivity of chromatographic separations [72–74], because the separation process is more selective than that between the eluent and the stationary phase alone. In HPLC, the application of the CDs has achieved a spectacular success. Their incorporation into the mobile phase allows improving the separations, since they are soluble in water and provide reversible and selective complexation. In addition, they are stable and show no absorption in the UV-visible region of the electromagnetic spectrum. These characteristics mean that CDs are generally used in reverse phase separation processes, achieving the separation of isomers, diastereoisomers, and enantiomers [75–78]. The high resolution obtained is due to the differences in the stability constants of the complexes in the mobile phase and the different adsorption of these complexes in the stationary phase. CDs may also be incorporated as support for the stationary phases. Capillary electrophoresis has also found use in chiral analytical separations [79–82].

4. A primer on pharmaceutical, food and cosmetic cyclodextrin studies

4.1. Bioavailability

CDs have mainly been used as complexing agents to improve the aqueous solubility of molecules. This allows the use of CDs to reduce or prevent gastrointestinal or ocular irritation by lowering the local concentration of the free drug below the irritancy threshold. Also, unpleasant odor or taste of drugs can be hidden by complexation of the functional groups that produce them with CDs, occulting them from the sensory receptors [83–85], furthermore, reducing their hydrophobicity using CDs. Finally, CDs can increase percutaneous or rectal absorption of drugs and their derivatives can increase the guest molecule bioavailability [84]. Recently, CDs and their derivatives have been used in dispersed vehicle systems such as emulsions, microcapsules, microspheres, nanospheres, nanocapsules, liposomes, and beads [86]. Additionally, the host-guest property allows CDs to be used as building blocks in supramolecular chemistry [7]. Suvarna et al. [87] explain an insight in the use of CDs to increase the bioavailability to resolve the problem of solubility and stability of phytochemicals. The authors describe that some chemicals as quercetin, curcumin, arteminsinin, resveratrol or naringenin increased their bioavailability due to the inclusion complexes with CDs. Authors concluded that CDs need to be more explored to cover some molecules that have potential biological activity but have not been approached.

4.2. Encapsulation

The encapsulation with CDs is gaining interest in different industries; this is reflected in the large number of publication and products related with it, such as drug delivery systems [7, 35]. This capacity of encapsulating compounds is used for a wide variety of things, among them is to protect the compounds, or to transport them to a target. This ability is due to the toroidal shape of CDs which makes possible to encapsulate hydrophobic molecules fully or partially in their cavity [14, 35]. This characteristic let the CDs being used for oral, sublingual, ocular, nasal, rectal, pulmonary, dermal, and other drug delivery systems, especially in systems of type 1/1 (one molecule per CD). The encapsulation with CDs enhanced the bioavailability of lipophilic drugs, as they are 17β-estradiol, androstenediol, clomipramine, and others. A limitation of CD in sublingual route is that the quantity used for a proper formulation is too large to be considered. This increase in the bioavailability is also observed in the oral route for drugs such as diltiazem, flufenamic acid, molsidomine, salbutamol, having all of them a sustained release [88].

4.3. Controlled release

In order to optimize pharmacotherapy, drug release should be controlled in accordance with the therapeutic purpose and the pharmacological properties of the active substances. In recent years, the interest regarding the control of rate or time of delivery has significantly increased [88]. The multifunctional characteristics of CDs allow them to be used in most drug delivery systems [84]. The design process of drug delivery systems is currently more focused on the oral route, in which the release of the drug can be controlled by dissolution, diffusion, osmosis, density or pH. Challa et al. [89] give several examples of different uses in oral delivery. The use of β-CD increased the bioavailability of ketoprofen, terfenadine, and griseofulvin; but, the same CD, also demonstrated higher intensity or longer duration of therapeutic activity in tolbutamide or terfenadine. Although there are different effects depending on the modified CD used, for example, the solubility and dissolution rate can be increased using HP-β-CD, for drugs as albendazole, ketoprofen, phenytoin, and gliclazide; or an improvement of hydrolysis stability γ-CD, for drugs as digoxin, camptothesin and paclitaxel. For oral administration, all CDs can be used because they are not toxic.

4.4. Nano

The improvement of the efficacy and bioavailability of poorly soluble drugs can be achieved by nanoparticles, which are stable systems that are used to create drug delivery systems [83]. Nanoparticles are 100–10,000 times smaller than human cells and their uses revolutionize diagnosis, treatment, therapeutic efficacy, and patient compliance [83, 90]. However, nanoparticles are limited by their low drug loading and entrapment ability, which compromises their safety and efficacy [84]. The use of CDs as a polymer increases the loading capacity of nanoparticle systems [89]. Furthermore, the optimal drug bioavailability and biodistribution can be achieved with a proper manipulation of physico-chemical and biological mechanisms, which can be provided by the hybrid functionalities of CD nanosystems [91]. A new class of colloidal polymer is nanosponges, which consist of solid nanoparticles with colloidal shape and nanocavities. Examples of nanosponges are those based on CDs. It should be noted that the type, number, and position of the substituent on the CD affect the complexation ability of nanosponges. Thus, it is crucial to know which CD derivative to use. Tejashri et al. [92] expose the use of CD to make nanosponge, and the use of it to load drugs and use as carriers. The crosslinking of CDs with compounds, as carbonyl or dicarboxylate, creates the different types of nanosponge, polyamide, carbonate, etc. Authors concluded that this novel class of

CD-based nanosponge let drugs to be released in a controlled form at the target place, and its spherical shape let nanosponge to be administered as parental, aerosol, topical, tablets, and capsules forms.

4.5. Food

In last years the application of CDs in the food-industry have increased mainly due to the use of them as a protective agent against oxygen, to protect flavor of volatile compounds, to enriched food with vitamins and color components (such as anthocyanins) or to stabilize them [93, 94]. Another advantage for the food industry is that CD are tasteless, odorless, and non-caloric saccharides, and that they have an antidiabetic effect due to their low glycemic index and their capability to decrease the glycemic index of the food, and also to improve the cholesterol index. Human gastrointestinal enzymes cannot digest them, so it can be used as a dietary fiber, which is fermented by microflora, what makes them a prebiotic compound. All these properties make them nutraceuticals and bioactive food supplements [95, 96]. López-Nicolás et al. [97] analyzed the positive effects of CDs in the encapsulation of antioxidant, and the repercussion on important factors as K_F or pH values. They also reviewed the antioxidant capacity of CDs, but they concluded that there is a necessity of more studies in this aspect.

4.6. Cosmetic

The cosmetic industry is looking for products with a good biological activity and adequate delivery on the skin [98]. The applications of CDs in cosmetics are similar to the pharmaceutical ones, e.g., stabilizing substances or increasing their solubility [99–101]. Centini et al. [98] associated ferulic acid, which is a photoprotector agent and an antioxidant compound, and CD. However, ferulic acid is not too much used due to the instability of it in the presence of air, UV-light, and heat; so, the aim of the work was to enhance the physico-chemical stability. The authors concluded that the complex ferulic acid/CD have a better photostability and do not generate degradation products. Buschmann and Schollmeyer [99] explained the use of CD against the vaporization of slow release of the volatile compounds in perfumes; or the opposite, they also explained the use of CD to eliminate undesired odors, such as mercapto derivate used in waving lotion. More applications will become possible when CDs price decreases. CDs can also be used in the textile industry as depots of cosmetic molecules providing new cosmetic formulations.

4.7. Miscellaneous applications of cyclodextrins: tabular form

A more detailed picture of most recent selected applications in various areas, ranging from general reviews to inclusion complexes, metal and organometallic complexes, food, pharmaceutical, pharmacological, medical and biomedical, environmental chemistry, personal care and toiletry, industrial, nanotechnological, industrial and analytical applications to enzyme, biomimetic, bioactive assembles and recognition, as well as miscellaneous applications is compiled in **Table 1**, which gives an idea of the importance and relevance of the CDs field. **Figure 5** shows the number of publications cited per year, whereas in **Figure 6**,

Content	Authors	Refs.
General reviews		
Overview about the work carried out on CDs concerning with: the general characteristics of CDs and derivatives, the preparation and evaluation of inclusion complexes, the use of CDs in the preparation of drug delivery systems, and their use for the preparation of biomaterials and nanoparticles.	Duchêne and Bochot (2016)	[14]
Comprehensive overview on the methods used for analysis of CDs and CD-derivatives. The paper intends to act as a guide in looking around the classical and modern instrumental analytical methods suitable for identification, characterization and determination of CDs themselves, CDs in finished products or even in biological samples.	Szente et al. (2016)	[2]
Current review on various aspects of CDs with regard to their chemical characteristics, properties, approaches used for complexation, characterization techniques, uses along with and future potential.	Khan and Durakshan (2013)	[7]
Pharmaceutical applications of CDs with an emphasis on their solubilizing properties, their tendency to self-assemble to form aggregates, CD ternary complexes, and their metabolism and pharmacokinetics.	Kurkov and Loftsson (2013)	[3]
Overview about several aspects related to the physico-chemical properties of CDs and their potential applications illustrated by recent examples.	Venturini et al. (2008)	[102]
Inclusion complexation and CDs: physicochemical parameters of the guest molecule and improvements in the molecule's solubility, stability, taste, safety, bioavailability, etc.	Mosher and Thompson (2007)	[103]
CDs and their use in industrial products, technologies and analytical methods.	Martin del Valle (2004)	[93]
Overview about past, present, and future of CD research. Potential uses of CDs in pharmaceuticals, foods, cosmetics, and chemical products and technologies.	Szejtli (2004)	[26]
CDs: structure, complex formation, drug solubility and non-conventional CD complexes.	Loftsson (2002)	[9]
Scientific and technological aspects of CDs: from computational chemistry to industrial uses of CDs.	D'souza and Lipkowitz (1998)	[104]
History (the three stages in the development of CD chemistry), fundamentals of CD chemistry and future trends.	Szejtli (1998)	[105]
The properties and potential uses of CD derivatives: dimethyl- and dimethyl-βCD (DIMEB and TRIMEB).	Szejtli (1992)	[106]
Catalyses by CDs leading to practical usages: covalent, non-covalent and asymmetric catalyses by CDs.	Bender and Komiyama (1978)	[107]
Inclusion complexes		
The inclusion complex of oxyresveratrol in modified CDs: a thermodynamic, structural, physicochemical, fluorescent and computational study.	Matencio et al. (2017)	[108]

Content	Authors	Refs.
Summary of method for inclusion complex formation of CD with its guests and its applications.	Cheirsilp and Rakmai (2016)	[42]
Literature review to characterize the formation of inclusion complexes by different techniques in the solid and in the solution state complexation.	Maazaoui and Abderrahim (2015)	[44]
Use of CDs as complexing agents to enhance the solubility of poorly soluble drugs and hence to resolve the many issues associated with developing and commercializing poorly water-soluble drugs.	Chaudhary and Patel (2013)	[16]
Survey of crystal structures of pure CD hosts and CD inclusion compounds carried out during the last six years. The entries range from simple alkylated derivatives to elegant multi-substituted target CD molecules, with and without included guests.	Caira (2011)	[40]
CD inclusion of four phenylurea herbicides: determination of complex stoichiometries and stability constants using solution ^{1}H NMR spectroscopy.	Smith et al. (2010)	[69]
Comparison of the inclusion complexation between host and guest in CD chemistry with the coordination interaction between central ion (M^{x+}) and ligands in coordination chemistry.	Song et al. (2009)	[36]
Threading CDs molecules onto polymer chains to form crystalline inclusion complexes organized by non-covalent interactions.	Martinez and Gomez (2007)	[109]
Practical considerations in development of solid dosage forms that contain CD.	Miller et al. (2007)	[110]
CD inclusion complexes with a solvatochromic fluorescent probe: an undergraduate physical chemistry lab experiment to establish the solvatochromic nature of PRODAN and then use the changes in the emission spectra upon inclusion in β- or γ-CD to determine stoichiometry and formation constants for the complexes.	Baker et al. (2002)	[111]
Some applications of CD/ substrate inclusion complexes.	Crini et al. (2001)	[8]
Determination of thermodynamic parameters of the CD inclusion processes: an undergraduate physical chemistry lab experiment.	Valero et al. (1999)	[112]
Complexation thermodynamics of CDs.	Rekharsky and Inoue (1998)	[113]
Applications of CDs to pharmaceutical industry and chemical catalysis. Analytical applications are also considered, since CDs inclusion improves the sensitivity and selectivity of most analytical methods.	Muñoz-Botella et al. (1995)	[37]
β-CD inclusion complexes with iodine: an advanced and inexpensive undergraduate chemistry experiment.	Diaz et al. (1994)	[114]
Critical overview about past, present and future of CDs: properties, studies on CD inclusion compounds and its applications.	Davies et al. (1983)	[115]
Metal and organometallic complexes		
Synthesis, reactivity and structural diversity of well-defined metal complexes derived essentially from native CDs. Structural motifs for metal complexes based on CDs: from monomeric species, dinuclear systems, homo- and heterometallic sandwich-type complexes to cylindrical, extended structures.	Prochowic et al. (2016)	[116]

Content	Authors	Refs.
Overview of recent advances of CD catalyzed reactions, which is organized in the order of the following reaction types: the modified CD catalyzed organic reaction, CD catalyzed organic reaction of metal ion present, CD catalyzed organic reaction without metal ion, and CD catalyzed organic reactions in application of asymmetric synthesis and photochemical reactions.	Hong et al. (2015)	[117]
Research and application of CDs and their derivatives in asymmetric and stereospecific syntheses, with their division into three main groups: (1) CDs promoting asymmetric and stereospecific catalysis in water; (2) CDs' complexes with transition metals as asymmetric and stereospecific catalysts; and (3) CDs' non-metallic derivatives as asymmetric and stereospecific catalysts.	Macaev and Boldescu (2015)	[118]
Preparation and analysis of CD-based metal–organic frameworks: laboratory experiments adaptable for high school students.	Smith et al. (2015)	[49]
Selectively functionalized CDs and their metal complexes: recent applications as chiral receptors and catalytic center in the mimicking of metalloenzymes.	Bellia et al. (2009)	[119]
Metal complexing properties of native CDs (including deprotonation in alkaline medium) and a report on some recent results on composition and stability of metal–CD complexes.	Norkus (2009)	[120]
CDs as supramolecular hosts for organometallic complexes.	Hapiot et al. (2006)	[121]
Food applications		
Complexation of poorly water-soluble phytochemicals (flavonoids, phenolic derivatives, coumestans to triterpenes) with CDs to improve their aqueous solubility, stability, rate of dissolution and bioavailability.	Suvarna et al. (2017)	[87]
CDs in food technology and human nutrition: benefits and limitations. The recent applications of CDs for reducing unwanted components, such as trans-fats, allergens, mycotoxins, acrylamides, bitter compounds, as well as in smart active packaging of foods are also overviewed.	Fenyvesi et al. (2016)	[95]
History, chemistry, methods of complexation and application of CDs into different areas, particularly in the pharmaceutical and food industry.	Maazaoui and Abderrahim (2015)	[44]
Properties, enzymatic production, and food applications of α-CD, as well as its differences with β- and γ-CDs.	Li et al. (2014)	[122]
Studies on the complexes formed between several important types of antioxidant compounds and CDs.	López-Nicolás et al. (2014)	[97]
Applications of CDs as food additives and in food processing: transport of previously nontransportable foods and prevention of the spread of microbial infections.	Martina et al. (2013)	[123]
CDs as novel solutions for the food industry concerning with their role as dietary fiber, in food and drink with health-promoting additives, protect sensitive ingredients, improve taste and odor, or their positively influence to the texture and consistency of food.	Zipp (2012)	[96]
Factors controlling flavors binding constants to CDs and their applications in foods.	Astray et al. (2010)	[124]

Content	Authors	Refs.
CD encapsulation of essential oils and volatiles: methods for the preparation of inclusion complexes, analytical techniques and applications.	Cabral Marques (2010)	[33]
Practical aspects of the utilization of CDs and CD inclusion compounds to food manufacture, focusing on the technical advantages of their use in food processing and as food additives.	Moreira da Silva (2009)	[125]
Use of CDs in the food industry: properties from a technological point of view, such as solubility and their capability to form inclusion complexes are described.	Astray et al. (2009)	[94]
Isolation and identification of native and branched-type (glucosylated and maltosylated) CDs in different enzyme- and heat-processed starch-containing food products.	Szente et al. (2006)	[35]
Practical aspects of the utilization of CDs and CD complexes in the food industry: molecular encapsulation of lipophilic food ingredients, long-term storage stability and technological advantages and food processing technologies.	Szente and Szejtli (2004)	[126]
CDs: application to food processing.	Yoshii (2004)	[127]
Pharmaceutical applications		
Reviews		
CDs: history, chemical structure, synthesis, physicochemical properties, uses, complexation phenomenon, approaches for making inclusion complexes, and its characterisation, advantages of inclusion complexes, mechanism of drug release, regulatory status and its applications.	Kanaka Durga Devi et al. (2010)	[83]
Basic science information and data on the development of drugs in CD-containing formulations.	Loftsson and Brewster (2010)	[128]
Critical review about experimental methods for determination of the binding constant between CD and a guest molecule.	Funasaki et al. (2008)	[10]
Historical development of CDs with emphasis on their use in pharmaceutical formulations.	Loftsson and Duchêne (2007)	[129]
CD-based pharmaceutics: past, present and future applications.	Davis and Brewster (2004)	[130]
CDs: structure, complex formation and drug solubility and non-conventional CD complexes.	Loftsson (2002)	[9]
Main impetus for the research into CD-drug combinations.	Frömming and Szejtli (1994)	[131]
Delivery release		
Application of CD nanosystems for oral drug delivery: strategies for the synthesis of these nanosystems, and their potential for the intelligent navigation of the gastrointestinal tract for optimal bioavailability and biodistribution.	Adeoye et al. (2017)	[91]
CD-mediated hierarchical self-assembly and its potential in drug delivery applications.	Antoniuk and Amiel (2016)	[132]
Supramolecular nanostructures based on CD and poly(ethylene oxide): syntheses, structural characterizations and applications for drug delivery.	Zheng and Wyman (2016)	[133]

Content	Authors	Refs.
Recent advances in drug delivery techniques utilizing CDs, and cyclic oligosaccharides consisting of α-1,4-linked α-D-glucopyranose units. Especially, drug delivery system consisting of combination systems of CDs and functional materials such as dendrimer, liposome and PEG are introduced.	Arima et al. (2015)	[134]
Relationship between CDs structure and physicochemical characteristics: self assembly and drug delivery. Importance of the nanoparticle technology preparation for the stability and application of this nanodevice.	Bonnet et al. (2015)	[135]
CD-based carriers for delivery of chemotherapeutic cytotoxic anticancer drugs.	Gidwani and Vyas (2015)	[90]
CD-based delivery systems for arthritic diseases: from development to experimental therapeutics.	Nascimento et al. (2015)	[136]
A vision for CD nanoparticles in drug delivery systems and pharmaceutical applications.	Lakkakula and Krause (2014)	[137]
CD containing biodegradable particles: from preparation to drug delivery applications.	Zafar et al. (2014)	[138]
State of the art and recent advances in the construction of CD-based assemblies and their applications for controlled drug delivery.	Zhang and Ma (2013)	[139]
CD in drug delivery: complexing agents, bioavailability and industrial applications.	Chordiya Mayur and Senthilkumaran (2012)	[86]
Recent developments of CDs in drug delivery using various routes of administration.	Laza-Knoerr et al. (2010)	[140]
Advantages of CD inclusion complexation, effects on important drug properties in formulation and applications in delivery systems (oral drug, rectal dug, nasal drug, transdermal drug, ocular drug, controlled and targeted drug, peptide and protein delivery, gene and oligonucleotide delivery, dermal and transdermal delivery, brain drug delivery or brain targetting).	Tiwari et al. (2010)	[84]
CD-based supramolecular architectures: syntheses, structures, and applications for drug and gene delivery.	Li and Loh (2008)	[141]
Applications of CDs and their derivatives in different areas of drug delivery: parenteral, oral, ophthalmic and nasal drug delivery. Other routes including dermal, rectal, sublingual and pulmonary delivery are also briefly addressed.	Rasheed et al. (2008)	[88]
Applications and comparative benefits of use of CDs and their derivatives in the design of novel delivery systems like liposomes, microspheres, microcapsules, nanoparticles, CD grafted cellulosic fabric, hydrogels, nano- sponges, beads, nanogels/nanoassemblies and CD-containing polymers.	Vyas et al. (2008)	[142]
The utility of CDs for enhancing oral bioavailability.	Carrier et al. (2007)	[143]
CDs as cosmetic delivery systems: study of ferulic acid/ CD association complexes at the light of its possible use as sunscreen.	Centini et al. (2007)	[98]
Effects of hydrophilic CDs on drug permeation through membranes and possible mechanism of action based on the current knowledge of the structural characteristics of water and the unstirred water layer juxtaposed to the membrane of interest.	Loftsson et al. (2007)	[144]

http://dx.doi.org/10.5772/intechopen.72736

Content	Authors	Refs.
Interesting findings and applications of CDs and their derivatives in different areas of drug delivery, particularly in protein and peptide drug delivery and gene delivery. Applications in the design of various novel delivery systems like liposomes, microspheres, microcapsules, and nanoparticles.	Challa et al. (2005)	[89]
CDs in drug delivery. Pharmaceutical products worldwide containing drug/CD complexes on the market.	Loftsson et al. (2005)	[145]
Recent findings and applications of both unmodified and modified CDs for in vivo drug delivery. Use of CDs for parenteral, oral, ophthalmic, and nasal drug delivery. Other routes including dermal, rectal, and pulmonary delivery are also briefly addressed.	Rajewski and Stella (1996)	[54]
Carrier		
Potential therapeutic application of dendrimer/CD conjugates with targeting ligands as advanced carriers for gene and oligonucleotide drugs.	Arima et al. (2017)	[146]
Drug carrier systems based on CD supramolecular assemblies and polymers: present and perspectives.	González-Gaitano et al. (2017)	[147]
CD-based polymeric nanoparticles as efficient carriers for anticancer drugs.	Duchêne et al. (2016)	[148]
Potential use of chemically modified CDs as high-performance drug carriers in drug delivery systems with emphasis on the more recent developments.	Rasheed et al. (2008)	[88]
CD drug carrier systems: characteristics, improvements of drug properties by CD complexation and CD-based drug delivery systems.	Uekama et al. (1998)	[149]
Solubilization and permeation		
CDs in pharmaceutical formulations: solubilization, binding constant, and complexation efficiency.	Jambhekar and Breen (2016)	[18]
Use of CDs as complexing agents to enhance the solubility of poorly soluble drugs and issues associated with developing and commercializing poorly water soluble drugs.	Chaudhary and Patel (2014)	[16]
Pharmaceutical applications of CDs: effects on drug permeation through biological membranes.	Loftsson and Brewster (2011)	[150]
General background to the use of CD as solubilizers as well as highlight kinetic and thermodynamic tools and parameters useful in the study of drug solubilization by CDs.	Brewster and Loftsson (2007)	[15]
CDs as solubilizers as well as highlight kinetic and thermodynamic tools and parameters useful in the study of drug solubilization	Loftsson and Brewster (1996)	[19]
Protein		
Use of CDs and their derivatives as antiaggregant agents in a number of proteins and some multimeric enzymes.	Oliveri and Vecchio (2016)	[151]
CD-based multivalent glycodisplays: covalent and supramolecular conjugates to assess carbohydrate–protein interactions.	Martínez et al. (2013)	[152]
CD interactions with protein-like structures in order to describe their possible applications in the formulation of pharmaceutical proteins.	Varca et al. (2010)	[153]

Content	Authors	Refs.
Encapsulation		
Encapsulation of CD/drug inclusion complex into conventional, deformable and double loaded liposomes: characteristics of these systems and advantages and disadvantages of each one.	Gharib et al. (2015)	[154]
Encapsulation of biocides by CDs: toward synergistic effects against pathogens.	Nardello-Rataj and Leclercq (2014)	[34]
Use of CDs as encapsulating agents for bioactive plant molecules in the pharmaceutical field.	Pinho et al. (2014)	[35]
Excipients		
Background review for CDs used as excipients.	EMA/CHMP/333892/ (2013)	[155]
CDs as functional excipients: methods to enhance complexation efficiency.	Loftsson and Brewster (2012)	[156]
Formulations		
CDs in pharmaceutical formulations: structure and physicochemical properties, formation of complexes, and types of complex.	Jambhekar and Breen (2016)	[17]
Evaluation of CDs drug complexes in pharmaceutical formulation: preparation of sodium valproate phenytoin sodium/ β-CD inclusion complex in a trial to stabilize the drug against moisture absorption and forming non-hygroscopic powders and preparation of phenytoin sodium/β-CD inclusion complex in a trial to stabilize the drug against moisture absorption and mask its bitter taste.	Akasha et al. (2014)	[157]
CDs in topical drug formulations: drug delivery from aqueous CD solutions by diffusion and membrane controlled.	Loftsson and Masson, (2001)	[158]
Miscellaneous		
Types of fluorophores which have been used for CD tagging: synthetic strategies used for the conjugation and pharmaceutical applications of these 'visualized' macrocycles including their use in photodynamic therapy.	Benkovics et al. (2017)	[159]
CDs' legacy as complexing agents and future prospects of this class of chemical entities in pharmaceutics as new active pharmaceutical ingredients.	di Cagno and Pio (2017)	[160]
Use of CD in the different routes of drug administration.	Shimpi et al. (2005)	[161]
Recent findings on the safety profiles of three natural CDs and several chemically modified CDs: stability against non-enzymatic and enzymatic degradations in various body fluids and tissue homogenates and their pharmacokinetics via parenteral, oral, transmucosal, and dermal routes of administration.	Irie and Hekama (1997)	[162]
Pharmacology		
Production, physiochemical properties, pharmacokinetics, toxicity and applications of γ-CD and its derivatives.	Saokham et al. (2017)	[15]
Interactions between CDs and cellular components: medical applications.	Leclercq (2016)	[163]
Inclusion of terpenes in CDs: preparation, characterization and pharmacological approaches.	Lima et al. (2016)	[164]

Content	Authors	Refs.
Self-assembly of CDs and their complexes in aqueous solutions.	Ryzhakov et al. (2016)	[48]
Diagnostic utility of flow cytometry and improvement of rocuronium-induced anaphylaxis with the use of sugammadex.	Takazawa et al. (2016)	[165]
Sugammadex for reversal of rocuronium-induced neuromuscular blockade in pediatric patients: a systematic review and meta-analysis.	Won et al. (2016)	[166]
Improving the therapeutic response of analgesic drugs by CDs.	De Oliveira et al. (2015)	[167]
Types of CDs, and their efficacy, physicochemical properties and transformation into nanoparticles with interesting in vitro and in vivo applications.	Lakkakula and Krause (2014)	[137]
Potential therapeutic use of CDs and CD nanoparticles in neurodegenerative diseases, stroke, neuroinfections and brain tumors.	Vecsernyés et al. (2014)	[168]
Basic and clinical pharmacology of sulfobutylether-β-CD.	Loftsson and Brewster (2010)	[128]
Basic and clinical pharmacology of sulfobutylether-β-CD.	Luke et al. (2010)	[169]
CD introduction to anesthesia practice: form, function, and application.	Welliver (2007)	[170]
Findings on the safety profiles of three natural CDs and several chemically modified.	Irie and Uekama (1997)	[162]
Medical and biomedical		
Key features of the CDs therapeutic discovery. Application of computational chemistry approaches such as QSAR/QSPR, molecular docking, and molecular/quantum mechanics for modeling of CD-drug system.	Abdolmaleki et al. (2017)	[171]
Recent development of copolymeric delivery system for anticancer agents based on CD derivatives.	Feng et al. (2016)	[172]
General features and applications of CDs and their interactions with isolated biomolecules leading to the formation of inclusion or exclusion complexes: potential medical applications.	Leclercq (2016)	[163]
Data on the general properties and complexing ability of CDs and assessment methods (phase solubility, DSC tests and X-ray diffraction, FTIR spectra).	Radu et al. (2016)	[173]
CD interactions with protein-like structures: possible applications in the formulation of pharmaceutical proteins.	Vecsernyés et al. (2014)	[168]
Amphiphilic CDs and their applications: preparation of nanoparticles based on amphiphilic CDs for biomedical applications.	Parrot-Lopez et al. (2010)	[174]
A supramolecular approach to medicinal chemistry: essential roles played by intermolecular forces in mediating the interactions between chemical molecules and biological systems.	Smith (2005)	[175]
Medicinal applications of CDs: improvement of drug properties, use of drug/CD complexes, CDs in tabletting and direct treatment with CDs.	Szejtli (1994)	[176]
Environmental Chemistry and Applications		
Nanosponge CD polyurethanes and their modification with nanomaterials for the removal of pollutants from waste water.	Leudjo Taka et al. (2017)	[177]

Content	Authors	Refs.
Progress in the immobilization of β-CD and their application in adsorption of environmental pollutants.	Han et al. (2016)	[178]
Interactions of CDs and their derivatives with toxic organophosphorus compounds.	Letort et al. (2016)	[179]
CD inclusion of four phenylurea herbicides: determination of complex stoichiometries and stability constants using solution ^{1}H NMR spectroscopy.	Smith et al. (2010)	[69]
Fluorescence spectroscopy as a tool to study the properties of CD host-guest complexes. Overview of recent studies concerned with exploiting the properties of CDs and their inclusion complexes to study energy transfer through the use of photochemical antennas and the development of chemical and environmental sensors.	Fakayode et al. (2007)	[27]
Synthesis and applications of adsorbents containing CDs in the field of chromatographic separations and in waste water treatment.	Crini and Morcellet (2002)	[180]
Personal care and toiletry		
CDs as cosmetic delivery system: study of ferulic acid/CD association complexes.	Centini et al. (2007)	[98]
Inclusion complex formation of CD with its guest and their applications in foods and flavors, personal care and toiletry, environment protection, pharmaceuticals among others.	Cheirsilp and Rakmai (2016)	[42]
Possible applications of CDs in cosmetic products and some examples of their present uses.	Buschmann and Schollmeyer (2002)	[99]
Industrial applications		
Enabling technologies and green processes in CD chemistry: microwaves, ultrasound and ball mills have become irreplaceable tools in the synthesis of CD derivatives. Examples of sonochemical selective modification of native α-, β- and γ-CDs including heterogeneous phase Pd- and Cu-catalysed hydrogenations and couplings.	Cravotto et al. (2016)	[181]
Major fields of enzyme application and overview on previous protein engineering studies wherein natural enzymes were modified to meet the operational conditions required for industrial application.	Jemli et al. (2016)	[24]
Applications of CDs in medical textiles: general data properties and complexing ability of CDs and assessment methods (phase solubility, DSC tests and X-ray diffraction, FTIR spectra, analytical method).	Radu et al. (2016)	[173]
Applications of CDs in various industrial products, technologies, analytical and chemical processes and recent industrial advancements.	Sharma and Baldi (2016)	[182]
General features of β-CD and their applications in the textile industry: attachment technique of β-CD to the textile's surface.	Bhaskara-Amrit et al. (2011)	[183]
CDs in pharmaceutics, cosmetics, and biomedicine: current and future industrial applications.	Bilensoy (2011)	[184]
Role of CDs in the textile chemical technology: remove the surfactants from the material or to inactivate them in liquid phase, to intensify the enzyme processes or as balancers in dyeing with reactive pigments.	Grigoriu and Popescu (2011)	[185]
Amphiphilic CDs and their applications. Preparation of nanoparticles based on amphiphilic CDs for biomedical applications.	Parrot-Lopez et al. (2010)	[174]

Content	Authors	Refs.
Applications of CDs in pharmaceuticals with a major emphasis on drug delivery systems. Utility in a variety of foods, flavors cosmetics, packaging and textiles.	Singh et al. (2002)	[186]
Applications of CDs in pharmaceuticals, foods and flavours, cosmetics, chemical industry, agricultural industry and adhesives, coatings and other polymers.	Arenskötter et al. (2001)	[187]
Industrial applications of CDs. Production and analysis of complexes.	Hedges (1998)	[188]
Utilization of CDs in industrial products and processes: (i) textiles, fibers and papers; (ii) foods and cosmetics; (iii) plastics and rubber; (iv) photographic and recording materials; (v) biotechnology and (vi) environmental protection.	Szejtli (1997)	[189]
Overview about industrial uses of CDs and their derivatives.	Duchêne and Wouessidjewe (1992)	[190]
CD inclusion compounds in research and industry: production of pharmaceuticals, pesticides, foodstuffs, and toilet articles among others.	Saenger (1980)	[191]
Nano		
General overview of CDs and pharmaceutical nanotechnology in oral delivery systems. Strategies for the synthesis of these nanosystems, and their potential for the intelligent navigation of the gastrointestinal tract for optimal bioavailability and biodistribution.	Adeoye and Cabral-Marques (2017)	[91]
Nanosponge CD polyurethanes and their modification with nanomaterials for the removal of pollutants from waste water.	Leudjo et al. (2017)	[177]
CD-based supramolecular host–guest interactions for engineering supramolecular nanoparticles: biomedical applications.	Mejia-Ariza et al. (2017)	[192]
CD-based polymeric nanoparticles as efficient carriers for anticancer drugs.	Duchene et al. (2016)	[148]
CD-based nanosponges: a versatile platform for cancer nanotherapeutics development.	Swanimathan et al. (2016)	[193]
Supramolecular nanostructures based on CD and poly(ethylene oxide): syntheses, structural characterizations and applications for drug delivery.	Zheng and Wyman (2016)	[133]
Nano-sized CD-based molecularly imprinted polymer adsorbents for perfluorinated compounds.	Karoyo and Wilson (2015)	[194]
Overall view of the diversity of designs of CD-based supramolecular nanosystems with a special focus on the advances materialized in the last five years, including clinical trials.	Simoes et al. (2015)	[195]
Recent advances in the construction of nanoassemblies driven by CD-based inclusion complexation and their application in biomedical and biomimetic fields.	Kang et al. (2014)	[196]
A vision for CD nanoparticles in drug delivery systems and pharmaceutical applications.	Lakkakula and Krause (2014)	[137]
Approaches tested to synthesize nano- to macro-size covalently cross-linked CD networks: (i) direct cross-linking through condensation with di- or multifunctional reagents, (ii) copolymerization of CD derivatives with acrylic/vinyl monomers, and (iii) grafting of CDs to preformed medical devices.	Concheiro and Alvarez-Lorenzo (2013)	[197]

Content	Authors	Refs.
Development of nanosponges as drug delivery systems, with special reference to CD based nanosponges.	Tejashri et al. (2013)	[92]
Preparation, characterization and advantages for pharmaceutical and biomedical applications of CD-based nanogels.	Moya-Ortega et al. (2012)	[198]
Formation and applications of CD nanoaggregates induced by guest molecules, the concerned thermodynamics behind the process and the effect of concentration of the guest molecules on the morphology of the aggregates.	Purkayastha et al. (2012)	[199]
Approaches employed in delivering drugs to the central nervous system. Changes in blood-brain barrier function in several neurological disorders.	Martín-Banderas et al. (2011)	[200]
Fabrication technologies of supramolecular systems including nanoplatforms and hydrogels as well as their applications in nanomedicine and pharmaceutical sciences.	Zhang and Ma (2013)	[139]
Classification, physicochemical properties, efficacy and safety of nanoparticles prepared from different amphiphilic CDs are discussed in light of the current literature work with in vitro and in vivo findings.	Bilensoy and Hincal (2009)	[201]
Analytical and physicochemical applications		
Reviews		
Classical and modern instrumental analytical methods suitable for identification, characterization and determination of CDs themselves, CDs in finished products or even in biological samples.	Szente et al. (2016)	[2]
CDs in in sample preparation, sensitivity and selectivity improvement, enantio-separation, creating single-molecule sensors, and automatizing DNA sequencing.	Szente and Szeman (2013)	[28]
CDs: from molecular recognition to CDs as enzyme models. Reactivity and chemistry, chromatography, X-ray, NMR plus other physicochemical methods, as well as model calculations, rotaxane and catenane structures, and applications in the pharmaceutical industry are overviewed.	Dodziuk (2006)	[202]
Use of CDs in major areas of analytical chemistry such as chromatography, electrophoresis, spectroscopy, electrochemistry and as analytical sensors.	Mosinger et al. (2001)	[203]
Role of CDs in three of the major areas of modern instrumental analysis: separations, spectroscopy and electrochemical analysis.	Armstrong (1998)	[204]
Chirality		
CD-functionalized monolithic capillary columns: preparation and chiral applications.	Adly et al. (2016)	[205]
Recent developments in CD functionalized monolithic columns for the enantioseparation of chiral drugs.	Guo et al. (2016)	[206]
Advances on the use of CDs in the chiral analysis of drugs by capillary electrophoresis.	Saz and Marina (2016)	[80]

Content	Authors	Refs.
Recent contributions to the understanding of the binding mechanism between chiral selectors and selectands in analytical enantioseparations including polysaccharide derivatives, CDs, cyclofructans, macrocyclic glycopeptides, proteins, brush-type selectors, ion-exchangers, polymers, crown ethers, ligand-exchangers, molecular micelles, ionic liquids, metal-organic frameworks and nucleotide-derived selectors.	Scriba (2016)	[77]
Development of cationic CDs for chiral separation. Update of the research endeavors of synthetic and analytical chemists in evaluating enantioselectivity of cationic CDs using different analytical methods and the study of the chiral recognition mechanism.	Zhou and Scriba (2016)	[75]
Advances in enantiomeric resolution on monolithic chiral stationary phases in liquid chromatography and electrochromatography.	Al-Othman et al. (2014)	[207]
Recent examples of mechanistic aspects of capillary enantioseparations with regard to mathematical modeling of enantioseparations, investigations of the analyte-complex structures as well as new chiral selectors and applications of chiral analyses by CE and CEC.	Jac and Scriba (2013)	[208]
Review of the latest advances in developing modified CDs as chiral selectors for various chromatographic and electromigration techniques.	Tang et al. (2013)	[76]
Chiral analysis of amphetamines, methadone and metabolites in biological samples by electrodriven methods.	Mandrioli et al. (2011)	[209]
The growth and applications of CDs as chiral discriminator.	Pathak and Pathak (2008)	[210]
CDs in capillary electrophoresis enantioseparations: recent developments and applications.	Scriba (2008)	[90]
Separation of enantiomeric barbiturates by capillary electrophoresis using a CD containing run buffer: a laboratory experiments for degree students.	Contradi et al. (1997)	[82]
Complexes characterization		
Physicochemical characterization of CD-drug interactions in the solid state and the effect of water on these interactions.	Ogawa and Takahashi (2015)	[52]
Analytical techniques for characterization of CD complexes in the solid state.	Mura (2015)	[51]
Analytical tools which can be employed for the characterization of drug-CD inclusion complexes in solution, with emphasis on their respective potential merits, disadvantages and limits.	Mura (2014)	[47]
Surfactant-CD host-guest association: fundamentals, drawbacks and advantages of techniques commonly used to obtain insights on the structural and bulk solutions changes resulting from host-guest association mechanism, and corresponding methods for binding quantification.	Valente and Söderman (2014)	[30]
CD inclusion complexes probed by NMR techniques.	Pessine et al. (2012)	[45]
A literature review of CD inclusion complexes characterization: X-ray diffraction, infrared spectroscopy and nuclear magnetic resonance.	Takahashi et al. (2012)	[211]

Content	Authors	Refs.
A literature review of CD inclusion complexes characterization: differential scanning calorimetry and thermogravimetry.	Takahashi et al. (2012)	[68]
A bilogarithmic method for the spectrophotometric evaluation of stability constants of 1:1 weak complexes from mole ratio data.	Boccio et al. (2006)	[212]
NMR studies of CDs and CD complexes. Comprehensive overview about the most important approaches to structural problems with CDs, mainly in solution.	Schneider et al. (1998)	[213]
The stability of CD complexes in solution: binding equilibria and kinetics, strengths and structures of CD complexes, the sources of CD complex stability and prediction of CD complex stability.	Connors (1997)	[46]
Separation Methods		
State-of-the-art applications of CDs as functional monomers in molecular imprinting techniques.	Lay et al. (2016)	[214]
CDs in capillary electrophoresis: recent contributions, practical uses (e.g. solute-CD binding constant estimation and further potentials), developments and applications (mainly chiral and achiral analysis).	Escuder-Gilabert et al. (2014)	[79]
Recent developments and new trends.		
Separation processes in the presence of CDs using molecular imprinting technology and ionic liquid cooperating approach.	Zhang et al. (2011)	[59]
Role of CDs in chromatography. Influence of The formation the physicochemical parameters of the guest molecule (adsorption capacity, polarity, hydrophobicity, etc.).	Cserhat and Forgaes (2003)	[71]
Summary of the information concerning the synthesis of materials containing CDs and general overview of the different possible applications of CDs as sorbents in the field of separation techniques.	Crini and Morcellet (2002)	[180]
CDs as a versatile tool in separation science. The techniques examined include gel electrophoresis, isotachophoresis, isoelectric focusing, preparative scale electrophoretic techniques, thin-layer chromatography, electrochemically modulated liquid chromatography, use of monolithic media in liquid chromatography, microdialysis, separation on hollow fibers, foam flotation enrichment, solid- and liquid-phase extractions, countercurrent chromatography, separation through liquid and composite membranes, and CD applications in molecularly imprinted polymers.	Schneiderman and Stalcup (2000)	[61]
Utilization of CDs and their derivatives in gas-liquid and gas-solid-, gel-, inclusion-, thin-layer-, affinity-, and high performance liquid chromatography.	Szejtli (1987)	[215]
Applications of CDs in chromatographic separations and purification methods.	Hinze (1981)	[73]
Spectrofluorometric Methods		
Spectrofluorometric analytical applications of CDs based on host-inclusion complex.	Elbashir et al. (2014)	[64]
Room temperature phosphorescence in CDs: analytical applications.	Muñoz de la Peña et al. (2000)	[66]

Content	Authors	Refs.
Electrochemical Methods		
Advantages and detecting mechanism of electrochemical sensors based on CDs functionalized materials, and recent advances for CDs-based materials (including CDs/carbon nanotubes, CDs/graphene, CDs/conducting polymers and other CDs-based nanomaterials) in electrochemical sensing.	Zhu et al. (2016)	[70]
Substrate/analyte solubilization and stabilization to the development of CD based sensors and detectors.	Szente and Szejtli (1998)	[216]
State of the art of the electrochemistry of α-, β-, and γ-CDs and CD inclusion complexes and their polarographic and voltammetric assay.	Bersier et al. (1991)	[71]
Enzyme—Biomimetic-Bioactive assemblies recognition		
General overview of three different categories of CD-based artificial enzymes including metal free CD-based artificial enzymes, CD-based artificial metalloenzymes and CD-based artificial enzymes with computational design, focusing on their rate acceleration factor.	Aghahosseini and Ramazani (2016)	[21]
Major fields of enzyme application and overview on previous protein engineering studies wherein natural enzymes were modified to meet the operational conditions required for industrial application.	Jemli et al. (2016)	[24]
Macromolecules based on recognition between CD and guest molecules: synthesis, properties and functions.	Liu et al. (2015)	[217]
Representative contributions in the construction and the structural characteristics of CD-based supramolecular assemblies and their interactions with biologically important substrates.	Chen and Liu (2010)	[218]
New chemistry based on the principles used by Nature: biomimetic chemistry.	Breslow (2009)	[23]
Literature overview on reactions in which CDs bind substrates and then either catalyze their reactions or mimic a step in an enzymatic catalytic sequence.	Breslow and Dong (1998)	[22]
Adjusting the lock and adjusting the key in CD chemistry. An introduction in biomimetic chemistry.	Breslow (1980)	[219]
Miscellaneous		
Functioning via host–guest interactions: achievement of selective molecular adhesion, self-healing, toughness, and actuation properties. These functions have been achieved by reversible bond formation with CDs.	Takachima and Harada (2017)	[29]
Qualitative and quantitative analysis of research outputs on molecular modeling in CDs.	Zhao et al. (2017)	[220]
Supramolecular polymer assembly in aqueous solution arising from CD host-guest complexation. Effects of such complexation on properties at the molecular and macroscopic levels.	Wang et al. (2016)	[32]
Superstructures with CDs: chemistry and applications.	Wenz and Monflier (2016)	[221]

Content	Authors	Refs.
Synthesis of CD half-channels derived by per-functionalization of the CD primary positions and their activity as channels assessed by the bilayer clamp technique.	Chui and Fyles (2014)	[222]
Construction of supramolecular structures of CDs with some polymers (polyrotaxanes) and formation of supramolecular oligomers and polymers formed by CD derivatives.	Harada et al. (2009)	[223]
Systematic analysis of methods that are available for modification of CDs. The focus is on methods for transformation where the number and the exact positions of modifications are ascertained and pure compounds with unambiguous structures are obtained.	Khan et al. (1998)	[224]
Applications of computational chemistry to the study of CDs: molecular modeling, structural features of CDs, dynamical aspects of CD structure and computational studies of host–guest complexation.	Lipkowitz (1998)	[225]
CD-based catenanes and rotaxanes.	Nepogodiev et al. (1998)	[226, 227]
Organic reactions mediated by CDs: effect in solid CD complexes.	Takahashi (1998)	[228]

Table 1. Selected papers on food, pharmaceutical, pharmacology, cosmetic, industrial, and analytical applications of cyclodextrins (CDs).

the number of papers cited by journal for the most cited journal (number of references ≥2) appears. Emphasis is stressed on reviews and taking into account the high number of references available, the authors apologize for those they may have overlooked or inadvertently omitted.

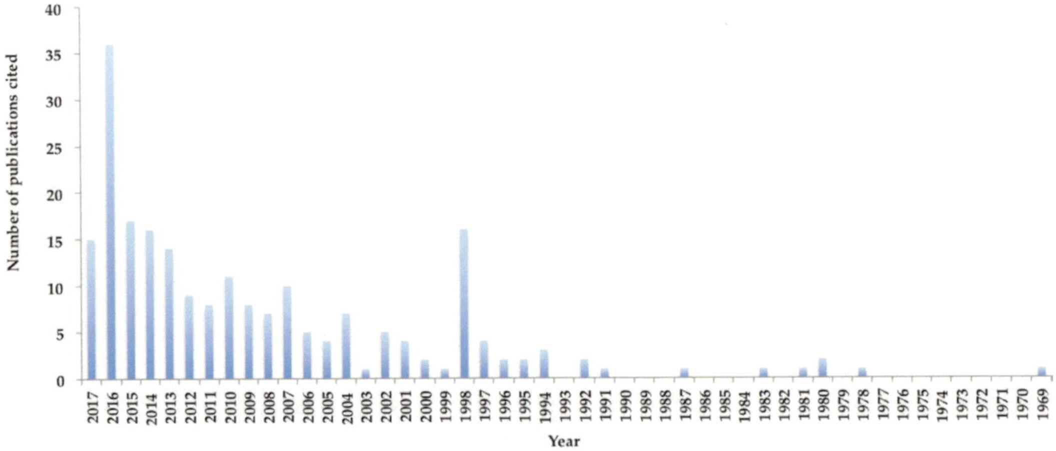

Figure 5. Number of publications cited per year.

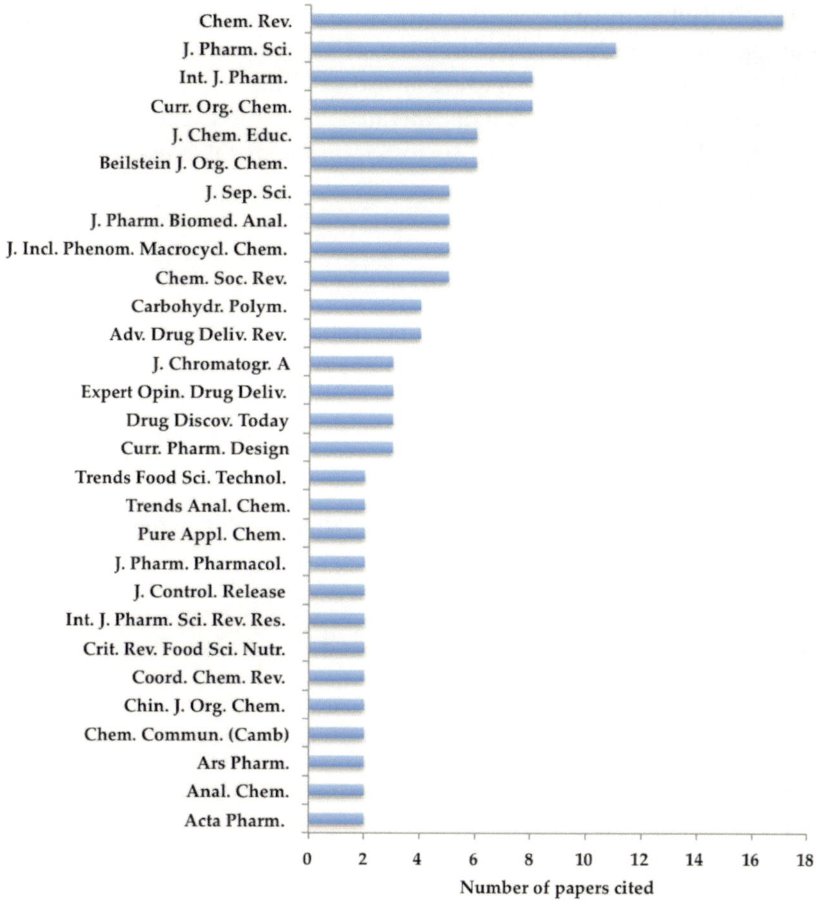

Figure 6. Number of papers cited by journal.

5. Conclusion

Currently, there are a large number of drugs with poor solubility, bioavailability, permeability issues, undesirable properties as taste and odor, and irritation potential, and CDs can become an useful tool for optimizing drugs problematic [84]. Additionally, new uses of cyclodextrins are being explored, in different fields as nanoparticles, liposome and microsphere. The ability of making inclusion complexes with drugs makes CDs have a great future, as reflected by the rising number of publications and patents having been filed. Some researchers also believe that there will be more a still wider use for CDs as the knowledge about their properties increase [7]. The studies of CD-based nanosystems have recently increased, as they become platforms providing pharmacokinetic and formulation design efficiency without posing security problems [91]. CDs are also generating interest for gene therapy and exploration of non-viral methods, probably for the difficulties in viral gene delivery [7]. A new area which is going to increase, is the study of the effects of the environment in the reactivity between CD-guest molecules [115]. The creation of new types of CD is going to enhance due to the wide range of possibilities in the treatments of atherosclerosis, cancer, and degenerative brain disease that are considered lethal disease [160]. CDs will surprise us in the future with not predictable uses [184].

Author details

Julia Martin[1]*, Enrique Jacobo Díaz-Montaña[2] and Agustín G. Asuero[2]

*Address all correspondence to: jbueno@us.es

1 Department of Analytical Chemistry, Escuela Politécnica Superior, University of Seville, Seville, Spain

2 Department of Analytical Chemistry, Faculty of Pharmacy, University of Seville, Seville, Spain

References

[1] Crini G. Review: A history of cyclodextrins. Chemical Reviews. 2014;**114**(21):10940-10975

[2] Szente L, Szeman J, Sohajda T. Analytical characterization of cyclodextrins: History, official methods and recommended new techniques. Journal of Pharmaceutical and Biomedical Analysis. 2016;**130**:347-365

[3] Kurkov SV, Loftsson T. Cyclodextrins. International Journal of Pharmaceutics. 2013; **453**(1):167-180

[4] Berthelot M. Ciclodextrinas. In: Khymos FF, editor. Química y Medida. Madrid, Spain: Alhambra; 1969. pp. 312-438

[5] Cramer F, Hettler H. Inclusion compounds of cyclodextrins. Die Naturwissenschaften. 1967;**54**:625-632

[6] Gattuso G, Nepogodiev SA, Stoddart JF. Synthetic cyclic oligosaccharides. Chemical Reviews. 1998;**98**(5):1919-1958

[7] Khan NA, Durakshan M. Cyclodextrin: An overview. International Journal of Bioassays. 2013;**2**:858-865

[8] Crini G, Morcellet M, Morin N. Quelques applications des complexes d´inclusion cyclodextrine/substrat. L'Actualité Chimique. 2001;**1**:18-25

[9] Loftsson T. Cyclodextrins. Journal of Inclusion Phenomena and Macrocyclic Chemistry. 2002;**44**:213-218

[10] Funasaki N, Ishikawa S, Neya S. Advances in physical chemistry and pharmaceutical applications of cyclodextrins. Pure and Applied Chemistry. 2008;**80**(7):1511-1524

[11] Saenger W, Jacob J, Gessler K, Steiner T, Hoffmann D, Sanbe H, Koizumi K, Smith SM, Takaha T. Structures of the common cyclodextrins and their larger analogues beyond the doughnut. Chemical Reviews. 1998;**98**(5):1787-1802

[12] Saokham P, Loftsson T. γ-Cyclodextrin. International Journal of Pharmaceutics. 2017; **516**(1-2):278-292

[13] Maistrenko VN, Gusakov VN, Sangalov EY. Voltammetry of host-guest complexes. Journal of Analytical Chemistry (Engl. Trasl.). 1995;**50**:528-533

[14] Duchêne D, Bochot A. Thirty years with cyclodextrins. International Journal of Pharmaceutics. 2016;**514**(1):58-72

[15] Brewster ME, Loftsson T. Cyclodextrins as pharmaceutical solubilizers. Advanced Drug Delivery Reviews. 2007;**59**(7):645-666

[16] Chaudhary VB, Patel JK. Cyclodextrin inclusion complex to enhance solubility of poorly water soluble drugs: A review. International Journal of Pharmaceutical Sciences and Research. 2013;**4**(1):68-76

[17] Jambhekar SS, Breen P. Cyclodextrins in pharmaceutical formulations I: Structure and physicochemical properties, formation of complexes, and types of complex. Drug Discovery Today. 2016;**21**(2):356-362

[18] Jambhekar SS, Breen P. Cyclodextrins in pharmaceutical formulations II: Solubilization, binding constant, and complexation efficiency. Drug Discovery Today. 2016;**21**(2):363-368

[19] Loftsson T, Brewster ME. Pharmaceutical applications of cyclodextrins. 1. Drug solubilization and stabilization. Journal of Pharmaceutical Sciences. 1996;**85**(10):1017-1025

[20] Shieh WJ, Hedges AR. Properties and applications of cyclodextrins. Journal of Macromolecular Science Part A. 2006;**33**(5):673-683

[21] Aghahosseini H, Ramazani A. General overview on cyclodextrin-based artificial enzymes' activity. Current Organic Chemistry. 2016;**20**(26):2817-2836

[22] Breslow R, Dong SD. Biomimetic reactions catalyzed by cyclodextrins and their derivatives. Chemical Reviews. 1998;**98**(5):1997-2012

[23] Breslow R. Biomimetic chemistry. The Journal of Biological Chemistry. 2009;**284**:1337-1342

[24] Jemli S, Ayadi-Zouari D, Hlima HB, Bejar S. Biocatalysts: Application and engineering for industrial purposes. Critical Reviews in Biotechnology. 2016;**36**(2):246-258

[25] Jabbari A, Sadeghian H. Amphiphilic cyclodextrins, synthesis, utilities and application of molecular modeling in their design. In: Demir Sezer A, editor. Recent Advances in Novel Drug Carrier Systems. Chapter 12. Rijeka: InTech; 2012. pp. 331-354. ISBN: 978-953-51-0810-8

[26] Szejtli J. Past, present, and future of cyclodextrin research. Pure and Applied Chemistry. 2004;**76**(10):1825-1845

[27] Fakayode SO, Lowry M, Fletcher KA, Huang X, Powe AM, Warner IM. Cyclodextrins host-guest chemistry in analytical and environmental chemistry. Current Analytical Chemistry. 2007;**3**(3):171-181

[28] Szente L, Szemán J. Cyclodextrins in analytical chemistry: Host-guest type molecular recognition. Analytical Chemistry. 2013;**85**:8024-8030

[29] Takashima Y, Harada A. Functioning via host-guest interactions. Journal of Inclusion Phenomena and Macrocyclic Chemistry. 2017;**87**(3-4):313-330

[30] Valente AJM, Söderman O. The formation of host-guest complexes between surfactants and cyclodextrins. Advances in Colloid and Interface Science. 2014;**205**:156-176

[31] Loftsson T, Másson M, Brewster ME. Self-association of cyclodextrins and cyclodextrin complexes. Journal of Pharmaceutical Sciences. 2004;**93**(5):1091-1099

[32] Wang J, Qiu ZQ, Wang YM, Li L, Guo XH, Pham DT, Lincoln SF, Prud'homme RK. Supramolecular polymer assembly in aqueous solution arising from cyclodextrin host-guest complexation. Beilstein The Journal of Organic Chemistry 2016;**12**:50-72

[33] Cabral Marques HM. A review on cyclodextrin encapsulation of essential oils and volatiles. Flavour Fragrance Journal. 2010;**25**(5):313-326

[34] Nardello-Rataj V, Leclerq L. Encapsulation of biocides by cyclodextrins: Toward synergistic effects against pathogens. Beilstein. The Journal of Organic Chemistry. 2014;**10**:2603-2622

[35] Pinho E, Grootveld M, Soares G, Henriques M. Cyclodextrins as encapsulation agents for plant bioactive compounds. Carbohydrate Polymers. 2014;**101**(1):121-135

[36] Song LX, Bai L, XM X, He J, Pan SZ. Inclusion complexation, encapsulation interaction and inclusion number in cyclodextrin chemistry. Coordination Chemistry Reviews. 2009;**253**(9-10):1276-1284

[37] Munoz-Botellas S, Del Castillo B, Martin MA. Cyclodextrins: Properties and applications of inclusion complex formation. Ars Pharmaceutica. 1995;**36**(2):187-198

[38] Li S, Purdy WC. Cyclodextrins and their applications in analytical chemistry. Chemical Reviews. 1992;**92**:1457-1470

[39] Szente L, Harangi J, Greiner M, Mandel F. Cyclodextrins found in enzyme- and heat-processed starch-containing foods. Chemistry & Biodiversity. 2006;**3**(9):1004-1014

[40] Caira MR. Structural aspects of crystalline derivatized cyclodextrins and their inclusion complexes. Current Organic Chemistry. 2011;**15**(6):815-830

[41] Iacovino R, Caso JV, Di Donato C, Malgieri G, Palmieri M, Russo L, Isernia C. Cyclodextrins as complexing agents: Preparation and applications. Current Organic Chemistry. 2017;**21**(2):162-176

[42] Cheirsilp B, Rakmai J. Inclusion complex formation of cyclodextrin with its guest and their applications. Biology, Engineering and Medicine. 2016;**2**(1):1-6

[43] Liu Y, Wang K. Thermodynamics of resulting complexes between cyclodextrins and bile salts. In: Morales-Rodriguez R, editor. Thermodynamics–Fundamentals and Its Application in Science. Chapter 12. Rijeka: InTech; 2012. pp. 305-318. ISBN: 978-953-51-0779-8

[44] Maazaoui R, Abderrahim R. Applications of cyclodextrins: Formation of inclusion complexes and their characterization. International Journal of Advanced Research. 2015;**3**(2):1030-1030

[45] Pessine FBT, Calderini A, Alexandrino GL. Review: Cyclodextrin inclusion complexes probed by NMR techniques. In: Kim D, editor. Magnetic Resonance Spectroscopy. Chapter 12. Rijeka: InTech; 2012. pp. 237-264. ISBN: 978-953-51-0065-2

[46] Connors KA. The stability of cyclodextrin complexes in solution. Chemical Reviews. 1997;**97**:1325-1357

[47] Mura P. Analytical techniques for characterization of cyclodextrin complexes in aqueous solution: A review. Journal of Pharmaceutical and Biomedical Analysis. 2014;**101**:238-250

[48] Ryzhakov A, Do Thi T, Stappaerts J, Bertoletti L, Kimpe K, Sá Couto AR, Saokham P, Van den Mooter G, Augustijns P, Somsen GW, Kurkov S, Inghelbrecht S, Arien A, Jimidar MI, Schrijnemakers K, Loftsson T. Self-assembly of cyclodextrins and their complexes in aqueous solutions. Journal of Pharmaceutical Sciences. 2016;**105**(9):2556-2569

[49] Smith MK, Angle SR, Northrop BH. Preparation and analysis of cyclodextrin-based metal-organic frameworks: Laboratory experiments adaptable for high school through advanced undergraduate students. Journal of Chemical Education. 2015;**92**(2):368-372

[50] Harata K. Structural aspects of stereodifferentiation in the solid state. Chemical Reviews. 1998;**98**(5):1803-1828

[51] Mura P. Analytical techniques for characterization of cyclodextrin complexes in the solid state: A review. Journal of Pharmaceutical and Biomedical Analysis. 2015;**113**:226-238

[52] Ogawa N, Takahashi C, Yamamoto H. Physicochemical characterization of cyclodextrin-drug interactions in the solid state and the effect of water on these interactions. Journal of Pharmaceutical Sciences. 2015;**104**(3):942-954

[53] Gelb RI, Schwartz LM, Radeos M, Edmonds RB, Laufer AA. Cyclohexaamylose complexation with organic solvent molecules. Journal of the American Chemical Society. 1982;**104**:6283-6288

[54] Rajewski RA, Stella VJ. Pharmaceutical applications of cyclodextrins. 2. In vivo drug delivery. Journal of Pharmaceutical Sciences. 1996;**85**(11):1142-1169

[55] Pendergast DD. Stability constants of α-cyclodextrin complexes of disubstituted benzenes: Interpretation of structural effects in terms of a binding. [PhD thesis]. University of Wisconsin-Madison; 1983

[56] Innocenti A. Stoichiometry and Research – The Importance of Quantity in Biomedicine. Rijeka: InTech; 2012. pp. 388. ISBN: 978-953-51-0198-7

[57] Martina K, Cravotto, G. Cyclodextrin as a food additive in food processing. In: Cirillo G, Spizzirri UG, Iemma F, editors. Functional Polymers in Food Science: From Technology to Biology. Hoboken, NJ, USA: John Wiley & Sons, Inc., 2015

[58] SzejtliB,ZsadonT,CserhatiT.Cyclodextrinuseinseparations.In:HinzeWL,ArmstrongDW, editors. Ordered Media in Chemical Separations. Washington, DC: ACS; 1987. pp. 200-217

[59] Zhang J, Shen X, Chen Q. Separation processes in the presence of cyclodextrins using molecular imprinting technology and ionic liquid cooperating approach. Current Organic Chemistry. 2011;**1**:74-85

[60] Khalafi L, Rafiee M. Cyclodextrin based spectral changes. In: Rádis Baptista G, editor. An Integrated View of the Molecular Recognition and Toxinology – From Analytical Procedures to Biomedical Applications. Chapter 19. Rijeka: InTech; 2013. pp. 471-493. ISBN: 978-953-51-1151-1

[61] Schneiderman E, Stalcup AM. Cyclodextrins: A versatile tool in separation science. Journal of Chromatography. B, Biomedical Sciences and Applications. 2000;**745**(1):83-102

[62] Singh V, He YP, Wang CF, JH X, Xu XN, Li HY, Singh P, York P, Sun LX, Zhang JW. A comparison report of three advanced methods for drug-cyclodextrin interaction measurements. Journal of Pharmaceutical and Biomedical Analysis. 2017;**134**:252-258

[63] Tablet C, Matei I, Hillebrand M. The determination of the stoichiometry of cyclodextrin inclusion complexes by spectral methods: Possibilities and limitations. In: Innocenti A, editor. Stoichiometry and Research – The Importance of Quantity in Biomedicine. Chapter 3. Rijeka: InTech; 2012. pp. 47-76. ISBN: 978-953-51-0198-7

[64] Elbashir AA. Supramolecular Analytical Applications of Cyclodextrins and their Derivatives Using Fluorescence Spectroscopy. New York: Nova Science Publishers; 2017. pp. 209-232

[65] Elbashir AA, Dsugi NFA, Mohmed TOM, Aboul-Enein HY. Spectrofluorometric analytical applications of cyclodextrins. Luminescence. 2014;**29**(1):1-7

[66] Muñoz de la Peña A, Mahedero MC, Bautista Sánchez A. Room temperature phosphorescence in cyclodextrins. Analytical applications. Analusis. 2000;**28**(8):670-678

[67] Gunshefski M, Santana JJ, Stephenson J, Stephenson J, Winefordnet JD. Solid-phase room-temperature phosphorescence. Applied Spectroscopy Reviews. 1992;**27**(2): 143-192

[68] Takahashi AI, Veiga FJB, Ferraz HG. A literature review of cyclodextrin inclusion complexes characterization – Part III: Differential scanning calorimetry and thermogravimetry. International Journal of Pharmaceutical Sciences Review and Research. 2012;**12**(1):16-20

[69] Smith VJ, Bogdan D, Caira MR, Bogdan M, Bourne SA, Fărcaş SI. Cyclodextrin inclusion of four phenylurea herbicides: Determination of complex stoichiometries and stability constants using solution 1H NMR spectroscopy. Supramolecular Chemistry. 2010;**22**(3):172-177

[70] Zhu G, Yi Y, Chen J. Recent advances for cyclodextrin-based materials in electrochemical sensing. Trends in Analytical Chemistry. 2016;**80**:232-241

[71] Bersier PM, Bersier J, Klingert B. Electrochemistry of cyclodextrins and cyclodextrin inclusion complexes. Electroanalysis. 1991;**3**(6):443-455

[72] Cserhati T, Forgacs E. Cyclodextrins in chromatography. In: Smith RM, editor. RSC Chromatography Monographs. United Kingdom: Series; 2003. p. 170. DOI: 10.1039/1757-7063

[73] Hinze WL. Applications of cyclodextrins in chromatographic separations and purification methods. Separation and Purification Methods. 1981;**10**:159-237

[74] Zhu Q, Scriba GKE. Advances in the use of cyclodextrins as chiral selectors in capillary electrokinetic chromatography: Fundamentals and applications. Chromatographia. 2016;**79**(21-22):1403-1435

[75] Zhou J, Tang J, Tang W. Recent development of cationic cyclodextrins for chiral separation. Trends in Analytical Chemistry. 2015;**65**:22-29

[76] Tang W, Ng S, Sun D. In: Tang W, editor. Modified Cyclodextrins for Chiral Separation. Berlin Heidelberg: Springer-Verlag; 2013. pp. 258. ISBN: 978-3-642-37647-4

[77] Scriba GKE. Chiral recognition in separation science – An update. Journal of Chromatography. A. 2016;**1467**:56-78

[78] Guttman A, Aumateli A, Brunet S, Cooke N. Cyclodextrin array chiral analysis. American Laboratory. 1995;**27**(18):18-22

[79] Escuder-Gilabert L, Martín-Biosca Y, Medina-Hernández MJ, Sagrado S. Cyclodextrins in capillary electrophoresis: Recent developments and new trends. Journal of Chromatography. A. 2014;**1357**:2-23

[80] Saz JM, Marina ML. Recent advances on the use of cyclodextrins in the chiral analysis of drugs by capillary electrophoresis. Journal of Chromatography. A. 2016;**1467**:79-94

[81] Scriba GKE. Cyclodextrins in capillary electrophoresis enantioseparations – Recent developments and applications. Journal of Separation Science. 2008;**31**(11):1991-2011

[82] Contradi S, Vogt C, Rohde E. Separation of enantiomeric barbiturates by capillary electrophoresis using a cyclodextrin containing run buffer. Journal of Chemical Education. 1997;**74**(9):1122

[83] Kanaka Durga Devi N, Prameela Rani A, Muneer Aved M, SaiKumar K, Kaushik J, Sowjanya V. Cyclodextrins in pharmacy – An overview. Pharmacophore. 2010;**1**(3):155-165

[84] Tiwari G, Tiwari R, Rai AK. Cyclodextrins in delivery systems: Applications. Journal of Pharmacy & Bioallied Sciences. 2010;**2**(2):72-79

[85] Das SK, Rajabalaya R, David S, Gani N, Khanam J, Nanda A. Cyclodextrins-the molecular container. Research Journal of Pharmaceutical, Biological and Chemical Sciences. 2013;**4**(2):1694-1720

[86] Chordiya Mayur A, Senthilkumaran K. Cyclodextrin in drug delivery: A review. Research & Reviews In Pharmacy & Pharmaceutical Sciences. 2012;**1**(1):19-20

[87] Suvarna V, Gujar P, Murahari M. Complexation of phytochemicals with cyclodextrin derivatives – An insight. Biomedicine & Pharmacotherapy. 2017;**88**:1122-1144

[88] Rasheed A, Kumar CKA, Sravanthi VVNSS. Cyclodextrins as drug carrier molecule: A review. Scientia Pharmaceutica. 2008;**76**(4):567-598

[89] Challa R, Ahuja A, Ali J, Khar RK. Cyclodextrins in drug delivery: An updated review. AAPS PharmSciTech. 2005;**6**(2):E329-E357

[90] Gidwani B, Vyas A. A comprehensive review on cyclodextrin-based carriers for delivery of chemotherapeutic cytotoxic anticancer drugs. BioMed Research International. 2015;**2015**:1-15

[91] Adeoye O, Cabral-Marques H. Cyclodextrin nanosystems in oral drug delivery: A mini review. International Journal of Pharmaceutics 2017. DOI: http://dx.doi.org/doi:10.1016/j.ijpharm.2017.04.050

[92] Tejashri G, Amrita B, Darshana J. Cyclodextrin based nanosponges for pharmaceutical use: A review. Acta Pharmaceutica. 2013;**63**(3):335-358

[93] Martín del Valle EM. Cyclodextrins and their uses: A review. Process Biochemistry. 2004;**39**(9):1033-1046

[94] Astray G, Gonzalez-Barreiro C, Mejuto JC, Rial-Otero R, Simal-Gandara J. A review on the use of cyclodextrins in foods. Food Hydrocolloids. 2009;**23**(7):1631-1640

[95] Fenyvesi É, Vikmon MA, Szente L. Cyclodextrins in food technology and human nutrition: Benefits and limitations. Critical Reviews in Food Science and Nutrition. 2016;**56**:1981-2004

[96] Zipp H. Cyclodextrins – Novel solutions for the food industry. Food Ingredients Brasil. 2012;**22**:56

[97] López-Nicolás JM, Rodríguez-Bonilla P, García-Carmona F. Cyclodextrins and antioxidants. Critical Reviews in Food Science and Nutrition. 2014;**54**(2):251-276

[98] Centini M, Maggiore M, Casolaro M, Andreassi M, Maffei Facino R, Anselmi C. Cyclodextrins as cosmetic delivery systems. Journal of Inclusion Phenomena and Macrocyclic Chemistry. 2007;**57**(1-4):109-112

[99] Buschmann H-J, Schollmeyer E. Applications of cyclodextrins in cosmetic products: A review. Journal of Cosmetic Science. 2002;**53**(3):185-191

[100] Voncina B, Vivod V. Cyclodextrins in textile nishing. In: Günay M, editor. Eco- Friendly Textile Dyeing and Finishing. Chapter 3. Rijeka: InTech; 2013. pp. 53-75. ISBN: 978-953-51-0892-4

[101] Voncina B. Application of cyclodextrins in textile dyeing. In: Hauser PJ, editor. Textile Dyeing. Chapter 17. Rijeka: InTech; 2013. pp. 373-392

[102] Venturini CDG, Nicolini J, Machado C, Machado VG. Properties and recent applications of cyclodextrins. Quimica Nova. 2008;**31**(2):360-368

[103] Mosher G, Thompson DO. Complexation: Cyclodextrins. In: Encyclopedia of Pharmaceutical Technology. New York: Marcel Dekker, Inc.; 2007. pp. 671-696

[104] D'Souza VT, Lipkowitz KB. Cyclodextrins: Introduction. Chemical Reviews. 1998;**98**(5):1741-1742

[105] Szejtli J. Introduction and general overview of cyclodextrin chemistry. Chemical Reviews. 1998;**98**(97):1743-1753

[106] Szejtli J. The properties and potential uses of cyclodextrin derivatives. Journal of Inclusion Phenomena and Molecular Recognition in Chemistry. 1992;**14**(1):25-36

[107] Bender ML, Komiyama M. Cyclodextrin chemistry. In: Hafner K, Rees CW, Trost BM, Lehn J-M, von Rague Schleyer P, Zahradnik R, editors. Reactivity and Structure Concepts in Organic Chemistry. Vol. 6. Berlin Heidelberg: Springer-Verlag; 1978. p. 269. ISBN-13: 978-3-642-66844-9. Available from: http://link.springer.com/10.1007/978-3-642-66842-5

[108] Matencio A, García-Carmona F, López-Nicolás JM. The inclusion complex of oxyresveratrol in modified cyclodextrins: A thermodynamic, structural, physicochemical, fluorescent and computational study. Food Chemistry. 2017;**232**:177-184

[109] Martínez G, Gómez MA. Ciclodextrinas: Complejos de inclusión con polímeros. Revista Iberoamericana de Polímeros. 2007;**8**(4):300-312

[110] Miller LA, Carrier RL, Ahmed I. Practical considerations in development of solid dosage forms that contain cyclodextrin. Journal of Pharmaceutical Sciences. 2007;**96**(7):1691-1707

[111] Baker GA, Crane NJ, Mayrhofer RC, Betts TA. Cyclodextrin inclusion complexes with a solvatochromic fluorescent probe. Journal of Chemical Education. 2002;**79**(10):1261

[112] Valero M, Rodriguez LJ, Velázquez MM. Determination of thermodynamic parameters of the cyclodextrin inclusion processes: An undergraduate physical chemistry lab experiment. Journal of Chemical Education. 1999;**76**(3):418

[113] Rekharsky MV, Inoue Y. Complexation thermodynamics of cyclodextrins. Chemical Reviews. 1998;**98**(5):1875-1918

[114] Diaz D, Vargas-Baca I, Gracia-Mora J. Beta-cyclodextrin inclusion complexes with iodine: An advanced and inexpensive undergraduate chemistry experiment. Journal of Chemical Education. 1994;**71**(8):708

[115] Davies JED, Kemula W, Powell HM, Smith NO. Inclusion compounds – past, present, and future. Journal of Inclusion Phenomena. 1983;**1**(1):3-44

[116] Prochowicz D, Kornowicz A, Justyniak I, Lewiński J. Metal complexes based on native cyclodextrins: Synthesis and structural diversity. Coordination Chemistry Reviews. 2016;**306**:331-345

[117] Hong SB, Liu MY, Zhang W, Deng W. Organic reactions catalyzed by cyclodextrin and its derivatives. Chinese Journal of Organic Chemistry. 2015;**35**(2):325-336

[118] Macaev F, Boldescu V. Cyclodextrins in asymmetric and stereospecific synthesis. Symmetry-Basel. 2015;**7**(4):1699-1720

[119] Bellia F, La Mendola D, Pedone C, Rizzarelli E, Saviano M, Vecchio G. Selectively functionalized cyclodextrins and their metal complexes. Chemical Society Reviews. 2009;**38**(9):2756-2781

[120] Norkus E. Metal ion complexes with native cyclodextrins: An overview. Journal of Inclusion Phenomena and Macrocyclic Chemistry. 2009;**65**(3):237-248

[121] Hapiot F, Tilloy S, Monflier E. Cyclodextrins as supramolecular hosts for organometallic complexes. Chemical Reviews. 2006;**106**(3):767-781

[122] Li Z, Chen S, Gu Z, Chen J, Wu J. Alpha-cyclodextrin enzymatic production and food applications. Trends in Food Science and Technology. 2014;**35**(2):151-160

[123] Martina K, Binello A, Lawson D, Jicsinszky L, Cravotto G. Recent applications of cyclodextrins as food additives and in food processing. Current Nutrition & Food Science. 2013;**9**(3):167-179

[124] Astray G, Mejuto JC, Morales J, Rial-Otero R, Simal-Gándara J. Factors controlling flavors binding constants to cyclodextrins and their applications in foods. Food Research International. 2010;**43**(4):1212-1218

[125] Moreira Da-Silva A. Cyclodextrins as food additives and ingredients: Nutraceutical applications. In: Conference Paper. May, 2009. Available from: https://www.researchgate.net/publication/246546437_CYCLODEXTRINS_AS_FOOD_ADDITIVES_AND_INGREDIENTS_NUTRACEUTICAL_APPLICATIONS

[126] Szente L, Szejtli J. Cyclodextrins as food ingredients. Trends in Food Science and Technology. 2004;**15**(3-4):137-142

[127] Yoshii H. Application to food processing of cyclodextrin. Nippon Shokuhin Kagaku Kogaku Kaishi. 2004;**51**(12):647-655

[128] Loftsson T, Brewster ME. Pharmaceutical applications of cyclodextrins: Basic science and product development. The Journal of Pharmacy and Pharmacology. 2010; **62**(11):1607-1621

[129] Loftsson T, Duchêne D. Cyclodextrins and their pharmaceutical applications. International Journal of Pharmaceutics. 2007;**329**(1-2):1-11

[130] Davis ME, Brewster ME. Cyclodextrin-based pharmaceutics: Past, present and future. Nature Reviews. Drug Discovery. 2004;**3**:1023-1035

[131] Frömming K-H, Szejtli J. Cyclodextrins in pharmacy. In: Davies JED, editor. Topics in Inclusion Science. Vol. 5. Dordrecht-Boston-London: Kluwer Academic Publishers; 1994. p. 233

[132] Antoniuk I, Amiel C. Cyclodextrin-mediated hierarchical self-assembly and its potential in drug delivery applications. Journal of Pharmaceutical Sciences. 2016;**105**(9):2570-2588

[133] Zheng Y, Wyman IW. Supramolecular nanostructures based on cyclodextrin and poly(ethylene oxide): Syntheses, structural characterizations and applications for drug delivery. Polymer. 2016;**8**(5):198

[134] Arima H, Hayashi Y, Higashi T, Motoyama K. Recent advances in cyclodextrin delivery techniques. Expert Opinion on Drug Delivery. 2015;**12**(9):1425-1441

[135] Bonnet V, Gervaise C, Djedaini-Pilard F, Furlan A, Sarazin C. Cyclodextrin nanoassemblies: A promising tool for drug delivery. Drug Discovery Today. 2015;**20**(9):1120-1126

[136] Nascimento MHM, Akkari ACS, Mariano KCF, Braz ASK, Lombello CB, de Araujo DR. Cyclodextrin-based delivery systems for arthritic diseases: From development to experimental therapeutics. Current Pharmaceutical Design. 2015;**21**(33):4907-4916

[137] Lakkakula JR, Krause RWM. A vision for cyclodextrin nanoparticles in drug delivery systems and pharmaceutical applications. Nanomedicine (London, England). 2014;**9**(6):877-894

[138] Zafar N, Fessi H, Elaissari A. Cyclodextrin containing biodegradable particles: From preparation to drug delivery applications. International Journal of Pharmaceutics. 2014;**461**(1-2):351-366

[139] Zhang J, Ma PX. Cyclodextrin-based supramolecular systems for drug delivery: Recent progress and future perspective. Advanced Drug Delivery Reviews. 2013;**65**(9):1215-1233

[140] Laza-Knoerr AL, Gref R, Couvreur P. Cyclodextrins for drug delivery. Journal of Drug Targeting. 2010;**18**(9):645-656

[141] Li J, Loh XJ. Cyclodextrin-based supramolecular architectures: Syntheses, structures, and applications for drug and gene delivery. Advanced Drug Delivery Reviews. 2008;**60**(9):1000-1017

[142] Vyas A, Saraf S, Saraf S. Cyclodextrin based novel drug delivery systems. Journal of Inclusion Phenomena and Macrocyclic Chemistry. 2008;**62**(1-2):23-42

[143] Carrier RL, Miller LA, Ahmed I. The utility of cyclodextrins for enhancing oral bioavailability. Journal of Controlled Release. 2007;**123**(2):78-99

[144] Loftsson T, Vogensen SB, Brewster ME, Konradsdottir F. Effects of cyclodextrins on drug delivery through biological membranes. Journal of Pharmaceutical Sciences. 2007;**96**(10):2532-2546

[145] Loftsson T, Jarho P, Másson M, Järvinen T. Cyclodextrins in drug delivery. Expert Opinion on Drug Delivery. 2005;**2**(2):335-351

[146] Arima H, Motoyama K, Higashi T. Potential therapeutic application of dendrimer/cyclodextrin conjugates with targeting ligands as advanced carriers for gene and oligonucleotide drugs. Therapeutic Delivery. 2017;**8**(4):215-232

[147] Gonzalez-Gaitano G, Isasi JR, Velaz I, Zornoza A. Drug carrier systems based on cyclodextrin supramolecular assemblies and polymers: Present and perspectives. Current Pharmaceutical Design. 2017;**23**(3):411-432

[148] Duchene D, Cavalli R, Gref R. Cyclodextrin-based polymeric nanoparticles as efficient carriers for anticancer drugs. Current Pharmaceutical Biotechnology. 2016;**17**(3):248-255

[149] Uekama K, Hirayama F, Irie T. Cyclodextrin drug carrier systems. Chemical Reviews. 1998;**98**(81):2045-2076

[150] Loftsson T, Brewster ME. Pharmaceutical applications of cyclodextrins: Effects on drug permeation through biological membranes. The Journal of Pharmacy and Pharmacology. 2011;**63**(9):1119-1135

[151] Oliveri V, Vecchio G. Cyclodextrins as protective agents of protein aggregation: An overview. Chemistry – An Asian Journal. 2016;**11**(11):1648-1657

[152] Martínez Á, Ortiz Mellet C, García Fernández JM. Cyclodextrin-based multivalent glycodisplays: Covalent and supramolecular conjugates to assess carbohydrate-protein interactions. Chemical Society Reviews. 2013;**42**(11):4746-4773

[153] Varca GHC, Andréo-Filho N, Lopes PS, Ferraz HG. Cyclodextrins: An overview of the complexation of pharmaceutical proteins. Current Protein & Peptide Science. 2010;**11**(4):255-263

[154] Gharib R, Greige-Gerges H, Fourmentin S, Charcosset C, Auezova L. Liposomes incorporating cyclodextrin-drug inclusion complexes: Current state of knowledge. Carbohydrate Polymers. 2015;**129**:175-186

[155] European Medicines Agency. Background Review for Cyclodextrins Used as Excipients – In the Context of the Revision of the Guideline on 'Excipients in the Label and Package Leaflet of Medicinal Products for Human Use' (CPMP/463/00 Rev. 1). 2014. pp. 1-17. Available from: www.ema.europa.eu/contact

[156] Loftsson T, Brewster ME. Cyclodextrins as functional excipients: Methods to enhance complexation efficiency. Journal of Pharmaceutical Sciences. 2012;**101**(9):3019-3032

[157] Akasha AA, Elwahedi MA, Eldeeb AM. Cyclodextrins and their pharmaceutical applications. PharmaTutor Magazine. 2014;**2**(7):40-46

[158] Loftsson T, Masson M. Cyclodextrins in topical drug formulations: Theory and practice. International Journal of Pharmaceutics. 2001;**225**(1-2):15-30

[159] Benkovics G, Malanga M, Fenyvesi É. The 'Visualized' macrocycles: Chemistry and application of fluorophore tagged cyclodextrins. International Journal of Pharmaceutics. 2017;**531**(2):689-700

[160] di Cagno M, Pio M. The potential of cyclodextrins as novel active pharmaceutical ingredients: A short overview. Molecules. 2016;**22**(1):1-14

[161] Shimpi S, Chauhan B, Shimpi P. Cyclodextrins: Application in different routes of drug administration. Acta Pharmaceutica. 2005;**55**(2):139-156

[162] Irie T, Uekama K. Pharmaceutical applications of cyclodextrins. III. Toxical issues and safety evaluation. Journal of Pharmaceutical Sciences. 1997;**86**(2):147-162

[163] Leclercq L. Interactions between cyclodextrins and cellular components: Towards greener medical applications? Beilstein Journal of Organic Chemistry. 2016;**12**:2644-2662

[164] Lima PSS, Lucchese AM, Araujo HG, Menezes PP, Araujo AAS, Quintans LJ, Quintans JSS. Inclusion of terpenes in cyclodextrins: Preparation, characterization and pharmacological approaches. Carbohydrate Polymers. 2016;**151**:965-987

[165] Takazawa T, Mitsuhata H, Mertes PM. Sugammadex and rocuronium-induced anaphylaxis. Journal of Anesthesia. 2016;**30**(2):290-297

[166] Won YJ, Lim BG, Lee DK, Kim H, Kong MH, Lee IO. Sugammadex for reversal of rocuronium-induced neuromuscular blockade in pediatric patients: A systematic review and meta-analysis. Medicine. 2016;**95**(34):4678

[167] de Oliveira MGB, Guimaraes AG, Araujo AAS, Quintans JSS, Santos MRV, Quintans LJ. Cyclodextrins: Improving the therapeutic response of analgesic drugs: A patent review. Expert Opinion on Therapeutic Patents. 2015;**25**(8):897-907

[168] Vecsernyés M, Fenyvesi F, Bácskay I, Deli MA, Szente L, Fenyvesi É. Cyclodextrins, blood-brain barrier, and treatment of neurological diseases. Archives of Medical Research. 2014;**45**(8):711-729

[169] Luke DR, Tomaszewski K, Damle B, Schlamm HT. Review of the basic and clinical pharmacology of sulfobutylether-beta-cyclodextrin (SBECD). Journal of Pharmaceutical Sciences. 2010;**99**(8):3291-3301

[170] Welliver M. Cyclodextrin introduction to anesthesia practice: Form, function, and application. AANA Journal Course. 2007;**75**(4):289-296

[171] Abdolmaleki A, Ghasemi F, Ghasemi JB. Computer-aided drug design to explore cyclodextrin therapeutics and biomedical applications. Chemical Biology and Drug Design. 2017;**89**(2):257-268

[172] Feng RL, Deng PZ, Teng FF, Song ZM. Recent development of copolymeric delivery system for anticancer agents based on cyclodextrin derivatives. Anti-Cancer Agents in Medicinal Chemistry. 2016;**16**(3):299-308

[173] Radu CD, Parteni O, Ochiuz L. Applications of cyclodextrins in medical textiles – Review. Journal of Controlled Release. 2016;**224**:146-157

[174] Parrot-Lopez H, Perret F, Bertino-Ghera B. Amphiphilic cyclodextrins and their applications. Preparation of nanoparticles based on amphiphilic cyclodextrins for biomedical applications. Annales Pharmaceutiques Françaises. 2010;**68**(1):12-26

[175] Smith DKA. supramolecular approach to medicinal chemistry: Medicine beyond the molecule. Journal of Chemical Education. 2005;**82**(3):393

[176] Szejtli J. Medicinal applications of cyclodextrins. Medicinal Research Reviews. 1994;**14**(3):353-386

[177] Leudjo Taka A, Pillay K, Yangkou Mbianda X. Nanosponge cyclodextrin polyurethanes and their modification with nanomaterials for the removal of pollutants from waste water: A review. Carbohydrate Polymers. 2017;**159**:94-107

[178] Han YQ, Zhou WJ, Shen HM, Liu QP, WY Y, Ji HB, She YB. Progress in the immobilization of beta-cyclodextrin and their application in adsorption of environmental pollutants. Chinese Journal of Organic Chemistry. 2016;**36**(2):248-257

[179] Letort S, Balieu S, Erb W, Gouhier G, Estour F. Interactions of cyclodextrins and their derivatives with toxic organophosphorus compounds. Beilstein Journal of Organic Chemistry. 2016;**12**:204-228

[180] Crini G, Morcellet M. Synthesis and applications of adsorbents containing cyclodextrins. Journal of Separation Science. 2002;**25**(13):789-813

[181] Cravotto G, Caporaso M, Jicsinszky L, Martina K. Enabling technologies and green processes in cyclodextrin chemistry. Beilstein Journal of Organic Chemistry. 2016;**12**(1):278-294

[182] Sharma N, Baldi A. Exploring versatile applications of cyclodextrins: An overview. Drug Delivery. 2016;**23**(3):729-747

[183] Bhaskara-Amrit UR, Agrawal PB, Warmoeskerken MMCG. Applications of beta-cyclodextrins in textiles. Autex Research Journal. 2011;**11**(4):94-101

[184] Bilensoy E. Cyclodextrins in pharmaceutics, cosmetics, and biomedicine: Current and future industrial applications. Hoboken, New Jersey: John Wiley & Sons, Inc.; 2011. ISBN: 9780470926819

[185] Grigoriu A, Popescu O. Applications of cyclodextrines in textiles – A review. Buletinul Institutului Politehnic din Iaşi. 2011;**11**:94-101

[186] Singh M, Sharma R, Banerjee UC. Biotechnological applications of cyclodextrins. Biotechnology Advances. 2002;**20**(5-6):341-359

[187] Arenskötter M, Folmer F, Llewellyn C, Pardo A, Reinecke F, Trebbi G. Cyclodextrins. An overview. Intensive Program Agriculture: Source of Raw Materials for Industry. Germany: Gent. 2001:1-7. Available from: http://www.oocities.org/florecitafolmer/paper04a.htm

[188] Hedges AR. Industrial applications of cyclodextrins. Chemical Reviews. 1998;**98**(5): 2035-2044

[189] Szejtli J. Utilization of cyclodextrins in industrial products and processes. Journal of Materials Chemistry. 1997;**7**(4):575-587

[190] Duchene D, Wouessidjewe D. Industrial uses of cyclodextrins and their derivatives. Journal of Coordination Chemistry. 1992;**27**(1-3):223-236

[191] Saenger W. Cyclodextrin inclusion compounds in research and industry. Angewandte Chemie (International Ed. in English). 1980;**19**(5):344-362

[192] Mejia-Ariza R, Grana-Suarez L, Verboom W, Huskens J. Cyclodextrin-based supramolecular nanoparticles for biomedical applications. Journal of Materials Chemistry. 2017;**5**(1):36-52

[193] Swaminathan S, Cavalli R, Trotta F. Cyclodextrin-based nanosponges: A versatile platform for cancer nanotherapeutics development. Wiley Interdisciplinary Reviews. Nanomedicine and Nanobiotechnology. 2016;**8**(4):579-601

[194] Karoyo AH, Wilson LD. Nano-sized cyclodextrin-based molecularly imprinted polymer adsorbents for perfluorinated compounds – A mini-review. Nanomaterials. 2015;**5**(2):981-1003

[195] Simoes SM, Rey-Rico A, Concheiro A, Alvarez-Lorenzo C. Supramolecular cyclodextrin-based drug nanocarriers. Chemical Communications (Camb). 2015;**51**(29):6275-6289

[196] Kang Y, Guo K, Li BJ, Zhang S. Nanoassemblies driven by cyclodextrin-based inclusion complexation. Chemical Communications (Camb). 2014;**50**(76):11083-11092

[197] Concheiro A, Alvarez-Lorenzo C. Chemically cross-linked and grafted cyclodextrin hydrogels: From nanostructures to drug-eluting medical devices. Advanced Drug Delivery Reviews. 2013;**65**(9):1188-1203

[198] Moya-Ortega MD, Alvarez-Lorenzo C, Concheiro A, Loftsson T. Cyclodextrin-based nanogels for pharmaceutical and biomedical applications. International Journal of Pharmaceutics. 2012;**428**(1-2):152-163

[199] Purkayastha P, Jaffer SS, Ghosh P. Physicochemical perspective of cyclodextrin nano and microaggregates. Physical Chemistry Chemical Physics. 2012;**14**(16):5339-5348

[200] Martín-Banderas L, Holgado MA, Venero JL, Álvarez-Fuentes J, Fernández-Arévalo M. Nanostructures for drug delivery to the brain. Current Medicinal Chemistry. 2011;**18**(34):1-19

[201] Bilensoy E, Hincal AA. Recent advances and future directions in amphiphilic cyclodextrin nanoparticles. Expert Opinion on Drug Delivery. 2009;**6**(11):1161-1173

[202] Dodziuk H. Molecules with holes – Cyclodextrins. In: Dodziuk H, editor. Cyclodextrins and Their Complexes: Chemistry, Analytical Methods, Applications. Weinheim: Wiley-VCH Verlag GmbH & Co., KGaA; 2006. pp. 1-30. ISBN: 3-527-31280-3

[203] Mosinger J, Tomankova V, Nemcova I, Zyka J. Cyclodextrins in analytical chemistry. Analytical Letters. 2001;**34**(12):1979-2004

[204] Armstrong DW. Cyclodextrin in analytical chemistry. In: Huber O, Szejtli J, editors. Proceedings of the Fourth International Symposium on Cyclodextrins. Germany: Springer; 1998. pp. 437-439

[205] Adly FG, Antwi NY, Ghanem A. Cyclodextrin-functionalized monolithic capillary columns: Preparation and chiral applications. Chirality. 2016;**28**(2):97-109

[206] Guo JL, Lin YJ, Xiao Y, Crommen J, Jiang ZJ. Recent developments in cyclodextrin functionalized monolithic columns for the enantioseparation of chiral drugs. Journal of Pharmaceutical and Biomedical Analysis. 2016;**130**:110-125

[207] Al-Othman ZA, Al-Warthan A, Ali I. Advances in enantiomeric resolution on monolithic chiral stationary phases in liquid chromatography and electrochromatography. Journal of Separation Science. 2014;**37**(9-10):1033-1057

[208] Jáč P, Scriba GKE. Recent advances in electrodriven enantioseparations. Journal of Separation Science. 2013;**36**(1):52-74

[209] Mandrioli R, Mercolini L, Raggi MA. Chiral analysis of amphetamines, methadone and metabolites in biological samples by electrodriven methods. Electrophoresis. 2011;**32**(19):2629-2639

[210] Pathak D, Pathak K. Cyclodextrins as chiral discriminators. Indian Drugs. 2008;**45**(10): 765-774

[211] Takahashi AI, Veiga FJB, Ferraz HG. A literature review of cyclodextrin inclusion complexes characterization – Part II: X-ray diffraction, infrared spectroscopy and nuclear magnetic resonance. International Journal of Pharmaceutical Sciences Review and Research. 2012;**12**(1):8-15

[212] Boccio M, Sayago A, Asuero AGA. bilogarithmic method for the spectrophotometric evaluation of stability constants of 1:1 weak complexes from mole ratio data. International Journal of Pharmaceutics. 2006;**318**(1-2):70-77

[213] Schneider H-J, Hacket F, Rüdiger V, Ikeda H. NMR studies of cyclodextrins and cyclodextrin complexes. Chemical Reviews. 1998;**98**(5):1755-1786

[214] Lay S, Ni X, Yu H, Shen S. State-of-the-art applications of cyclodextrins as functional monomers in molecular imprinting techniques: A review. Journal of Separation Science. 2016;**39**(12):2321-2331

[215] Szejtli J. Application of cyclodextrins in the chromatography. Starch-Stärke. 1987;**39**(10): 357-362

[216] Szente L, Szejtli J. Non-chromatographic analytical uses of cyclodextrins. Analyst. 1998;**123**(4):735-741

[217] Liu BW, Zhou H, Zhou ST, Yuan JY. Macromolecules based on recognition between cyclodextrin and guest molecules: Synthesis, properties and functions. European Polymer Journal. 2015;**65**:63-81

[218] Chen Y, Liu Y. Cyclodextrin-based bioactive supramolecular assemblies. Chemical Society Reviews. 2010;**39**(2):495-505

[219] Breslow R. Adjusting the lock and adjusting the key in cyclodextrin chemistry. An introduction in biomimetic chemistry. In: Dolphin D, et al., editors. Advances in Chemistry. Washington, DC: American Chemical Society; 1980. pp. 1-15

[220] Zhao QQ, Zhang WX, Wang RM, Wang YT, Ouyang DF. Research advances in molecular modeling in cyclodextrins. Current Pharmaceutical Design. 2017;**23**(3):522-531

[221] Wenz G, Monflier E. Superstructures with cyclodextrins: Chemistry and applications III. Beilstein Journal of Organic Chemistry. 2016;**12**:937-938

[222] Chui JKW, Fyles TM. Cyclodextrin ion channels. Organic & Biomolecular Chemistry. 2014;**12**:3622-3634

[223] Harada A, Takashima Y, Yamaguchi H. Cyclodextrin-based supramolecular polymers. Chemical Society Reviews. 2009;**38**:875-882

[224] Khan AR, Forgo P, Stine KJ, D'Souza VT. Methods for selective modifications of cyclodextrins. Chemical Reviews. 1998;**98**(5):1977-1996

[225] Lipkowitz KB. Applications of computational chemistry to the study of cyclodextrins. Chemical Reviews. 1998;**98**(5):1829-1874

[226] Nepogodiev SA, Kurkov SV, Loftsson T. Cyclodextrins. International Journal of Pharmaceutics. 2013;**453**(1):167-180

[227] Nepogodiev SA, Stoddart JF. Cyclodextrin-based catenanes and rotaxanes. Chemical Reviews. 1998;**98**(5):1959-1976

[228] Takahashi K. Organic reactions mediated by cyclodextrins. Chemical Reviews. 1998;**98**(5):2013-2034

Chapter 2

Aggregation of Cyclodextrins: Fundamental Issues and Applications

Tânia F.G.G. Cova, Sandra M.A. Cruz,
Artur J.M. Valente, Paulo E. Abreu,
Jorge M.C. Marques and Alberto A.C.C. Pais

Additional information is available at the end of the chapter

http://dx.doi.org/10.5772/intechopen.73532

Abstract

The aggregation of cyclodextrins (CD) in aqueous solution is an old, yet still vastly unexplored topic that has been studied at least since the 1980s. At that time, few authors took into consideration the possibility of formation of aggregates for the interpretation of thermodynamic and thermophysical properties of CDs in aqueous solution. The aggregates appear at quite low CD concentrations and seem to encompass only a small number of CD molecules. They also occur in water in the presence of hydrophobic or amphiphilic moieties, including surfactants, assuming a preassembled state with the hydrophobic chains threading through one or two CDs. After a long period in which it has been neglected, CD aggregation is now a hot topic and one far from gathering consensus. In this chapter, a timely and critical review on the phenomenon of CD aggregation and the respective supramolecular properties, including some computational rationales, will be presented. A comprehensive summary of CD aggregates studied to date, indicating the formation conditions, characterization techniques, and applications, is also provided.

Keywords: cyclodextrins, self-aggregation, molecular dynamics simulations

1. Introduction

The aggregation of amphiphilic molecules in solution, including cyclodextrins (CD), can be considered as a microphase separation between polar and nonpolar phases. This involves both the hydrophobic and hydrophilic moieties of such molecules and induces the formation and growth of the so-called self-assembled structures. This is a spontaneous process in which the system components form ordered aggregates, typically involving an enthalpy gain in solvation, due to hydrogen bond formation, and possibly a gain in the entropy of bulk water (hydrophobic effect) [1–3]. The process is promoted by balanced attractive and repulsive interactions between molecules, or specific moieties, occurring from a less organized state (e.g., a solution or a disordered aggregate) to a final ordered state (e.g., a crystal). These soft interactions are generally weak, comparable to thermal energies, and noncovalent, with typical strengths varying from less than 5 kJ mol^{-1} for van der Waals forces to ca. 120 kJ mol^{-1} for hydrogen bonds [4]. Other relevant factors, such as the reversibility and flexibility, the interaction with the respective environment, and the mobility of the system components, determine the success of the molecular self-assembly process.

Despite a substantial amount of data is available on the thermodynamic aspects of assembly processes, the information on the CD aggregation behavior and detailed mechanisms of formation and dissociation is scarce. In particular, little is known about the cooperative nature of individual contributions of components for the CD aggregation/assembly process. This is clearly a topic requiring fundamental research, with direct implications for improving the supramolecular assembly design and fostering new formulation opportunities in this type of systems.

Table 1 summarizes some relevant physical-chemical parameters of the three most common natural cyclodextrins (α-, β-, and γ-CDs). CDs are able to solubilize, protect, or transport a large variety of compounds, e.g., inorganic nonassociated salts [5, 6], surfactants [7–9], and nonpolar drugs, by forming inclusion and noninclusion complexes. These CD-guest interactions are facilitated by the unique structure of CD, which combines a hydrophilic outer surface with an internal lipophilic cavity. The extensive development of CD-based formulations has resulted in a strongly increasing amount of studies, in which the exceptional features of CDs have been described (see e.g., [10–16]). In the last decades, CDs have also been used for functionalization of organic [17] and inorganic [18, 19] materials and for the synthesis of nanosized and microsized aggregates [20–22]. This process is, in general, driven by hydrophobic interactions and is affected by a number

Property	**α-CD**	**β-CD**	**γ-CD**
N° glucose unites	6	7	8
MW (g mol^{-1})	972	1135	1297
Solubility in water (g L^{-1})	145	18.5	232
Outer diameter cavity, wide end (Å)	13.7	15.3	16.9
Inner diameter cavity, wide end (Å)	5.7	7.8	9.5
Volume cavity ($Å^3$)	174	262	427
No. water molecules* inside cavity	5.8	8.7	14.2

*Assuming a volume of a water molecule equal to 30 $Å^3$.

Table 1. Some physical properties of α-, β- and γ-cyclodextrins [26, 33].

of factors such as the concentration of reactants, temperature, nature of solvent, and the addition of neutral or ionic co-solutes [23–25].

One of the most intriguing properties of CDs is the respective solubility. It should be noted that the solubility in water does not follow a common trend, with β-CD being much less soluble than the other two [26]. Such behavior has attracted researchers for several years, and different explanations have been proposed, based on the rigidity of the CD ring and intramolecular H-bonding or CD self-aggregation [27, 28]. The formation of aggregates can be enhanced upon formation of inclusion complexes, but in general, negligible amounts of aggregates (of ca. 200–300 nm in size) are formed in pure CD solutions. The extent of aggregation may also increase with increasing CD concentration [29].

A wide range of experimental data generated on CD aggregation have been reviewed (see e.g., Refs. [2, 3, 30]). However, the mechanisms of CD aggregation and CD complex aggregation have not yet been fully understood.

CD aggregation was originally recognized based on random observations and deviations from theoretical expectations. Several analytical techniques are now being optimized to specifically detect and characterize CD aggregates. In CD solutions, dynamic light scattering (DLS) [31, 32] and transmission electron microscopy (TEM) [32] have provided reliable evidence of the aggregation phenomenon. The latter has been recognized by several research groups that identified aggregates of different shapes, including spherical and elongated particles, welded fibers, and rods [1, 2, 19, 29].

A significant amount of microscopic data [1] and information from dynamic light scattering (DLS), nuclear magnetic resonance spectroscopy (NMR), and dialysis membrane permeation methods have been reported [1, 3]. However, the full characterization of the CD aggregation behavior by these techniques has been hampered by the poor stability of the aggregates, which makes them sensitive to disruptive side effects [2, 32].

Computational methods, including molecular dynamics (MD) simulations, have been increasingly recognized as a valuable tool for the interpretation of experimental data [34–39]. The simulation of host-guest systems involving CD molecules and aggregates is already offering ways to explore a set of favorable events at different scales. The aggregation behavior and the interaction patterns of CDs, modified with both hydrophobic and hydrophilic groups, have been investigated by atomistic simulations in nonpolar solvents and in water [34, 35]. Although most studies involving amphiphilic CDs have been conducted in aqueous solutions, the use of nonpolar solvents is also important in preparative and characterization procedures, as suggested in Ref. [34]. In the latter, dynamic light scattering observations demonstrated that CDs can form well-defined aggregates in nonpolar solvents (e.g., dichloromethane), displaying a hydrodynamic radius of ca. 80 nm and low polydispersity [34]. Electrostatic, dipolar, and dispersion forces have been recognized as playing a definite role in the formation and stabilization of these aggregates. The formation of even larger structures (such as micelles, vesicles, membranes, and nanoparticles) based on these smaller aggregates and their potential uses in pharmaceutical formulations have also been suggested.

2. General aspects

2.1. Controversial experimental evidence

The aggregation of single CDs and CD-guest complexes, in water, has been described in a number of recent studies [2, 3, 13, 31, 37, 40, 41]. Despite the efforts for understanding the factors that govern inclusion/binding with guest molecules, the precise manner in which CD molecules aggregate and the cooperative effects underlying this phenomenon are much less studied and still far from consensual.

The classical assumption states that inclusion complexes between CDs and guest molecules are always formed in "ideal" solutions, with individual complexes, independent of each other. However, the treatment of the interaction between system components is oversimplified, since CDs can form both inclusion and noninclusion complexes and water-soluble aggregates. In some cases, the aggregation results in the opalescence of aqueous CD solutions. However, the reduced diameter of the aggregates is smaller than the wavelength of visible light, and more often, the formation of clear solutions is promoted.

The first reference to the occurrence of self-aggregation dates back to 1983, when Koichiro and co-workers [42], based on viscosity and activity coefficient data of aqueous solutions of α- and γ-cyclodextrins, suggested the occurrence of dimers or larger aggregates. These authors have also pointed out that such aggregates are acting as "structure making" [43] by tightening water-water hydrogen bonding. Ten years later, Häusler and Müller-Goymann [44] have observed that at concentrations above 50% (w/w), hydroxypropyl (HP)-β-CD self-aggregates, leading to an increase in the solution viscosity. They also found that the addition of chaotropic solutes (e.g., urea and NaCl) tends to decrease the viscosity of solutions as a consequence of CD disaggregation. A similar observation was found for γ-CD (10% m/v) solutions [45], at physiological pH. In both works, the self-aggregation of CD monomers is justified by intermolecular H-bonding interactions promoted by the hydrophilic rims [44, 45].

Coleman et al. [46] extended their studies to β-CD, the less soluble of all native CDs, and found that the addition of water structure-breaking solutes or an increase of the solution pH (at values higher than 12.5) in order to ionize the —OH groups leads to an increase in solubility. This evidence led them to hypothesize that the solubility of CDs is related with interactions between CD aggregates and the water through the formation of two chains of hydrated β-CD molecules forming a rod-like aggregate. These authors also argue that the structure of water has an important role on the aggregation/disaggregation of CDs.

The quantification of size and mass percentage of aggregates was only made possible, in a systematic manner, during the last decade. It has been concluded, by using photon correlation spectroscopy (PCS), that in a 12 mM CD aqueous solution, large polydisperse aggregates of 200–300 nm in size were formed [29]; however, the respective mass percentage is quite small when compared to that of free CD. For example, the mass contribution of α-CD aggregates is ca. 0.8% (0.096 mM) assuming coils or 0.001% (0.12 μM) considering spheres [29]. For other CDs, the mass contribution of aggregates is also residual, i.e., 0.0011% (0.13 μM) for β-CD [47] and 0.02% (0.154 mM) for γ-CD solutions [45]. He et al. [48] found a bimodal distribution in dynamic light scattering (DLS) data of α-, β-, and γ-CD aqueous solutions, with mean hydrodynamic radii of less than 1 nm and higher than 60 nm for the fast and slow components (attributed to monomer and aggregated CDs, respectively). This has been revisited by Puskás et al. [49] by using the same technique. However, contrarily to previous work, only γ-CD aggregates have been confirmed. In other studies [50, 51], the formation of CD aggregates of globular shape, at [β-CD] = 3 mM, with a minimum hydration radius of ca. 90 nm, was reported, based on data from different techniques, including DLS, cryo-TEM, and electron spin resonance (ESR) probe spectroscopy. However, at higher CD concentrations, these particles coexist with other structures as large as a few micrometers. These aggregates tend to increase with CD concentration, suggesting cooperative aggregate-aggregate interactions. In addition, Rao and Geckeler [52] have concluded that the formation of β-CD supramolecules can also occur in aqueous solutions at room temperature. By increasing the stirring time and the concentration from 4 to 10 mM, it was possible to follow the formation of cage-like structures (with an average size of 7.5 nm), after 5 h stirring, evolving to channeled structures (with 260 nm length) after 72 h stirring.

The thermodynamic parameters of β-CD self-assembly have been studied, assuming that CDs behave as colloids [36]. For that, the scattering intensity of a set of β-CD solutions of different concentrations has been measured; by plotting the scattering intensity as a function of β-CD concentration, two different regimes were observed below and above the critical aggregation concentration (cac = 1.6 mM): at c < cac, and only monomers (the occurrence of small aggregates cannot be also neglected) exist in solution. However, at c > cac, the scattering intensity increases linearly with β-CD concentration, indicating the presence of two different set of aggregates of 60.0 nm and 120 nm in size. Using the pseudo-phase model, the free energy of aggregation was calculated to be −15.95 kJ mol^{-1} at 298.15 K. From the dependence of the cac value on the temperature, the variation of the aggregation enthalpy and, consequently, the aggregation entropy (at 298.15 K) were estimated to be −26.48 kJ mol^{-1} and −35.32 J K^{-1} mol^{-1}, respectively. These values for thermodynamic parameters suggest that (a) aggregation is enthalpy driven, and (b) aggregates are formed by noncovalent interactions [53]. Following the same approach, the cac values of α-, β-, and γ-CDs were also measured by using a permeability technique, based on dialysis membranes with molecular weight cut off higher than 2 kDa [3, 54]. Inspecting the flux of CDs as a function of concentration, two different regimes have been found which were attributed to the formation of aggregates. The calculated values for the cac of α-, β-, and γ-CDs are 12.2, 6.1, and 7.2 mM, respectively.

Recently, a thermodynamic study [55] using isodesmic and K2-K self-assembly models was performed on CD derivatives (HP-β-CD, HP-γ-CD, and sulfobutylether(SBE)-β-CD). The isodesmic model assumes that the Gibbs energy and equilibrium constant (K) are equal in each addition of a monomer to an aggregate (for details see Ref. [55]). The K2-K model is a modified version of the former, in which the K value of the first step of the self-assembly is different from those of the remaining steps. The respective cac values and aggregate sizes were also determined, using DLS and ^{1}H NMR and TEM, respectively. These CDs displayed similar cac values of ca. 2% (m/v). Three different groups of particle sizes were also identified based on correlation functions: (i) a group with ca. 1 nm corresponds to a single CD molecule, (ii) a group with size values ranging from ca. 30 nm to 60 nm for HP-γ-CD and from ca. 10–70 nm for SBE-β-CD and from ca. 100 nm to 200 nm for HP-β-CD, and (iii) a small group with larger aggregates attaining 1 μm, for β-CD derivatives, and from ca. 140 nm to 1 μm for HP-γ-CD. It was shown that aggregation results from some cooperative contributions, with the first step of the aggregation being less favorable than the subsequent ones. A thermodynamic analysis indicated that the aggregation process was spontaneous, exothermic, and associated to an entropy loss. The calculated standard free energies range from −7.1 kJ mol^{-1} for SBE-β-CD to −10.6 for HP-γ-CD, and the enthalpy values were −20.6, −27.5, and −46.3 kJ mol^{-1} for HP-β-CD, SBE-β-CD, and HP-γ-CD, respectively [55].

The dynamics of CDs in aqueous solution has been fully assessed using different NMR techniques by Valente et al. [56, 57]. The analysis of ^{1}H NMR self-diffusion of deuterium solutions of CDs shows that, for all three natural CDs, diffusion coefficients depend linearly on the CD volume fraction, suggesting that these molecules are behaving as nonaggregate hard spheres. A similar conclusion was reached from the analysis of the dependence of mutual diffusion coefficients on CDs concentration [58–60].

A more sensitive parameter, related to the volume of the diffusing particle, is the transverse magnetization relaxation time T_2 or the spin–spin relaxation rate R_2 (=$1/T_2$). In a perfectly homogeneous magnetic field, the R_2 relaxation rate can be measured directly from the free induction decay in the time domain or the full width at half height of the resonance in the frequency domain. The dependence of R_2 as a function of CD volume showed that no aggregation occurs. However, the presence of more transient aggregates cannot be excluded for cases in which the lifetime of the aggregate is short compared to the respective tumbling time. The presence of very large aggregates, not visible in the NMR spectra on account of their slow rotational tumbling, cannot also be ruled out. A different but complementary aspect is related to the mechanism of interaction. For further insight, the aggregation of α-, β-, and γ-CDs in aqueous solutions was addressed by focusing on the CD-CD interactions using deuterium relaxation rates (R_1) for deuterium-labeled CDs. In this particular case, the dependence of T_1 (=$1/R_1$) on the CD concentration, for all CDs, was explained by the equilibrium between monomeric and dimeric CDs and, again, no evidence in favor of large aggregates of CDs involving a nonnegligible fraction was found [57].

The formation of aggregates in aqueous solutions containing CDs can be promoted by the presence of guest compounds, which upon inclusion can also contribute to understand and predict the CD aggregation behavior. The structure and nature of the guest molecule can thus affect the CD aggregation process. CD-guest complexes are, most often, simply formed by one guest molecule and one CD molecule. However, ternary complexes are also frequently described [61], where water-soluble polymers [62, 63], metal ions, or organic salts [64] are used to potentiate some CD effect.

The coexistence of inclusion and noninclusion complexes in aqueous solutions containing CDs has been documented and associated with the formation of aggregates based on these complexes. The first evidence for aggregation with complexes involving CDs and lipophilic guests was reported by Mele and co-workers [65] in 1998. Later studies [32], with contradictory or ambiguous results, fostered further investigation on this phenomenon.

The formation of CD-guest-based aggregates in the nanosize range has been confirmed by DLS and TEM analyses and associated to a negative deviation from linearity. This type of aggregation occurred for a CD concentration of ca. 10% (wt/vol) [32]. The self-association of CD complexes can explain the observed decrease in the activity coefficient with increasing CD concentration and the dependence of the complex stoichiometry on the method applied.

Several model compounds [32] have been recently used to investigate the effect of the physicochemical properties of the guest molecules on the CD aggregation behavior. For instance, the impact of a set of esters of para-hydroxybenzoic acid, differing in the side chain length, on HP-β-CD aggregation was recently evaluated resorting to permeation experiments, DLS and MS. The number and size of CD aggregates (<200 nm) increased in the presence of longer guests. However, no clear relation was found between the extent of aggregate formation and the CD concentration.

The premicellar association of inclusion complexes of cationic surfactants and β-CD followed by micellar association of the inclusion complexes has also been suggested based on NMR studies. In addition, micelle-like assemblies with diameters exceeding 200 nm have been observed in aqueous solutions containing trans-β-carotene and β-CD and γ-CD [2].

The structure and size of these CD aggregates are clearly affected by water molecules and hydration shells [2]. Although the ability of CDs to self-assemble to form aggregates is well documented, it has also been shown that the aggregates are very unstable. Attempts to stabilize nanosize self-assembled CD aggregates of the native CDs, and their hydrophilic and monomeric derivatives have not been successful so far.

2.2. Common arrangements in CD aggregates

There are two typical crystal structures for native CDs: cages and channels (see **Figure 1**) [22, 66]. The cage arrangement occurs when CDs are grouped crosswise, displaying a herringbone pattern (**Figure 1a**), or are aligned in adjacent layers leading to a brick-like pattern (**Figure 1b**). In both cases, the formation of inclusion complexes is prevented, as the CD cavities are blocked on both portals by neighboring CDs.

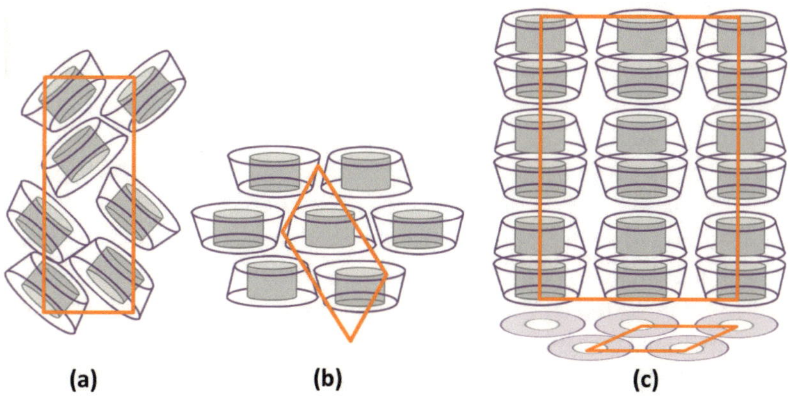

Figure 1. Schematic representation of packing structures of (a) cage-type; (b) layer-type CD; and (c) head-to-tail channel-type CD crystals. Adapted from Ref. [66].

The channel-type assembly is observed if the CDs are stacked in columns so that cavities are aligned to produce channels (**Figure 1c**). While the channel arrangement can be converted into a cage structure by water sorption/desorption cycles, with an intermediated amorphous state, the latter usually results from rapid recrystallization of CDs. After reaching sorption equilibrium, CD molecules undergo a slow rearrangement to the cage structure with defined water content [66]. Similar patterns are generally found in the crystal structures of CD inclusion complexes [66]. When the entire guest molecule is small enough to be included inside a single CD cavity, cage-type structures are formed. On the other hand, a channel-type structure is observed in the presence of long-chain molecules as guests (e.g., polymers). Rusa and co-workers [2, 67] have reported the encapsulation of poly(ε-caprolactone) and poly(L-lactic acid) into α- and γ-CDs. The inclusion of a polymeric structure inside the CD cavity induces the formation of channel-like structures, being those with α-CD more stable due to the increase of hydrophobic CD-polymer interactions. By using an appropriate experimental procedure, the authors were able to produce a solid-state channel packing of CDs containing only water molecules inside the cavities [68]. The same group has also observed the inclusion of poly(vinyl alcohol) (PVA) into CD by taking advantage of the freezing–thawing process for PVA gelation [69]. In this process, the gelation of PVA-containing composites occurs, taking into account two different types of cross-linking: (i) the hydrogel-bonding naturally observed during the freezing-thawing process and (ii) the CD-CD aggregation resulting from channel-type arrangements [69].

2.3. Applications of CD aggregates

As already mentioned in previous sections, a variety of CD-based aggregates can be formed under different conditions (e.g., concentration, solvent medium, and temperature). These include native and modified CDs, inclusion complexes and the respective aggregates and also rotaxanes and polyrotaxanes, nanotubes, and other high-order structures, such as nanospheres and network aggregates [70]. The potential uses of these self-assembled nanomaterials have been explored for advanced applications, ranging from drug solubilization and drug delivery [71], selective binding [72], and controlled adsorption [72]. In pharmaceutical and biomedical fields, it is expected that such applications may include (i) the nanoencapsulation of drugs in the hydrophobic interchain volumes and nanocavities of modified CDs, which can be used as drug carriers or pharmaceutical excipients, (ii) anticancer phototherapy, (iii) gene delivery, and (iv) protection of unstable active components through the formation of inclusion complexes [70]. Several interesting examples of these potential applications have been focused on amphiphilic CDs, which allow to easily modulate both hydrophobic/hydrophilic and self-assembly properties, by grafting different substituents on the portals of native CDs [70, 73]. For instance, supramolecular assemblies based on CD/porphyrin nanoassemblies have been studied *in vitro* [74] as potential nanotherapeutics in A375 human melanoma cells. Other micellar structures and spherical vesicles based on CD-perylene conjugates have been designed to be included in fluorescence sensory and photoresponsive materials, photoinduced electron transfer systems, and organic electronic devices [75]. The self-aggregation of amphiphilic CDs has also been explored for drug delivery applications, as they are able to capture selectively drug molecules, displaying enhanced solubilization capacity [71]. The affinity of amphiphilic CDs for incorporation in model and biological membranes has also been investigated and explored for preparation of functionalized lipid membranes and improved biomimetic systems [73, 76]. A broader range of potential applications of CD aggregates are compiled in recent publications such as Refs. [70, 73] and references therein.

3. Computational observations

Irrespective of the disparate observations, the self-aggregation capability must be affected by the type of CD present in the aqueous solution. While the aggregates of α-CD and γ-CD can be completely removed by standard filtering procedures, the formation of β-CD aggregates (at least as dimers or possibly larger aggregates) persists in solution, displaying fast aggregation kinetics. This suggests that the hydrophilic CD portals play a definite role in the aggregation process [29]. The disruption of hydrogen bond networks by ionization (or functionalization) increases solubility and may suppress aggregation.

The presence of aggregates in solutions containing structure-breaking solutes, in which the solubility of β-CD is enhanced, has provided new insights into this unusual behavior. The low solubility of β-CD (see **Table 1**) has been explained by the presence of aggregates and the respective unfavorable interaction with the hydrogen bond network of bulk water [46]. Note that aspects pertaining to the binding energy in the solid state cannot be disregarded. The relevance of understanding the mechanistic details of the CD aggregation phenomena encompasses either controlling/preventing the formation of aggregates (that preclude the development of specific formulations and the product development) or designing novel formulation strategies.

Computer simulations have been used to rationalize the experimental findings concerning recognition [77], inclusion [36, 77–79], and aggregation [34, 35, 37, 39, 80]. The cooperative binding of at least two CD monomers to a guest molecule has been considered the driving force responsible for self-assembly processes in the construction of CD-based nanoarchitectures [78]. For aggregation processes without using guests, the assembly is usually driven by the hydrophilic portals (in native CDs) and by interactions between substituent chains of CD derivatives with the neighboring cavities of other CDs [78].

The orientational patterns of inter-glucopyranose hydrogen bonds at the secondary portal of β-CD and the respective effect on the CD structure and dimer binding/stability in polar and nonpolar solvents have been explored by van der Spoel et al. [77] in the presence of various guest molecules. It was demonstrated, based on MD simulations and free energy calculations, that polar solvents with stronger hydrogen bond accepting abilities can easily disrupt intermolecular hydrogen bonds, resulting in less stable dimers. Also, the guest models included in the channel-type cavity increase the binding affinity between CD monomers, particularly in polar solvents [77]. Using a similar computational approach, the authors have explored the effect of three different dimerization modes of β-CD molecules and the presence of isoflavone drug analogues in the construction of CD-based nanostructured materials. It was demonstrated that the cooperative binding of CD cavities to guest molecules favors the dimerization process and, consequently, the overall stability and assembly of the CD nanostructures. It was also proved that the desolvation of CD dimers and entropy changes upon complexation cooperatively contributes to the binding process [78].

Another study [37] focused on the spontaneous adsorption of native CDs and the respective aggregates and the related dependence on temperature. It was found that the adsorption of both individual CDs and small CD aggregates (ca. 20 molecules) to the solution/air interface is negligible. The solute-solute interactions were significantly larger for β-CD than for α-CD at 298 K, and the dependence of these interactions on temperature was more relevant for the smaller CD, which displayed a more favorable aggregation at 283 K than at 298 K. The dynamic exchange of hydrogen bonds between the CD hydroxyl groups and the neighboring water molecules indicated a much larger occupancy for individual intramolecular H-bonds in β-CD.

In what follows, the CD-CD interactions [57], for deuterium labeled CDs, in aqueous solutions are further explored by atomistic simulations. Two types of systems are defined, one in which the β-CD is free in water and three others in which two β-CDs are present and may form dimers. In what concerns the latter, these include initial arrangements with proximity of one primary portal and one secondary portal (PS), two primary portals (PP), and two secondary portals (SS), as shown in **Figure 2**.

The molecular dynamics simulations were performed with Gromacs (version 4.6.5), using the all-atom amber99sb [81] force field and the TIP3P water model. The initial coordinates of the β-CD were extracted from the RCSB protein data bank (PDB code: 1DMB), and partial charges were generated using the R.E.D.D. Server [82].

In each system, the molecules were accommodated in a cubic box (7.5 nm edge-length) containing approximately 13,000 explicit TIP3P water molecules. To obtain a starting configuration, each system was firstly subjected to an energy minimization step. All the calculations were carried out in NPT ensemble with periodic boundary conditions at a constant temperature of 300 K and a pressure coupling of 1.0 bar, respectively, to V-rescale and Berendsen external baths. A standard time step of 2 fs was used for both equilibration and production runs. A cut-off of 0.9 nm was used for calculating the Lennard-Jones interactions. Electrostatic interactions were evaluated using the particle mesh Ewald method [83]. Constraints were applied for bond lengths with the LINCS algorithm [84].

Equilibrium properties, structure, and dynamics of β-CD systems were calculated for the simulation runs of 50 ns after the systems were equilibrated for 2 ns. Geometric clustering was performed to identify dominant CD-CD structures, sampled during the MD simulations. The algorithm for cluster analysis is based on the hierarchical (top-down) approach [85] and

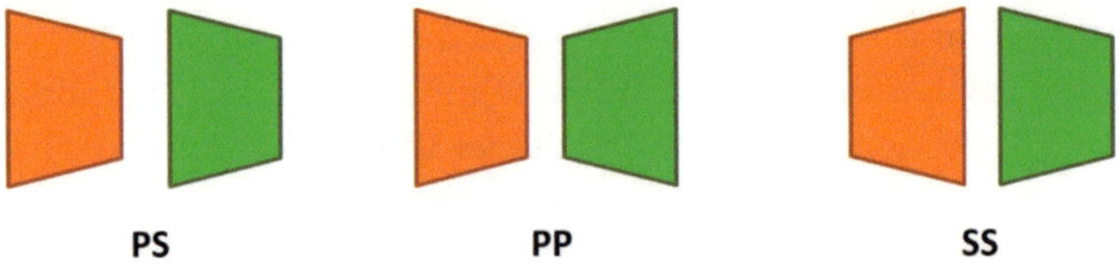

Figure 2. Schematic illustration of the pairwise initial arrangements of β-CDs, with facing primary and secondary portals (PS), two primary portals (PP), and two secondary portals (SS), with the colors used to distinguish β-CD molecules.

allows evaluating the conformational prevalence of each CD-CD structure, by determining dominant clusters based on the root mean square deviation of the atom positions between all pairs of structures. For each CD-CD arrangement, the number of neighboring structures is calculated for RMSD values of 0.35 nm. **Figure 3** presents the behavior of CD-CD structures in different simulation runs, each corresponding to a different initial arrangement. For each CD backbone, the center of mass (COM) of the oxygen atoms at the secondary portal (S) was defined as the reference point for evaluating aggregated and nonaggregated structures and the possible rotation or tilt of the CD molecules.

For PS and PP as initial arrangements, a significant evolution is observed in the relative positioning of the two molecules. The PS and PP initial arrangements display an almost complete rotation or a tilt of one molecule with respect to the other, leading to most favorable SS and PS arrangements, respectively. The "intermediate" PS arrangement increases the CD-CD interactions through partial inclusion of some P groups in the hydrophobic cavity of the other CD molecule. The initial SS arrangement prevails over the course of the simulation with a typical COM distance of 0.46 nm. In addition to intramolecular hydrogen bonds, the two CD molecules can form additional intermolecular hydrogen bonds, optimizing the CD-CD interaction. PS is an intermediate arrangement between the most (SS) and the least stable (PP) arrangements. The PP arrangement of the CD pair involves weaker interactions between P groups of the two molecules, producing a low prevalence, relatively open aggregate (COM distance of 0.8 nm), suggesting a relatively poor clustering of this dimeric aggregate.

The rotational autocorrelation functions (ACF) corresponding to the motion of each CD, free or mostly in dimer arrangement, were also inspected. Two alternatives were tested for the

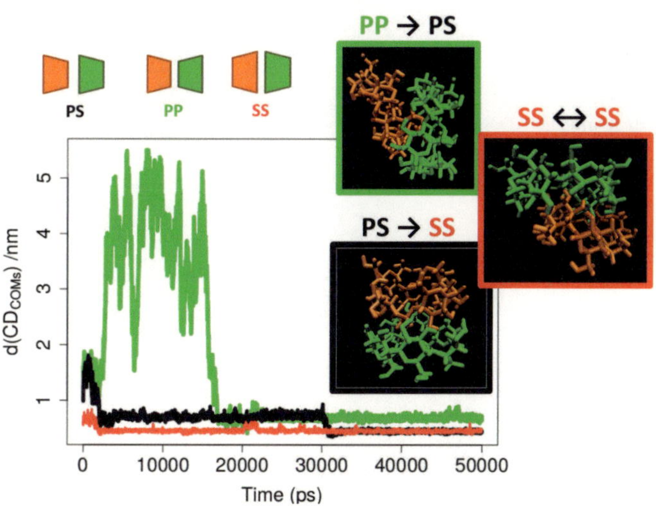

Figure 3. Distribution of distances between the centers of mass of β-CD molecules, defined by the oxygen atoms of the secondary portals. Right panels illustrate the final conformations for the imposed PS, PP, and SS initial arrangements of β-CD molecules, in aqueous solution, sampled during the MD simulations at 300 K and identified by geometric cluster analysis. The color codes for CD molecules are as in **Figure 1**, while the initial arrangements are represented in black, green, and red, for PS, PP, and SS, respectively.

definition of the molecule fixed rotating vector defined by the C2-D (see **Figure 4**) bond (vector) and resulting from the use of three atoms (triplet), those of the C2-D bond and the adjacent carbon C3, and defined as the cross product of vectors C3-C2 and C2-D.

The ACF curves are similar for these two cases and are represented in **Figure 5**. The curves were fitted from 0 up to 500.0 ps to a one-parameter exponential and reflect the slower decay for the dimer situation, $\tau_0 = 1250$ ps, much larger than for the free CD, $\tau_0 = 448$ ps. These values are of order of magnitude of those experimentally obtained.

The organization of CD molecules in aggregates when in the presence of guest-entities clearly deserves further attention. As an example of the behavior found for more complex systems, simulations were performed to study the inclusion complex of β-CD and poly(vinyl alcohol) (PVA) molecules in water. The importance of this polymer is related to the ability to form hydrogels exhibiting a high degree of swelling in water that has demonstrated a great potential to act as a matrix for many applications, including drug delivery [86], wound dressing [87], and sensors [88]. More recently, it has been found that such broad applications of PVA result from its ability to behave as an amphiphilic polymer [89]. This latter feature is relevant for studying the ability of CDs for forming host-guest or aggregate complexes. In this example, three molecular dynamics simulations were performed in systems containing two β-CD

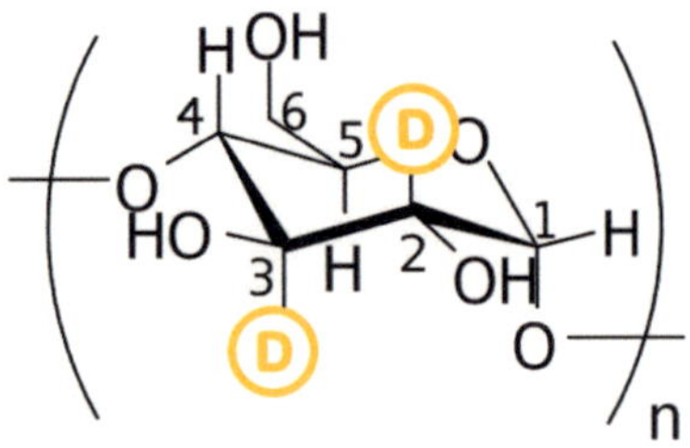

Figure 4. Cyclodextrin structures (α-CD, n = 6; β-CD, n = 7; and γ-CD, n = 8).

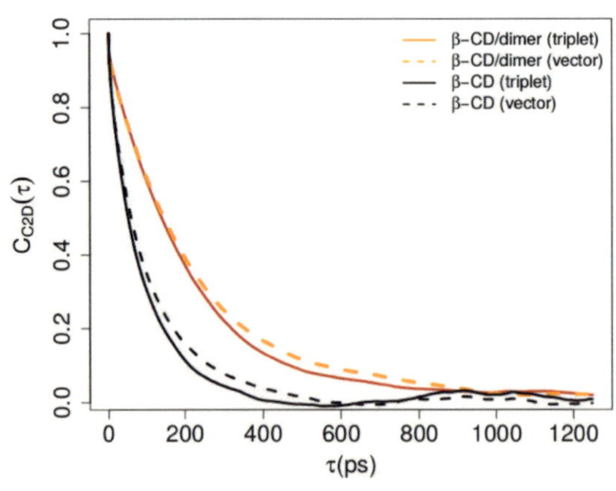

Figure 5. Average autocorrelation function of C2-D bond for CD molecules in the monomeric and dimeric states, in water at 300 K (black and orange curves, respectively).

(denoted as A and B) molecules with 1 PVA, 2 PVA (A and B), and 10 PVA (A to J) oligomers, respectively. The main results are illustrated in **Figure 6**.

CDs are able to form aggregates at early stages of the simulations, and PVA seems to promote the formation of CD dimers (Dm_{AB}). Indeed, PVA contains both hydrophilic and hydrophobic groups that may interact either with the outside part of the CDs or form inclusion complexes; the latter is shown by the snapshot in **Figure 6(b)**, while the former appears for simulations of

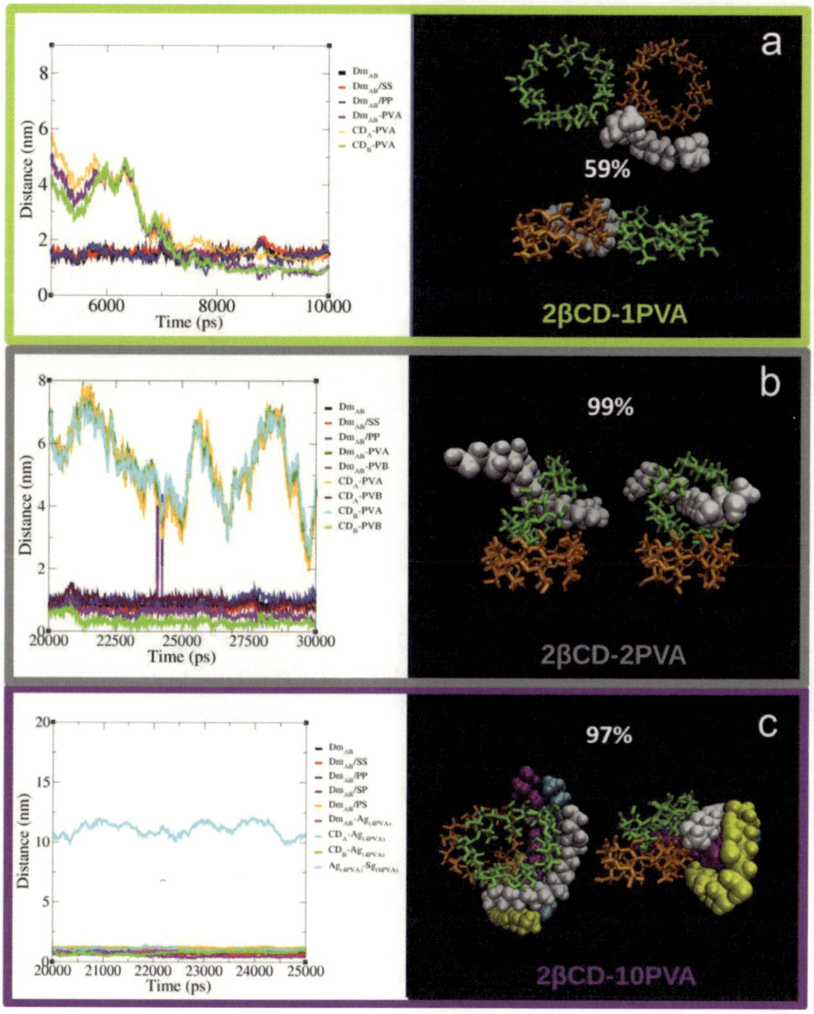

Figure 6. Summary of the MD simulations between β-CD and PVA molecules. Left panels present the measured distances between the center of mass of the groups formed by (i) each β-CD in the dimer (Dm_{AB}), (ii) the secondary-secondary portals (SS), the primary-primary portals (Dm_{AB}/PP), the secondary-primary portals (PS), Dm_{AB}-PVA, Dm_{AB}-PVB, CD_A-PVA, CD_A-PVB, CD_B-PVA, CD_B-PVB, Dm_{AB}-PVA aggregate ($Ag_{(4PVA)}$), CD_A-$Ag_{(4PVA)}$, CD_B-$Ag_{(4PVA)}$, and $Ag_{(4PVA)}$—6 separated PVA molecules ($Sg_{(6PVA)}$). Right panels illustrate the 3 dominant conformations (accounting for more than 50%, panel a, and 90%, panels b and c, of the occurrences) sampled during the equilibrated parts of the production runs, at 300 K, in aqueous solution. The force field F-85 [90] was used for the CDs. The electrostatic charges of PVA were calculated using the R.E.D. server and the remaining terms generated by ACPYPE [91].

2 CDs with 1 PVA (**Figure 6(a)**) and 2 CDs with 10 PVAs (**Figure 6(c)**). It is also noted that the prevalence of the type of configurations in **Figure 6(b)** and **Figure 6(c)** across the simulation is very high (above 95%), which is a strong indication that such complexes are stable; for 2 CDs with 1 PVA, even though the prevalence of the typical configuration represented in **Figure 6(a)** is above 50%, it is much less than the other two cases. Nonetheless, it is apparent from the distance versus time plot in **Figure 6(b)** that the PVA is able to exit the CD pocket in some instances and, then, reform the inclusion complex. In turn, it is particularly interesting to notice from the simulation of 2 CDs with 10 PVAs that the concomitant aggregation of the CDs with various molecules of PVA appears to be relatively stable complex and likely to delay or even prevent the formation of inclusion complexes (not observed during the course of the simulation).

4. Concluding remarks

CD aggregation is still a fertile ground for research, both from the experimental and computational points of view. It was seen that conflicting evidences are provided experimentally, calling for a comprehensive explanation of the involved phenomena, and that detailed, interaction-based, mechanistic rationales from computation are still much needed. It is expected that, in the next years, this subject will benefit from substantial efforts on both grounds.

Acknowledgements

The Coimbra Chemistry Centre is supported by the Fundação para a Ciência e a Tecnologia (FCT), Portuguese Agency for Scientific Research, through the Projects PEst-OE/QUI/UI0313/2014 and POCI-01-0145-FEDER-007630. Sandra M.A. Cruz thanks the postdoc grant from Coimbra Chemistry Centre (Contract DPDRH-56-15-401). Tânia F.G.G. Cova also acknowledge the PhD research Grant SFRH/BD/95459/2013 assigned by FCT.

Author details

Tânia F.G.G. Cova, Sandra M.A. Cruz, Artur J.M. Valente*, Paulo E. Abreu,
Jorge M.C. Marques and Alberto A.C.C. Pais

*Address all correspondence to: avalente@ci.uc.pt

CQC, Department of Chemistry, University of Coimbra, Coimbra, Portugal

References

[1] Kurkov SV, Loftsson T. Cyclodextrins. International Journal of Pharmaceutics. 2013;**453**: 167-180

[2] Ryzhakov A, Do Thi T, Stappaerts J, Bertoletti L, Kimpe K, Sá Couto AR, Saokham P, Van den Mooter G, Augustijns P, Somsen GW, Kurkov S, Inghelbrecht S, Arien A, Jimidar MI, Schrijnemakers K, Loftsson T. Self-assembly of cyclodextrins and their complexes in aqueous solutions. Journal of Pharmaceutical Sciences. 2016;**105**:2556-2569

[3] Saokham P, Sá Couto A, Ryzhakov A, Loftsson T. The self-assemble of natural cyclodextrins in aqueous solutions: Application of miniature permeation studies for critical aggregation concentration (cac) determinations. International Journal of Pharmaceutics. 2016;**505**:187-193

[4] Whitesides GM, Boncheva M. Beyond molecules: Self-assembly of mesoscopic and macroscopic components. Proceedings of the National Academy of Sciences. 2002;**99**:4769-4774

[5] Assaf KI, Ural MS, Pan F, Georgiev T, Simova S, Rissanen K, Gabel D, Nau WM. Water structure recovery in chaotropic anion recognition: High-affinity binding of dodecaborate clusters to γ-cyclodextrin. Angewandte Chemie International Edition. 2015;**54**:6852-6856

[6] Terekhova I, Chibunova E, Kumeev R, Kruchinin S, Fedotova M, Kozbiał M, Wszelaka-Rylik M, Gierycz P. Specific and nonspecific effects of biologically active inorganic salts on inclusion complex formation of cyclodextrins with aromatic carboxylic acids. Chemical Engineering Science. 2015;**122**:97-103

[7] Valente AJM, Söderman O. The formation of host–guest complexes between surfactants and cyclodextrins. Advances in Colloid and Interface Science. 2014;**205**:156-176

[8] Martín VI, Ostos FJ, Angulo M, Márquez AM, López-Cornejo P, López-López M, Carmona AT, Moyá ML. Host-guest interactions between cyclodextrins and surfactants with functional groups at the end of the hydrophobic tail. Journal of Colloid and Interface Science. 2017;**491**:336-348

[9] Shang C, Chen Y, Zhang C, Deng Z, Wu Z. Destruction of chitosan capsules based on host-guest interaction and controllable release of encapsulated dyes. Journal of Applied Polymer Science. 2017;**134**:45229

[10] Venuti V, Rossi B, Mele A, Melone L, Punta C, Majolino D, Masciovecchio C, Caldera F, Trotta F. Tuning structural parameters for the optimization of drug delivery performance of cyclodextrin-based nanosponges. Expert Opinion on Drug Delivery. 2017;**14**:331-340

[11] Liao R, Lv P, Wang Q, Zheng J, Feng B, Yang B. Cyclodextrin-based biological stimuli-responsive carriers for smart and precision medicine. Biomaterials Science. 2017; **5**:1736-1745

[12] Abdolmaleki A, Ghasemi F, Ghasemi JB. Computer-aided drug design to explore cyclodextrin therapeutics and biomedical applications. Chemical Biology & Drug Design. 2017; **89**:257-268

[13] Alsbaiee A, Smith BJ, Xiao L, Ling Y, Helbling DE, Dichtel WR. Rapid removal of organic micropollutants from water by a porous β-cyclodextrin polymer. Nature. 2016;**529**:190-194

[14] Peng L, Liu S, Feng A, Yuan J. Polymeric nanocarriers based on cyclodextrins for drug delivery: Host–guest interaction as stimuli responsive linker. Molecular Pharmaceutics. 2017;**14**:2475-2486

[15] Schmidt BVKJ, Barner-Kowollik C. Dynamic macromolecular material design—The versatility of cyclodextrin-based host–guest chemistry. Angewandte Chemie International Edition. 2017;**56**:8350-8369

[16] Harada A, Kobayashi R, Takashima Y, Hashidzume A, Yamaguchi H. Macroscopic self-assembly through molecular recognition. Nature Chemistry. 2011;**3**:34-37

[17] Medronho B, Andrade R, Vivod V, Ostlund A, Miguel MG, Lindman B, Voncina B, Valente AJM. Cyclodextrin-grafted cellulose: Physico-chemical characterization. Carbohydrate Polymers. 2013;**93**:324-330

[18] Krings JA, Vonhoren B, Tegeder P, Siozios V, Peterlechner M, Ravoo BJ. Light-responsive aggregation of [small beta]-cyclodextrin covered silica nanoparticles. Journal of Materials Chemistry A. 2014;**2**:9587-9593

[19] Polarz S, Smarsly B, Bronstein L, Antonietti M. From cyclodextrin assemblies to porous materials by silica Templating. Angewandte Chemie International Edition. 2001;**40**: 4417-4421

[20] Loftsson T. Self-assembled cyclodextrin nanoparticles and drug delivery. Journal of Inclusion Phenomena and Macrocyclic Chemistry. 2014;**80**:1-7

[21] Messner M, Kurkov SV, Jansook P, Loftsson T. Self-assembled cyclodextrin aggregates and nanoparticles. International Journal of Pharmaceutics. 2010;**387**:199-208

[22] He Y, Fu P, Shen X, Gao H. Cyclodextrin-based aggregates and characterization by microscopy. Micron. 2008;**39**:495-516

[23] Lo Nostro P, Lopes JR, Ninham BW, Baglioni P. Effect of cations and anions on the formation of polypseudorotaxanes. The Journal of Physical Chemistry B. 2002;**106**:2166-2174

[24] Becheri A, Lo Nostro P, Ninham BW, Baglioni P. The curious world of polypseudorotaxanes: Cyclodextrins as probes of water structure. The Journal of Physical Chemistry B. 2003;**107**:3979-3987

[25] De Sousa FB, Lima AC, Denadai AML, Anconi CPA, De Almeida WB, Novato WTG, Dos Santos HF, Drum CL, Langer R, Sinisterra RD. Superstructure based on [small beta]-CD self-assembly induced by a small guest molecule. Physical Chemistry Chemical Physics. 2012;**14**:1934-1944

[26] Sabadini E, Cosgrove T, Egídio FdC. Solubility of cyclomaltooligosaccharides (cyclodextrins) in H_2O and D_2O: A comparative study. Carbohydrate Research. 2006;**341**:270-274

[27] Naidoo KJ, Chen JY-J, Jansson JLM, Widmalm G, Maliniak A. Molecular properties related to the anomalous solubility of β-cyclodextrin. The Journal of Physical Chemistry B. 2004;**108**:4236-4238

[28] Cai W, Sun T, Shao X, Chipot C. Can the anomalous aqueous solubility of [small beta]-cyclodextrin be explained by its hydration free energy alone? Physical Chemistry Chemical Physics. 2008;**10**:3236-3243

[29] González-Gaitano G, Rodríguez P, Isasi JR, Fuentes M, Tardajos G, Sánchez M. The aggregation of cyclodextrins as studied by photon correlation spectroscopy. Journal of Inclusion Phenomena and Macrocyclic Chemistry. 2002;**44**:101-105

[30] Chen G, Jiang M. Cyclodextrin-based inclusion complexation bridging supramolecular chemistry and macromolecular self-assembly. Chemical Society Reviews. 2011;**40**: 2254-2266

[31] Zaman H, Bright AG, Adams K, Goodall DM, Forbes RT. Characterisation of aggregates of cyclodextrin-drug complexes using Taylor dispersion analysis. International Journal of Pharmaceutics. 2017;**522**:98-109

[32] Stappaerts J, Do Thi T, Dominguez-Vega E, Somsen GW, Van den Mooter G, Augustijns P. The impact of guest compounds on cyclodextrin aggregation behavior: A series of structurally related parabens. International Journal of Pharmaceutics. 2017;**529**:442-450

[33] Szejtli J. Introduction and general overview of cyclodextrin chemistry. Chemical Reviews. 1998;**98**:1743-1754

[34] Raffaini G, Mazzaglia A, Ganazzoli F. Aggregation behaviour of amphiphilic cyclodextrins: The nucleation stage by atomistic molecular dynamics simulations. Beilstein Journal of Organic Chemistry. 2015;**11**:2459-2473

[35] Raffaini G, Ganazzoli F, Mazzaglia A. Aggregation behavior of amphiphilic cyclodextrins in a nonpolar solvent: Evidence of large-scale structures by atomistic molecular dynamics simulations and solution studies. Beilstein Journal of Organic Chemistry. 2016; **12**:73-80

[36] Cova TFGG, Nunes SCC, Pais AACC. Free-energy patterns in inclusion complexes: The relevance of non-included moieties in the stability constants. Physical Chemistry Chemical Physics. 2017;**19**:5209-5221

[37] Mixcoha E, Campos-Terán J, Piñeiro Á. Surface adsorption and bulk aggregation of cyclodextrins by computational molecular dynamics simulations as a function of temperature: α-CD vs β-CD. The Journal of Physical Chemistry B. 2014;**118**:6999-7011

[38] López CA, de Vries AH, Marrink SJ. Molecular mechanism of cyclodextrin mediated cholesterol extraction. PLoS Computational Biology. 2011;**7**:e1002020

[39] Zhu X, Wu G, Chen D. Molecular dynamics simulation of cyclodextrin aggregation and extraction of anthracene from non-aqueous liquid phase. Journal of Hazardous Materials. 2016;**320**:169-175

[40] Lucio D, Irache JM, Font M, Martínez-Ohárriz MC. Nanoaggregation of inclusion complexes of glibenclamide with cyclodextrins. International Journal of Pharmaceutics. 2017; **519**:263-271

[41] Ghosh P, Maity A, Das T, Dash J, Purkayastha P. Modulation of small molecule induced architecture of cyclodextrin aggregation by guest structure and host size. The Journal of Physical Chemistry C. 2011;**115**:20970-20977

[42] Koichiro M, Masahiro S, Masayuki N. Viscosity B-coefficients, apparent molar volumes, and activity coefficients for α- and γ-cyclodextrins in aqueous solutions. Bulletin of the Chemical Society of Japan. 1983;**56**:3556-3560

[43] Marcus Y. Effect of ions on the structure of water: Structure making and breaking. Chemical Reviews. 2009;**109**:1346-1370

[44] Häusler O, Müller-Goymann CC. Properties and structure of aqueous solutions of hydroxypropyl-beta-cyclodextrin. Starch—Stärke. 1993;**45**:183-187

[45] Szente L, Szejtli J, Kis GL. Spontaneous opalescence of aqueous γ-cyclodextrin solutions: Complex formation or self-aggregation? Journal of Pharmaceutical Sciences. 1998;**87**:778-781

[46] Coleman AW, Nicolis I, Keller N, Dalbiez JP. Aggregation of cyclodextrins: An explanation of the abnormal solubility ofβ-cyclodextrin. Journal of Inclusion Phenomena and Molecular Recognition in Chemistry. 1992;**13**:139-143

[47] Wu A, Shen X, He Y. Micrometer-sized rodlike structure formed by the secondary assembly of cyclodextrin nanotube. Journal of Colloid and Interface Science. 2006;**302**:87-94

[48] Wu A, Shen X, He Y. Investigation on γ-cyclodextrin nanotube induced by N,N′-diphenylbenzidine molecule. Journal of Colloid and Interface Science. 2006;**297**:525-533

[49] Puskás I, Schrott M, Malanga M, Szente L. Characterization and control of the aggregation behavior of cyclodextrins. Journal of Inclusion Phenomena and Macrocyclic Chemistry. 2013;**75**:269-276

[50] Bonini M, Rossi S, Karlsson G, Almgren M, Lo Nostro P, Baglioni P. Self-assembly of β-cyclodextrin in water. Part 1: Cryo-TEM and dynamic and static light scattering. Langmuir. 2006;**22**:1478-1484

[51] Rossi S, Bonini M, Lo Nostro P, Baglioni P. Self-assembly of β-cyclodextrin in water. 2. Electron spin resonance. Langmuir. 2007;**23**:10959-10967

[52] Rao JP, Geckeler KE. Cyclodextrin supramacromolecules: Unexpected formation in aqueous phase under ambient conditions. Macromolecular Rapid Communications. 2011; **32**:426-430

[53] De M, You C-C, Srivastava S, Rotello VM. Biomimetic interactions of proteins with functionalized nanoparticles: A thermodynamic study. Journal of the American Chemical Society. 2007;**129**:10747-10753

[54] Saokham P, Loftsson T. A new approach for quantitative determination of γ-cyclodextrin in aqueous solutions: Application in aggregate determinations and solubility in hydrocortisone/γ-cyclodextrin inclusion complex. Journal of Pharmaceutical Sciences. 2015; **104**:3925-3933

[55] Do TT, Van Hooghten R, Van den Mooter G. A study of the aggregation of cyclodextrins: Determination of the critical aggregation concentration, size of aggregates and thermodynamics using isodesmic and K2–K models. International Journal of Pharmaceutics. 2017;**521**:318-326

[56] Valente AJM, Carvalho RA, Söderman O. Do cyclodextrins aggregate in water? Insights from NMR experiments. Langmuir. 2015;**31**:6314-6320

[57] Valente AJM, Carvalho RA, Murtinho D, Söderman O. Molecular dynamics of cyclodextrins in water solutions from NMR deuterium relaxation: Implications for cyclodextrin aggregation. Langmuir. 2017;**33**:8233-8238

[58] Ribeiro ACF, Leaist DG, Esteso MA, Lobo VMM, Valente AJM, Santos CIAV, Cabral AMTDPV, Veiga FJB. Binary mutual diffusion coefficients of aqueous solutions of β-cyclodextrin at temperatures from 298.15 to 312.15 K. Journal of Chemical & Engineering Data. 2006;**51**:1368-1371

[59] Ribeiro ACF, Santos CIAV, Valente AJM, Ascenso OS, Lobo VMM, Burrows HD, Cabral AMTDPV, Veiga FJB, Teijeiro C, Esteso MA. Some transport properties of γ-cyclodextrin aqueous solutions at (298.15 and 310.15) K. Journal of Chemical & Engineering Data. 2008;**53**:755-759

[60] Ribeiro ACF, Valente AJM, Santos CIAV, Prazeres PMRA, Lobo VMM, Burrows HD, Esteso MA, Cabral AMTDPV, Veiga FJB. Binary mutual diffusion coefficients of aqueous solutions of α-cyclodextrin, 2-hydroxypropyl-α-cyclodextrin, and 2-hydroxypropyl-β-cyclodextrin at temperatures from (298.15 to 312.15) K. Journal of Chemical & Engineering Data. 2007;**52**:586-590

[61] He J, Chipot C, Shao X, Cai W. Cooperative recruitment of amphotericin B mediated by a cyclodextrin dimer. The Journal of Physical Chemistry C. 2014;**118**:24173-24180

[62] Figueiras A, Sarraguça JMG, Carvalho RA, Pais AACC, Veiga FJB. Interaction of omeprazole with a methylated derivative of β-cyclodextrin: Phase solubility, NMR spectroscopy and molecular simulation. Pharmaceutical Research. 2007;**24**:377-389

[63] Figueiras A, Nunes SCC, Simões S, Pais AACC, Veiga F. Molecular interaction governing solubility and release profiles in supramolecular systems containing fenbufen, pluronics and cyclodextrins. Journal of Inclusion Phenomena and Macrocyclic Chemistry. 2015;**81**:395-407

[64] Li K, Wu X, Gu Q, Zhao X, Yuan M, Ma W, Ni W, Hou Z. Inclusion complexes of organic salts with [small beta]-cyclodextrin as organocatalysts for CO_2 cycloaddition with epoxides. RSC Advances. 2017;**7**:14721-14732

[65] Mele A, Mendichi R, Selva A. Non-covalent associations of cyclomaltooligosaccharides (cyclodextrins) with trans-β-carotene in water: Evidence for the formation of large aggregates by light scattering and NMR spectroscopy. Carbohydrate Research. 1998;**310**:261-267

[66] Hernández R, Rusa M, Rusa CC, López D, Mijangos C, Tonelli AE. Controlling PVA hydrogels with γ-cyclodextrin. Macromolecules. 2004;**37**:9620-9625

[67] Rusa CC, Fox J, Tonelli AE. Competitive formation of polymer–cyclodextrin inclusion compounds. Macromolecules. 2003;**36**:2742-2747

[68] Rusa CC, Bullions TA, Fox J, Porbeni FE, Wang X, Tonelli AE. Inclusion compound formation with a new columnar cyclodextrin host. Langmuir. 2002;**18**:10016-10023

[69] Papancea A, Valente AJM, Patachia S, Miguel MG, Lindman B. PVA–DNA cryogel membranes: Characterization, swelling, and transport studies. Langmuir. 2008;**24**:273-279

[70] Zerkoune L, Angelova A, Lesieur S. Nano-assemblies of modified cyclodextrins and their complexes with guest molecules: Incorporation in nanostructured membranes and amphiphile nanoarchitectonics design. Nanomaterials. 2014;**4**:741

[71] Auzély-Velty R, Djedaïni-Pilard F, Désert S, Perly B, Zemb T. Micellization of hydrophobically modified cyclodextrins. 1. Micellar structure. Langmuir. 2000;**16**:3727-3734

[72] Wang T, Chipot C, Shao X, Cai W. Structural characterization of micelles formed of cholesteryl-functionalized cyclodextrins. Langmuir. 2011;**27**:91-97

[73] Sliwa W, Girek T. Cyclodextrins: Properties and Applications. Weinheim, Germany: John Wiley & Sons; 2016

[74] Mazzaglia A, Bondì ML, Scala A, Zito F, Barbieri G, Crea F, Vianelli G, Mineo P, Fiore T, Pellerito C, Pellerito L, Costa MA. Supramolecular assemblies based on complexes of nonionic amphiphilic cyclodextrins and a meso-tetra(4-sulfonatophenyl)porphine tributyltin(IV) derivative: Potential nanotherapeutics against melanoma. Biomacromolecules. 2013;**14**:3820-3829

[75] Jiang B-P, Guo D-S, Liu Y. Self-assembly of amphiphilic perylene–cyclodextrin conjugate and vapor sensing for organic amines. The Journal of Organic Chemistry. 2010; **75**:7258-7264

[76] Bauer M, Charitat T, Fajolles C, Fragneto G, Daillant J. Insertion properties of cholesteryl cyclodextrins in phospholipid membranes: A molecular study. Soft Matter. 2012; **8**:942-953

[77] Zhang H, Tan T, Feng W, van der Spoel D. Molecular recognition in different environments: β-cyclodextrin dimer formation in organic solvents. The Journal of Physical Chemistry B. 2012;**116**:12684-12693

[78] Zhang H, Tan T, Hetényi C, Lv Y, van der Spoel D. Cooperative binding of cyclodextrin dimers to isoflavone analogues elucidated by free energy calculations. The Journal of Physical Chemistry C. 2014;**118**:7163-7173

[79] Zhang H, Tan T, Hetényi C, van der Spoel D. Quantification of solvent contribution to the stability of noncovalent complexes. Journal of Chemical Theory and Computation. 2013;**9**:4542-4551

[80] Zhang H, Tan T, van der Spoel D. Generalized born and explicit solvent models for free energy calculations in organic solvents: Cyclodextrin dimerization. Journal of Chemical Theory and Computation. 2015;**11**:5103-5113

[81] van der Spoel D, van Maaren PJ, Caleman C. GROMACS molecule & liquid database. Bioinformatics. 2012;**28**:752-753

[82] Vanquelef E, Simon S, Marquant G, Garcia E, Klimerak G, Delepine JC, Cieplak P, Dupradeau F-Y. R.E.D. Server: A web service for deriving RESP and ESP charges and building force field libraries for new molecules and molecular fragments. Nucleic Acids Research. 2011;**39**:W511-W517

[83] Wong-ekkabut J, Karttunen M. The good, the bad and the user in soft matter simulations. Biochimica et Biophysica Acta (BBA)—Biomembranes. 2016;**1858**:2529-2538

[84] Hess B, Bekker H, Berendsen HJC, Fraaije JGEM. LINCS: A linear constraint solver for molecular simulations. Journal of Computational Chemistry. 1997;**18**:1463-1472

[85] Daura X, Gademann K, Jaun B, Seebach D, van Gunsteren WF, Mark AE. Peptide folding: When simulation meets experiment. Angewandte Chemie International Edition. 1999;**38**:236-240

[86] Das S, Subuddhi U. Exploring poly(vinyl alcohol) hydrogels containing drug-cyclodextrin complexes as controlled drug delivery systems. Journal of Applied Polymer Science. 2014;**131**:40318(1-13)

[87] Zhou J, Wang Q, Lu H, Zhang Q, Lv P, Wei Q. Preparation and characterization of electrospun polyvinyl alcoholstyrylpyridinium/β-cyclodextrin composite nanofibers: Release behavior and potential use for wound dressing. Fibers and Polymers. 2016;**17**:1835-1841

[88] Ouyang L, Zhu L, Ruan Y, Tang H. Preparation of a native [small beta]-cyclodextrin modified plasmonic hydrogel substrate and its use as a surface-enhanced Raman scattering scaffold for antibiotics identification. Journal of Materials Chemistry C. 2015;**3**:7575-7582

[89] Knaapila M, Stewart B, Costa T, Rogers SE, Pragana J, Fonseca SM, Valente AJM, Ramos ML, Murtinho D, Pereira JC, Mallavia R, Burrows HD. Incorporation of a cationic conjugated polyelectrolyte CPE within an aqueous poly(vinyl alcohol) sol. Macromolecules. 2016;**49**:9119-9131

[90] Cezard C, Trivelli X, Aubry F, Djedaini-Pilard F, Dupradeau F-Y. Molecular dynamics studies of native and substituted cyclodextrins in different media: 1. Charge derivation and force field performances. Physical Chemistry Chemical Physics. 2011;**13**:15103-15121

[91] Sousa da Silva AW, Vranken WF. ACPYPE—AnteChamber PYthon parser interfacE. BMC Research Notes. 2012;**5**:367

Section 2

Applications of Cyclodextrins

Chapter 3

Interactions between Bio-Based Compounds and Cyclodextrins

Bruno Filipe Figueiras Medronho, Sandra Gonçalves, Raquel Rodríguez-Solana, Artur J.M. Valente and Anabela Romano

Additional information is available at the end of the chapter

http://dx.doi.org/10.5772/intechopen.73531

Abstract

Bio-based compounds, such as "green" surfactants and phytochemicals, are regarded as future sustainable resources for a vast range of applications in a modern society increasingly demanding economical, social, and environmental awareness. Natural compounds from plants (phytochemicals) are very sought by the pharmaceutical, cosmetic, and food industries. On the other hand, the growing interest in "green" surfactants (e.g., carbohydrate-based) is due to, inter alia, their preparation from renewable raw materials, ready biodegradability, and biocompatibility, among other reasons of fundamental, practical, economical, and environmental orders. Despite the wide range of potential applications of these bio-based compounds, their practical use is still limited due to many reasons such as poor aqueous solubility, volatility, reactivity, etc. Generally, when complexed with cyclodextrins, these bio-based compounds enhance considerably their performance and potential applications. Thus, this chapter aims at recalling some general fundamental aspects of phytochemicals and "green" surfactants, such as structure, function, and applications. In addition, their interactions with cyclodextrins are discussed from a physicochemical point of view with special focus on the techniques, mathematic modeling, and thermodynamic parameters (e.g., interactions, stoichiometries, association constants, etc.).

Keywords: sugar-based surfactants, phytochemicals, essential oils, polyphenols, cyclodextrins, host-guest complex

1. Cyclodextrins: general considerations

Due to its structure, cyclodextrins (CDs) readily form inclusion complexes through noncovalent interactions with molecular guests. The lipophilic cavity of CDs provides a microenvironment into which appropriately sized nonpolar moieties can enter. The hydrophobicity of the cavity enables the accommodation of a broad range of hydrophobic guests such as the alkyl chains of surfactants or different phytochemicals [1]. The hydrophilic exterior usually imparts CDs and their complexes, considerable solubility in water. The charge and polarity of the guest molecule play also an important role in the CD-substrate host-guest interaction [2]. However, this aspect is obviously less important than the geometric fitting. In the case of the charge, the complexation of neutral molecules is easier than the ionized counterpart. In general, molecules can be encapsulated by CDs when they are less hydrophilic or less polar than the solvent and when the formed complex is stable.

The main driving force for the formation of the complex is the release of enthalpy-rich water molecules from the cavity; water molecules are displaced by more hydrophobic guest molecules present in the solution to achieve the apolar-apolar interactions and decrease of CD ring strain resulting in a favorable lower energy state. The beneficial modification of guest molecular properties after the formation of the inclusion complex leads to a large number of applications in areas as diverse as encapsulation of active substances (i.e., flavoring agents, metallic cations, fragrances, and pesticides), enzymatic synthesis, catalysis, and energy transfer studies [3, 4]. Additionally, CDs also find important uses in cosmetics, environment protection, bioconversion, packing, textiles, and food domain [1, 5].

Less than 10% of all produced CDs and CD derivatives are used by the pharmaceutical industry. The largest CD users are the food and the cosmetic industry. CDs have a high level of biocompatibility, are absorbed in the gastrointestinal tract, and are completely metabolized by the colon microflora [6]. Some of them are approved by the Food and Drug Administration or have been accredited as being "generally recognized as safe" (GRAS) [7]. In the cosmetic area, CD performance stands out in the following: solubilize and stabilize specific sensitive components, stabilize emulsions, improve the absorption of active components onto the skin, reduce or eliminate bad aromas from certain components, and reduce the loss of the active components through volatilization, rapid oxidation, destruction by light, etc. [8].

2. Plant phytochemicals: brief introduction to polyphenols and essential oils

Plants are a remarkable source of biologically active compounds with potential applications in cosmetic, pharmaceutical, and food industries. Among the bioactive compounds synthesized as secondary metabolites by plants, phenolic compounds are probably the most relevant ones. Physiologically, they play a vital role in plant protection and contribute to plant odors, plant pigmentation, and/or their flavors. Structurally, phenolic compounds have, at least, one aromatic ring, with one or more hydroxyl groups attached [9]. The great diversity of phenolic compounds present in nature (i.e., more than 8000 different structures have been identified up to now) results from variations in the basic chemical skeleton (e.g., degree of oxidation, hydroxylation, methylation, glycosylation, and conjugation with further molecules, particularly lipids, proteins, other phenolics, and biomolecular metabolites) [9]. Phenolic compounds are grouped by the number of phenol rings they contain and the structural elements that bind these rings to another; flavonoids, phenolic acids, tannins, stilbenes, and lignans are examples of important representatives of these groups [10].

On the other hand, plant essential oils (EOs) are mixtures of numerous highly complex volatile compounds (hydrogenated and oxygenated monoterpenes, sesquiterpenes, phenols, simple alcohols, ketones, coumarins, etc.) present in variable concentrations whose aroma depends on the individual constituents present [11]. Due to their natural properties, EOs have been used as therapeutic remedies and flavoring agents since ancient times. In the last decades, many investigations showed that EOs have a wide range of valuable biological activities, such as antimicrobial, herbicidal, insecticidal, antioxidant, etc.

Although many investigations demonstrated the broad range of biological activities of many phytochemicals, they still have restricted applicability as pharmaceuticals or in food products due to their poor water insolubility and bioavailability, high volatility, rapid oxidation, or degradation when exposed to environmental factors. New approaches have been developed to overcome these drawbacks, and among them, CDs have been suggested as excellent vehicles for the protection of phytochemicals for food and drug delivery proposes [12–14].

2.1. Interaction between phytochemicals and cyclodextrins

According to a recent review by Suvarna et al., there are many phytochemicals whose solubility, bioavailability, or therapeutic activity is significantly improved by complexation with CDs (e.g., quercetin, curcumin, artemisinin, resveratrol, naringenin, etc.) [12]. The methods used for the formation of inclusion complexes between CDs and bioactive compounds are essentially neutralization, slurry, solution, coprecipitation, kneading, and grinding [15].

The encapsulation of phytochemicals with CDs usually involves the formation of 1:1 inclusion complexes with the most versatile CD, the β-CDs, and its derivatives. These derivatives can be classified according to their interaction with the water molecules in hydrophilic, hydrophobic, and ionizable derivatives [13]. Examples of used hydrophilic β-CDs are the methylated β-CDs—2,6-dimethyl-β-CD (DM-β-CD) and 2,3,6-trimethyl-β-CD (TM-β-CD)—the hydroxyalkylated β-CDs such as 2-hydroxypropyl-β-CD (HP-β-CD), and the branched β-CDs, glycosyl-β-CD (G-β-CD). These molecules are suitable for the formation of host-guest inclusion complexes with poor water-soluble compounds. On the other hand, the hydrophobic derivatives, such as the alkylated β-CD 2,6-diethyl-β-CD (DE-β-CD), are used to decrease and modulate the released rate of water-soluble molecules. Finally, the ionizable β-CD can enhance the dissolution rate and the inclusion capacity and even decrease the side effects of some molecules [16, 17]. Among the ionizable CDs, O-carboxymethyl-β-CD (CM-β-CD), O-carboxymethyl-O-ethyl-β-CD (CME-β-CD), and sulfate and sulfobutylether-β-CD (SBE-β-CD) should be highlighted.

Owing to their potential health promotion effects particularly the antioxidant, anti-inflammatory, and antimicrobial properties, one of the actual promising applications of phenolic compounds is their use in the food industry as additives, e.g., in the development of functional foods. Nevertheless, the efficacy of these natural compounds is dependent on the preservation or improvement of their stability, bioactivity, and bioavailability [18]. Inclusion complexation with CDs improves water solubility of phenolics and enhances their shelf life and biological activity [15]. Additionally, it has been shown that the inclusion of phenolic compounds (e.g., hydroxycinnamic and chlorogenic acids) with CD (β-CD) strongly limited their interactions with proteins, which is important regarding the use of phenolics as food additives [19, 20]. Note that the interactions of these compounds with proteins, frequently added to functional foods to improve nutritional value and proper texture characteristics, often decrease the bioavailability of both proteins and phenolics [21].

There are many examples showing that the complexation of β-CD, or some of its derivatives, increases the biological activity of phenolics. For instance, Shao et al. observed that the complexation of chlorogenic acid (CGA) with CDs (β-CD and HP-β-CD) improved its antioxidant activity [22]. Moreover, the addition of CGA-CD complexes to grape juice reduced the degradation of anthocyanins due to copigmentation effect with the CGA/HP-β-CD complex showing the superior activity and copigmentation effects. Gabaldon et al. also used HP-β-CD to increase the aqueous solubility of kaempferol, quercetin, and myricetin and to improve their antioxidant activity due to the protection toward free radical attack [23]. The complexation of curcumin with an ionizable β-CD (SBE-β-CD) enhanced its water solubility and, thus, improved the *in vitro* cytotoxic (on HepG-2 cells) and antioxidant activity of these compounds [24]. This β-CD derivative and the HP-β-CD are the most used derivatives on the pharmaceutical industry due to their low toxicity and high solubility [16, 25, 26].

EOs can be regarded as mixtures of phytochemicals, and there are several studies reporting the complexation of EOs or their components with CDs mainly to overcome problems related with EO water insolubility, high volatility, rapid oxidation, heat damage, and degradation on exposure to air [14]. Although many studies focus the complexation of EO components with β-CDs or its derivatives, e.g., eugenol/HP-β-CD [27] and linalool/HP-β-CD [28], it has been observed that sometimes γ-CD is a better complexing agent. Ciobanu et al. showed that menthol, menthone, and pulegone are capable to form stable 1:1 inclusion complexes with β-CD, but eucalyptol forms a more stable inclusion complex with γ-CD due to the size of its cavity [29]. Polymeric CDs, which can be synthesized using cross-linking agents such as epichlorohydrin, also revealed promising results in some specific cases but are not matter of discussion in this chapter [29].

The inclusion complexes of EOs (or their components) with CDs have been mainly tested for food and pharmaceutical applications, but they could be an efficient tool to improve the use of EOs in aromatherapy, cosmetic, and household cleaning products. An interesting application of EOs is related to their incorporation in food packaging systems or edible films due to their antimicrobial, antioxidant, and insect repellent capacity. However, this is often limited due to flavoring and organoleptic considerations. CD inclusion complexes could overcome these limitations allowing EOs to reach effective concentrations in the food matrices without exceeding organoleptically acceptable levels and even providing controlled release-rate kinetics. Recent studies encourage the use of CD-EO (e.g., β-CD/*Satureja montana* EO; γ-CD inclusion complex encapsulated electrospun zein nanofibrous webs/thymol) complexes as part of active packaging systems [30, 31] as well as promising candidates to be used as safe and effective antimicrobial agents (β-CD/eugenol) to control postharvest diseases in fruits [27].

There are scientific evidences that inclusion complexation with CDs improves the pharmacological effects of EOs or their components [28, 32, 33]. For instance, the complexation of *Hyptis pectinata* L. EO with β-CD improved its analgesic effect in a mice model [34]. Also Bomfim et al. observed that β-CD complexation increased *in vivo* tumor growth inhibition capacity of *Annona vepretorum* EO [32]. Recently, Lima et al. reviewed the preclinical and clinical studies published on complexes between CDs and terpenes [33]. These are the major components of EOs that exhibit a wide range of biological activities on the human body. Their survey shows that there is robust experimental evidence that CDs improve the oral absorption and pharmacological properties of terpenes. Nevertheless, more pharmacokinetic and clinical studies are required before they can be effectively used in clinical targets.

3. Natural surfactants: brief introduction to sugar-based amphiphiles

Despite the production of surfactants based on fats, oils, and carbohydrates, being a known area for several decades, on an industrial scale, this is a relatively new issue [35]. These amphiphilic molecules that have one of the main building blocks from a natural source are often called "natural surfactants" [36, 37]. For example, alkyl glycosides which are synthesized from a "natural" sugar unit and a "nonnatural" fatty alcohol are often regarded as natural surfactants. Considering their amphiphilic nature, it has been always a challenge to attach a carbohydrate molecule, such as the hydrophilic group (due to the numerous hydroxyl groups) to a fat and oil derivative, such as a fatty acid or a fatty alcohol. However, nowadays, several successful synthesis routes are well established, and numerous types of natural surfactants are known and available, even on a commercial scale [38]. Nowadays, carbohydrate-based surfactants (CBS) are among the most important classes of amphiphilic compounds [39–41]. Their structure results from the combination of sugar and lipids, naturally biosynthesized within living cells or, alternatively, synthetically prepared by sequential reactions using carbohydrate and fatty materials. The growing interest in such compounds is due to, inter alia, their preparation from renewable raw materials, biodegradability, mildness to the skin, and biocompatibility, among other reasons [42, 43]. In particular, CBS can be relatively easily prepared from the most abundant renewable vegetable raw materials (e.g., cellulose, pectin, hemicellulose, starch, etc.) in a wide range of structures and geometries by modular synthesis thanks to the presence of numerous reactive hydroxyl groups. Such structural diversity makes CBS excellent models to get insight on the surfactant mechanisms in modifying interfacial properties. This knowledge is crucial for the control of the formation and stability of diverse colloidal systems such as micelles, vesicles, foams, and emulsions [44]. An important structural feature of these surfactants is the typical sugar headgroup, a voluminous and relatively rigid moiety that can be functionalized by a myriad of reagents and synthetic schemes. Numerous properties and functionalities can be expected from such almost unlimited number of different compounds that can find specific applications in different industrial areas [45, 46]. CBS also present great advantages on the environmental side; their higher biodegradability and lower toxicity profile are important reasons to consider CBS as valid alternatives to the more common petrochemical-based surfactants.

3.1. Case study: alkyl polyglycosides

In recent years there has been a growing focus on three classes of surfactants with sugar or a polyol derivative as polar headgroup: alkyl polyglycosides (APGs), alkyl glucamides, and sugar esters. In this chapter, we will focus on APGs, which are regarded as "perfect amphiphilic structures" with excellent surface activity as well as solubility. APGs have been synthesized, for the first time, more than 100 years ago [47]. APGs are completely based on renewable resources and combine very good performance, multifunctionality, and competitive price with mildness. This explains why APGs are the most successful sugar-based surfactants nowadays. It is important to mention that not pure alkyl monoglucosides but rather a complex mixture of alkyl mono-, di-, tri-, and oligoglycosides is produced in the industrial processes. Because of this, the industrial products are called alkyl polyglycosides. The surfactants are thus characterized by the length of the alkyl chain and the average number of glucose units linked to it, the degree of polymerization (DP). Alkyl glycosides are stable at high pH and sensitive to low pH where they hydrolyze to sugar and fatty alcohol. The sugar unit is more water-soluble and less soluble in hydrocarbons than the corresponding polyoxyethylene unit; hence, APGs and other polyol-based surfactants are more lipophobic than their polyoxyethylene-based surfactant counterparts [48]. This makes the physicochemical behavior of APG surfactants in oil/water systems distinct from that of conventional nonionic surfactants. Moreover, APGs do not show the pronounced inverse solubility vs. temperature relationship that normal nonionics do [49]. This makes an important difference in solution behavior between APGs and polyoxyethylene-based surfactants. The critical micelle concentration (cmc) values of the pure alkyl monoglycosides and the technical APG are comparable with those of typical nonionic surfactants and decrease distinctly with increasing alkyl chain length. The alkyl chain length has a far stronger influence on the cmc than the number of glycoside groups of the APG. The influence of the DP of APGs on their phase behavior has been described by Fukuda et al. [50]. The region in which the liquid crystalline phases occur is only slightly dependent on the concentration with a greater expansion in the case of APGs with a higher DP.

Regarding their applications, APGs are mainly used in personal care products such as cosmetics, manual dishwashing, and detergents [51]. APGs have also been used in more advanced applications such as the extraction and purification of membrane proteins, which plays a major role in the determination of protein structures and functions [52]. This is because APGs have reduced protein-denaturing properties in comparison to conventional surfactants. Generally, the environmental fate of surfactants is inextricably linked with their biodegradation behavior. Thus, fast and complete biodegradability is the most important requirement for an environmentally compatible surfactant. The general environmental impact of chemicals lies mainly in their ecotoxicity, which is relatively high in the case of surfactants because of their surface activity and the resulting effects on biological membranes [47]. In the case of CBS (APGs in particular), they present a quite favorable environmental profile: the rate of biodegradation is usually high, and the aquatic toxicity is low. In addition, APGs exhibit favorable dermatological properties, being very mild when exposed to the skin and eye [53]. The interested reader can find excellent overviews on APGs elsewhere [45–47, 51, 54].

3.2. Interaction between alkyl polyglycosides and cyclodextrins

Saenger and Mullerfahrnow [55] and Casu et al. [56] were pioneers in the study of the interactions between APGs and CDs. By surface tension measurements, it was shown that the addition of CDs leads to an increase of surfactant critical micelle concentration. Furthermore, the interaction is most pronounced when the CD cavity and hydrophobic part of the surfactant exhibit the tightest fit. Such a view was also supported by ^{1}H and ^{13}C NMR spectroscopy. By the analysis of NMR chemical shifts of octyl- (C_8G_1) and dodecyl ($C_{12}G_1$) α- and β-D-glucopyranoside and octadecyl β-D-glucopyranoside ($C_{18}G_1$), in the absence and presence of α-CD, two main conclusions were taken: the presence of α-CD only affects the chemical shifts of the surfactant alkyl chain, and the chemical shift of the protons H-3 and H-5, located inside the CD cavity, increases by increasing the length of the alkyl chain [57]. Furthermore, it was observed that no chemical shift is observed for γ-CD/$C_{12}G_1$ mixed systems. These conclusions corroborate previous and subsequent studies, showing that the CD cavity is protruded just by the surfactant tail and the magnitude of the interaction is dependent on the relationship between the volumes of the CD cavity and the surfactant hydrocarbon chain [56, 58]. A better characterization of these complexes was accomplished by the quantification of the binding process. The effect of the length of the alkyl chain and surfactant head group on the association constant of APGs with different CDs was studied by ^{1}H NMR spectroscopy, NMR self-diffusion, and surface tension (**Table 1**) [58–60]. Although the comparison between association constants computed on the basis of different physical parameters is a difficult task [5], the analysis of the data allowed concluding that by increasing the alkyl chain length and decreasing the CD cavity volume (from β- to α-CD), the association constant increases. Furthermore, some authors stated that no interactions were observed when γ-CD was used [59, 58]. It is also worth noticing that for all mentioned systems, by subtracting the critical aggregation concentration (cac) of the surfactants from the CD concentration, the critical micelle concentration

	α-CD	β-CD	Obs.
β-C_8G_1	$3.68\ (\pm 1.6) \times 10^3$	$0.99\ (\pm 0.17) \times 10^3$	Self-diffusion NMR [58]
	$1.85\ (\pm 0.35) \times 10^3$		Surface tension [60]
β-C_9G_1	$76\ (\pm 750) \times 10^3$	$275\ (\pm 5300) \times 10^3$	Self-diffusion NMR [58]
β-$C_{10}G_1$		340 ± 30	[CD] = 1 mM, ^{1}H NMR [59]
β-$C_{12}G_1$		440 ± 40	[CD] = 2 mM, ^{1}H NMR [59]
		410 ± 40	[CD] = 1 mM, ^{1}H NMR [59]
β-$C_{12}G_2$		125 ± 10	[CD] = 1 mM, ^{1}H NMR [59]

Table 1. Binding constants, $K_{1,1}$ in $dm^3\ mol^{-1}$, for the inclusion complexes CD/APG.

value is obtained. This clearly suggests a 1:1 APG/CD complexation [4], and actually it finds support by the Job's plot [61] reported by Bernat et al. [60]. Another important issue is to understand the reliability of the binding constants reported in **Table 1**. Rymdén et al. found that an increase of a methylene group for a series of alcohols decreases the standard free energy of the alcohol: β-CD binding for ca. 3.0 kJ/mol [62]. This variation is similar to that observed for $K_{1,1}$ values obtained by self-diffusion coefficients (**Table 1**). On the other hand, comparing $K_{1,1}$ values for C_8G_1 with those from other monoalkyl surfactants, one can conclude that the APG/CD complex is more stable [63, 64]. This has been justified by the occurrence of hydrogen bonds between the sugar structure and the hydroxyl groups located at the rim of the CD. However, studying the effect of the number of sugar moieties in the surfactant head on the free energy of binding, an algebraic increase in the Gibbs free energy is observed. Indeed, comparing the binding constants for the interactions between β-$C_{12}G_1$ and β-$C_{12}G_2$ with β-CD, it is possible to conclude that the addition of an extra sugar moiety in the surfactant head decreases the *K* values for the supramolecular association. Thus, it can be hypothesized that no significant sugar-sugar interactions are involved in the interaction with CD, as it was previously discussed. Another hypothesis arises from the effect of carbohydrates on the water structure, for example, Ribeiro et al. have found that the presence of carbohydrates leads to an increase of the entropy in water [65], also called a structure "breaking effect" [66]. Consequently, an increase of the concentration of the sugar molecules in solution may contributes for a decrease in the binding entropy change and, consequently, to an increase in the binding Gibbs free energy.

Up to now, we have been discussing the binding process assuming a 1:1 APG/CD (α- and β-) binding stoichiometry. However, it should be stressed that from the study of the interactions between $C_{12}G_1$, and $C_{18}G_1$, and CDs, there are strong evidences for the occurrence of other species consistent with 1:2 complexes [56]. More recently, Haller and Kaatze, studying the interaction between C_8G_1 and α-CD by ultrasonic attenuation spectroscopy, concluded that besides 1:1 (APG/CD) complexes, the formation of 1:2 and 2:1 complexes (although in very low concentration) should not be ruled out [67].

4. On the methods to follow the interactions between cyclodextrins and phytochemicals and sugar-based surfactants

As it must have already been understood from the previous subsections, an accurate choice of the technique to follow the host-guest association is a key issue for a reliable thermodynamic and kinetic characterization of the association process. In general, the experimental techniques can be subdivided into two different categories, labeled as I and II [68]. Methods from group I (e.g., surface tension) are measuring changes in physical properties that are proportional, in some ways, to the extent of binding, while those from group II (e.g., ^{1}H NMR spectroscopy) rely on direct measurements of the free and bound ligand in a solution containing a known amount of the CD and guest molecule. Comments on such a division can be found in a couple of reviews (see, e.g., [69]), and it is outside of the scope of this chapter. The same is valid for computational techniques as relevant tools to infer on the structure of the supramolecular compounds [70–72].

In this section, the most relevant techniques used to study the interactions between CDs and sugar-based surfactants (e.g., APGs) and phytochemicals (e.g., polyphenols, EOs, and their components) will be highlighted (**Table 2**).

NMR spectrometry falls in the group II techniques, and it is used to determine association constants through the chemical shift changes noticed either by the guest or by the CD [73–75]. Focusing on EOs, and as previously discussed, they may present undesired features, such as volatility, poor aqueous solubility, and stability. Therefore, host-guest supramolecular complexes are often obtained by using solid-state-based methods [14] such as freeze-drying [76, 77], coprecipitation [78], and the saturated approach [27], improving the solubility of the EO and thus allowing the use of NMR techniques for the quantitative and qualitative assessment of the complexation process [74]. For example, the complexation of eugenol with β-CD was obtained by using the saturation method, and the obtained complex, in the solid state, was characterized by either ^{1}H, ^{13}C, or 2D NMR techniques, confirming the thread of CD's cavity by the aromatic ring of the eugenol [27]. Other techniques will be mentioned later since they fall on the so-called group II.

DOSY ^{1}H NMR has been used to study inclusion complexes between CD and different sugar-based substrates [79]. Kfoury et al. report a comprehensive study on the complexation between two phenol isomers (thymol and carvacrol) and CDs by using different NMR techniques, including DOSY [80]. The data allowed concluding that those isomers have a binding constant of 1344 M^{-1} and 1336 M^{-1}, respectively.

The self-diffusion measurements are, in principle, applicable to any systems as long as the free and complexed guests are soluble to an extent that allows for a good signal-to-noise ratio. It is important to note that on account of the rapid exchange on the NMR time scale, average diffusion coefficients for both the guest and for the CD are typically obtained. This method, as well as that involving chemical shift changes analysis, is also limited to systems where no overlapping of well-defined resonances is observed. The method relies on the fact that the self-diffusion coefficients of the uncomplexed guest are higher than the self-diffusion of the host-guest complex, as defined by the Stokes-Einstein equation. The change in the self-diffusion coefficient of the CD upon complexation is often small since the complex is often of the same size as the CD molecule, and thus the information from the CD self-diffusion is rather limited [58].

Ultrasonic relaxation technique falls into group II techniques and is based on the application of ultrasound to a given solution, with a frequency ranging from 20 kHz to several GHz, and subsequently measuring the molecular structural relaxation. The relaxation is sensitive to molecular volume changes [81], and thus, it may convey information on the stability constants of the host-guest complexes [82]. Furthermore, the use of a large frequency range allows to follow processes with relaxation times in the range from 20 ps to 20 μs [83], and thus the kinetics of the CD-surfactant association can be investigated. Haller and Kaatze, by using ultrasonic attenuation spectroscopy, were able to quantify the dynamics of unimer-micelle exchange of a sugar-based surfactant (i.e., octyl-β-D-glucopyranoside (C_8G_1)) in the presence of α-CD [67].

Also from group II, surface tension has also been used to follow the effect of CDs on the aggregation and interfacial properties of surfactants in CD-surfactant-containing solutions

Experimental methods	System	Obs.
NMR	Nerolidol + β-CD	[74]
UV-vis	Nerolidol + CDs[1]	[74]
Phase solubility studies	Cabreuva essential oil + HP-β-CD	[74]
Phase solubility studies	β-caryophyllene + HP-β-CD	[76]
UV-vis	Black pepper essential oil + HP-β-CD	[76]
Phase solubility studies	CDs[1] + PPs[2]	[77]
Phase solubility studies, NMR, TGA, DSC	β-CD + estragole	[78]
NMR	β-CD + eugenol	[27]
NMR	β-CD + rosmarinic acid	[75]
NMR	Cyclohexylacetic acid + β-CD	[79]
NMR	Cholic acid + β-CD	[79]
UV-vis, NMR	Thymol and carvacrol + CDs[1]	[80]
NMR	n-Octyl-β-D-glucoside and n-nonyl -β-D-glucoside + α-CD and β-CD	[58]
UAS[3]	Octyl-β-D-glucopyranoside + α-CD	[67]
Surface tension, NMR	APGs[4] + β-CD	[59]
Surface tension	Octyl-β-D-glucopyranoside + α-CD	[60]
Phase solubility studies, ITC, NMR	Nootkatone + β-CD and HP-β-CD	[71]
TGA	Cinnamon essential oil + β-CD	[90]
NMR, FTIR, release kinetics	Monochlorotriazinyl β-CD + EOs[5]	[91]
XRD, NMR, TGA	Thymol + γ-CD	[30]
DSC, TGA, FTIR, XRD, GC/MS, NMR	Isopulegol + α-CD and β-CD	[72]
Phase solubility studies, NMR, HPLC	Polymethoxyflavones + HP-β-CD	[93]
GC, total organic carbon, phase-solubility studies	SBE-β-CD, SBE-γ-CD and HP-β-CD + EOS[6]	[94, 95]

[1]CDs: alpha-cyclodextrin (α-CD), beta-cyclodextrin (β-CD), gamma-cyclodextrin (γ-CD), 2-hydroxypropyl-β-cyclodextrin (HP-β-CD), randomly methylated-beta-cyclodextrins (RAMEB), low methylated beta-cyclodextrin (CRYSMEB), and sulfobutylether β-cyclodextrin (SBE-β-CD)
[2]PPs: *trans*-anethole, estragole, eugenol, isoeugenol (phenylpropenes), caffeic acid, *p*-coumaric acid, and ferulic acid (hydroxycinnamic acids)
[3]UAS: ultrasonic attenuation spectroscopy
[4]APGs: glucopyranosides (octyl G8, decyl G10, dodecyl G12, tetradecyl G14) and two maltosides (decyl M10, dodecyl M12)
[5]EOs: essential oils of cedarwood, clove, eucalyptus, and peppermint
[6]EOS: essential oils of *Artemisia dracunculus, Citrus reticulata Blanco, Citrus aurantifolia, Melaleuca alternifolia, Melaleuca quinquenervia,* and *Rosmarinus officinalis cineoliferum*

Table 2. Compilation of the most relevant techniques used to study the interactions between CDs and bio-based compounds.

[59, 60]. Surface tension is a measure of cohesive forces between liquid molecules present at the surface, and it represents the quantification of force per unit length of free energy per unit area [84]. In general, the presence of CDs will increase the surface tension of an APG solution. Knowing that natural CDs are not surface-active, they cannot replace APG at the air interface [85]. Therefore, CD molecules contribute for the depletion of APG unimers from the interface due to the great interaction between these unimers and CDs. There are several examples where surface tension measurements have been used to assess the stoichiometry and stability constants of host-guest complexes [60, 86].

Isothermal titration calorimetry (ITC) is a sensitive and powerful technique to study host-guest interactions by measuring the enthalpy and the free energy of binding [86, 87]. There are also some cases where the kinetic constants of the binding process can be obtained by ITC (see, e.g., [88]). For example, the heat produced by a stepwise addition of HP-β- and β-CD solution to a nootkatone allowed to characterize the complexation process with a binding enthalpy and binding constant of -6.99 kJmol^{-1} and 4838 M^{-1}, and -14.38 kJmol^{-1}, and 5801 M^{-1}, respectively [71]. Unfortunately, the strict conditions required by this technique do not allow its routine implementation on a large scale [89].

As has been pointed out before, some phytochemicals (EOs, in particular) are, in general, poorly soluble in aqueous solutions; therefore, the formation of complexes with CD in solid state is a strategy for further applications. Consequently, there are several available techniques used to evaluate the complexation. Thermal techniques, such as thermal degradation analysis and differential scanning calorimetry, are classical examples of methods used to assess complexation. Moreover, thermal degradation also allows evaluating the thermal stability of the EO upon complexation [90, 78].

Other spectroscopic techniques, such as FTIR and XRD, which can be included in group II, have also been used, but the information is, in our opinion, rather qualitative [30, 72, 78, 91]. Another interesting approach to learn about the formation of host-guest complexes, in solid state, is to study the release kinetics of the EO. These studies, although do not allow to quantify the total amount of EO incorporated into the CDs, are of utmost importance to evaluate the presence of the EO in the complex as well as to provide hints on the release mechanism; the latter is quite relevant for EOs used as fragrances [91, 92].

For such poor soluble compounds, the complexation can also be evaluated by carrying out phase-solubility studies. These can be performed by using complexes in solid state or by checking the ability of increasing concentrations of CD to solubilize saturated solutions of EO. This allows assessing how much the solubility of the EO is improved upon complexation as well as the corresponding complex binding constant. The details on the quantitative determination of those parameters will be given in the next section. Different techniques can be applied to obtain the phase-solubility profiles. For instance, the solubility of black pepper EO in the presence of hydroxypropyl-beta-CD (HP-β-CD) was evaluated by UV-visible spectroscopy [76]. On the other hand, phase-solubility profiles for the encapsulation of polymethoxyflavones, obtained from mandarin EO, into HP-β-CD were obtained by using HPLC [93]. Kfoury et al. have used gas chromatography to study the ability of sulfobutylether-β- and sulfobutylether-γ-CD to encapsulate EOs components, such as limonene, estragole, and α- and β-pinene; their solubility in water was improved more than one order of magnitude [94]. It is worth noticing that, recently, a technique based on the total organic carbon determination has been reported and validated to follow the solubility improvement of EOs when increasing the concentration of CD [95].

5. Methods for computation of binding constants: the case of EO-CD association

In this section a rather simple and straight overview of the most used model equations to compute binding constants for host-guest association is provided. To do so, EO will be used as the guest compound.

As discussed before, the phase-solubility plots are a widely applied method to get knowhow on the improvement of the EO solubility driven by complexation and also for computation of the EO-CD association constant. Assuming that the most common type of EO/CD complex has a 1:1 stoichiometry, the corresponding reaction and equilibrium (binding) constant, K_1, can be written as

$$CD + G \overset{K_1}{\rightleftarrows} CD - G \tag{1}$$

$$K_1 = [CD - G]/\left([CD]_f[G]_f\right) \tag{2}$$

where G represents the guest (here the EO) and $[CD]_f$ and $[G]_f$ are the concentrations of uncomplexed (free) species in the system. Assuming that the change in the aqueous solubility of the EO (ΔS) is only due to the formation of the complex, we can write

$$\Delta S = S_T - S_0 = [CD - G] \tag{3}$$

where S_T is the measurable total solubility and S_0 is the solubility of the EO in water in the absence of CD (i.e., the intrinsic solubility). Thus, it follows that

$$[G]_f = S_0 \tag{4}$$

$$[CD]_f = [CD]_T - (S_T - S_0) \tag{5}$$

where $[CD]_T$ is the total concentration of CD in the solution.

Substituting Eqs. (3)–(5) in Eq. (2) and after algebraic manipulation, we obtain

$$S_T = S_0 + \frac{K_1 S_0}{1 + K_1 S_0}[CD]_T \tag{6}$$

Fitting Eq. (6) to experimental data of $S_T = f([CD]_T)$ allows the calculation of the intercept (S_0) and the association constant, K_1. As discussed by Loftsson et al., the determination of K is highly dependent on the intercept accuracy [96]. In order to overcome this drawback, the authors have established the concept of "complexation efficiency" (CE) which can be obtained independently of S_0, according to the following relation:

$$CE = \frac{[CD-G]}{[CD]_f} = \frac{(S_T - S_0)/[CD]_T}{1-(S_T - S_0)/[CD]_T} \tag{7}$$

where the term $((S_T - S_0)/[CD]_T)$ represents the slope of the phase-solubility profile.

The application of Eq. (7) is useful but limited by the dependence on S_0 and to 1:1 complexes.

Let us now assume the following reaction:

$$nCD + G \overset{K_n}{\rightleftarrows} CD_n - G \tag{8}$$

where n is a stoichiometry coefficient and K_n is the corresponding binding constant. Thus, from the conservation of mass equations:

$$[G]_f = [G]_T - [CD_n - G] \tag{9}$$

and

$$[CD]_f = [CD]_T - n[CD-G] \tag{10}$$

the binding constant equation can be written as

$$K_a = \frac{[CD_n - G]}{([CD]_T - n[CD-G])^n([S]_T - [CD-G])} \tag{11}$$

By measuring a physical parameter, known as ΔA, directly related with the formation of the complex $(CD_n - G)$, and performing the experiment in such a way that $[CD]_T >> [CD - G]$, Eq. (11) takes the form of the so-called "Benesi-Hildebrand" equation [97]:

$$K_a = \frac{\Delta A}{([CD]_T)^n([S]_T - \Delta A)} \tag{12}$$

or its linear form

$$\frac{1}{\Delta A} = \frac{1}{[S]_T} + \frac{1}{K_n[S]_T([CD]_T)^n} \tag{13}$$

However, it should be pointed out that nowadays, with the available software, such as Origin® and MatLab®, there is no need to linearize Eq. (12) once such methodology brings some restrictions to the computation of K and n.

It should be stressed that a key point in all these procedures is the accurate previous knowledge of the stoichiometry of complexation, but this is not always a simple task. The most common method for the determination of stoichiometry is the method of continuous variation or the Job plot; the virtues and limitations of this method were recently reviewed [98], and thus it is not our intention to further discuss it in the present chapter.

Going back to the determination of the binding constants, the most accurate way to compute K is by using the first principles. Here, for the sake of simplicity, the 1:1 and 1:2 (G/CD) stoichiometric ratios will be focused. Additionally, these examples also correspond to the large majority of complexes formed between CDs and EOs. For more complex stoichiometries, the computational treatment of the resulting equations (not shown) is not straightforward as a consequence of multicollinearity [99]. Multicollinearity causes larger standard errors in the quantities calculated and lowers statistical significance of the results. In limiting cases, several local minima may be obtained by iteration; these correspond to noticeably different combinations of the quantities calculated and may be the reason why different K values are reported for the same host-guest systems.

Assuming that a 1:1 complex (CD-S) is formed, the binding constant (Eq. (2)) can be rewritten as

$$K_1 = \frac{f}{(1-f)([CD]_T - f[G]_T)} \tag{14}$$

where f is defined as $[CD - G]/[G]_T$.

Despite the binding process being followed by ΔA (e.g., for ^{1}H NMR, ΔA will be equal to the chemical shift of a given ^{1}H resonance), the observed ΔA for a host molecule is expressed as

$$A_{obs} = (1-f)A_{CD,f} + fA_{CD-G} \tag{15}$$

where $A_{CD,f}$ and A_{CD-G} represent the measurable physical parameter related to CD in free and complexed states, respectively.

The variation of the physical parameter in the presence and absence of a guest molecule, $\Delta A_{obs} = \Delta A_{obs} - \Delta A_{CD}$, can be expressed as

$$\Delta A_{obs} = \frac{\Delta A_{CD-G}}{[CD]_T}[CD - G] \tag{16}$$

which, after some algebraic manipulation and simplification, results in [100, 101]:

$$\Delta A_{obs} = \frac{\Delta A_{CD-G}}{2\,[CD]_T}\left\{\left([G]_T + [CD]_T + \frac{1}{K_1}\right) - \left(\left([G]_T + [CD]_T + \frac{1}{K_1}\right)^2 - 4\left([G]_T[CD]_T\right)\right)^{1/2}\right\} \tag{17}$$

It should be stressed that the application of Eq. (17) shows some drawbacks when the total concentrations of CD and guest are low and/or the binding constant is very weak, i.e., for the simplest 1:1 case, when y is sufficiently small, $x - \sqrt{x^2 - y} \approx y/2x$, and, consequently, Eq. (17) reduces to

$$\Delta A_{obs} = \frac{\Delta A_{CD-G}}{T + \left(1/K_1\right)}[G]_T \tag{18}$$

where $T = [CD]_T + [S]_T$. If T is kept constant in the experiments, as is common practice when Job plots are used to obtain stoichiometries, the observed displacement varies linearly with $[S]_T$ or $[CD]_T$, but the fitting parameters are present in the form of a ratio that generates an infinite number of acceptable solutions. Consequently, it is suggested that T should be chosen in such a way that its value should be of the same order of magnitude than K_1^{-1} [102].

Another approach lays on the assumption of a 2:1 (CD/G) complexation, in a two-step mechanism. In these circumstances, the complexation process is defined by two binding constants K_1 and K_2, and the corresponding mass balances are defined as

$$[G]_f = [G]_T - [CD-G] - [CD_2-G] \tag{19}$$

and

$$[CD]_f = [CD]_T - [CD-G] - 2[CD_2-G] \tag{20}$$

From the equilibrium constants and Eqs. (19) and (20), we can write

$$A_{obs} = \frac{[CD]_f A_{CD} + [CD-G]A_{CD-G} + 2[CD_2-G]A_{CD_2-G}}{[CD]_f + [CD-G] + 2[CD_2-G]} \tag{21}$$

where A_{CD}, A_{CD-G}, and A_{CD2-G} are the contributions of the CD and 1:1 and 2:1 CD/G complexes, with concentrations $[CD]$, $[CD-G]$, and $[CD_2-G]$, respectively, for the observed (experimental) physical parameter A. Using a similar procedure to that used for a 1:1 complexation, it is possible to write Eq. (21) as a function of $[CD]$, that is

$$A_{obs} = \frac{A_{CD} + [CD]K_1 A_{CD-G} + K_1 K_2 [CD] A_{CD_2-G}}{1 + K_1[CD]_f + K_1 K_2 [CD]^2} \tag{22}$$

On the other hand, the free CD concentration is given by

$$[CD]^3 + \left(\frac{1}{K_{2,1}} - [CD]_T + 2[S]_T\right)[CD]^2 + \left(\frac{1}{K_{1,1}K_{2,1}} - \frac{[CD]_T}{K_{2,1}} + \frac{[S]_T}{K_{2,1}}\right)[CD] - \frac{[CD]_T}{K_{1,1}K_{2,1}} = 0 \tag{23}$$

One method for estimation of the free CD concentration is through an analytical solution of the real solution of a third-degree equation [103]:

$$f(x) = x^3 + ax^2 + bx + c \tag{24}$$

using a Cardin-Tartaglia formulae

$$x = r - \frac{1}{3}a - \frac{q}{r} \tag{25}$$

where

$$q = \frac{1}{3}b - \frac{1}{9}a^2 \quad (26)$$

and

$$r = \sqrt[3]{\frac{1}{6}ab - \frac{1}{2}c - \frac{1}{27}a^3 + \sqrt{\frac{1}{27}b^3 - \frac{1}{6}abc + \frac{1}{4}c^2 + \frac{1}{27}a^3c - \frac{1}{108}a^2b^2}} \quad (27)$$

6. Conclusions

Despite the huge potential of many bio-based compounds in several diverse areas, their use is still limited due to different reasons such as poor aqueous solubility, volatility, reactivity, etc. Therefore, advanced strategies have to be developed in order to minimize some of these weaknesses to make bio-based molecules usable on a larger scale. It became patent in this chapter that CD interaction with bio-based compounds, such as different phytochemicals or sugar-based surfactants, has generally a remarkable positive impact on their performance, improving their aqueous solubility and availability and decreasing their degradation rate, etc. Moreover, it is clear that this is not only interesting and beneficial from an application point of view but also very stimulating from a fundamental perspective where thermodynamics, modeling, and different experimental methodologies get together for a deep and challenging characterization of the systems. Eventually, such knowledge will be crucial for the future development of improved formulations and make use at full extent of the exciting properties of bio-based compounds.

Acknowledgements

The Portuguese Foundation for Science and Technology (FCT) is acknowledged through the projects PTDC/AGR-TEC/4814/2014, UID/QUI/UI0313/2013, and POCI-01-0145-FEDER-07630. FCT is also acknowledged through the researcher grants IF/01005/2014, SFRH/BPD/103086/2014, and SFRH/BPD/84112/2012 financed by POPH-QREN and subsidized by the European Science Foundation.

Author details

Bruno Filipe Figueiras Medronho[1]*, Sandra Gonçalves[1], Raquel Rodríguez-Solana[1], Artur J.M. Valente[2] and Anabela Romano[1]

*Address all correspondence to: bfmedronho@ualg.pt

1 Faculty of Sciences and Technology (MEDITBIO), University of Algarve, Faro, Portugal

2 CQC, Department of Chemistry, University of Coimbra, Coimbra, Portugal

References

[1] Del Valle EMM. Cyclodextrins and their uses: A review. Process Biochemistry. 2004; **39**(9):1033-1046. DOI: 10.1016/S0032-9592(03)00258-9

[2] Jiang LX, Yan Y, Huang JB. Versatility of cyclodextrins in self-assembly systems of amphiphiles. Advances in Colloid and Interface Science. 2011;**169**(1):13-25. DOI: 10.1016/j.cis.2011.07.002

[3] Medronho B, Andrade R, Vivod V, Ostlund A, Miguel MG, Lindman B, Voncina B, Valente AJM. Cyclodextrin-grafted cellulose: Physico-chemical characterization. Carbohydrate Polymers. 2013;**93**(1):324-330. DOI: 10.1016/j.carbpol.2012.08.109

[4] Nilsson M, Cabaleiro-Lago C, Valente AJM, Soderman O. Interactions between gemini surfactants, 12-s-12, and beta-cyclodextrin as investigated by NMR diffusometry and electric conductometry. Langmuir. 2006;**22**(21):8663-8669. DOI: 10.1021/la061220e

[5] Valente AJM, Soderman O. The formation of host-guest complexes between surfactants and cyclodextrins. Advances in Colloid and Interface Science. 2014;**205**:156-176. DOI: 10.1016/j.cis.2013.08.001

[6] Dias HMAM, Berbicz F, Pedrochi F, Baesso ML, Matioli G. Butter cholesterol removal using different complexation methods with beta-cyclodextrin, and the contribution of photoacoustic spectroscopy to the evaluation of the complex. Food Research International. 2010;**43**(4):1104-1110. DOI: 10.1016/j.foodres.2010.02.002

[7] Lakkakula JR, Krause RWM. A vision for cyclodextrin nanoparticles in drug delivery systems and pharmaceutical applications. Nanomedicine-UK. 2014;**9**(6):877-894. DOI: 10.2217/Nnm.14.41

[8] Szejtli J. Past, present, and future of cyclodextrin research. Pure and Applied Chemistry. 2004;**76**(10):1825-1845. DOI: 10.1351/pac200476101825

[9] Crozier A, Jaganath IB, Clifford MN. Dietary phenolics: Chemistry, bioavailability and effects on health. Natural Product Reports. 2009;**26**(8):1001-1043. DOI: 10.1039/b802662a

[10] Manach C, Scalbert A, Morand C, Remesy C, Jimenez L. Polyphenols: Food sources and bioavailability. The American Journal of Clinical Nutrition. 2004;**79**(5):727-747

[11] Vishwakarma GS, Gautam N, Babu JN, Mittal S, Jaitak V. Polymeric encapsulates of essential oils and their constituents: A review of preparation techniques, characterization, and sustainable release mechanisms. Polymer Reviews. 2016;**56**(4):668-701. DOI: 10.1080/15583724.2015.1123725

[12] Suvarna V, Gujar P, Murahari M. Complexation of phytochemicals with cyclodextrin derivatives—An insight. Biomedicine & Pharmacotherapy. 2017;**88**:1122-1144. DOI: 10.1016/j.biopha.2017.01.157

[13] Pinho E, Grootveld M, Soares G, Henriques M. Cyclodextrins as encapsulation agents for plant bioactive compounds. Carbohydrate Polymers. 2014;**101**:121-135. DOI: 10.1016/j.carbpol.2013.08.078

[14] Marques HMC. A review on cyclodextrin encapsulation of essential oils and volatiles. Flavour and Fragrance Journal. 2010;**25**(5):313-326. DOI: 10.1002/ffj.2019

[15] Esfanjani AF, Jafari SM. Biopolymer nano-particles and natural nano-carriers for nano-encapsulation of phenolic compounds. Colloids and Surfaces B: Biointerfaces. 2016;**146**: 532-543. DOI: 10.1016/j.colsurfb.2016.06.053

[16] Loftsson T, Duchene D. Cyclodextrins and their pharmaceutical applications. International Journal of Pharmaceutics. 2007;**329**(1–2):1-11. DOI: 10.1016/j.ijpharm.2006.10.044

[17] Matsuda H, Arima H. Cyclodextrins in transdermal and rectal delivery. Advanced Drug Delivery Reviews. 1999;**36**(1):81-99. DOI: 10.1016/S0169-409x(98)00056-8

[18] Fang ZX, Bhandari B. Encapsulation of polyphenols—A review. Trends in Food Science & Technology. 2010;**21**(10):510-523. DOI: 10.1016/j.tifs.2010.08.003

[19] Budryn G, Palecz B, Rachwal-Rosiak D, Oracz J, Zaczynska D, Belica S, Navarro-Gonzalez I, Meseguer JMV, Perez-Sanchez H. Effect of inclusion of hydroxycinnamic and chlorogenic acids from green coffee bean in beta-cyclodextrin on their interactions with whey, egg white and soy protein isolates. Food Chemistry. 2015;**168**:276-287. DOI: 10.1016/j.foodchem.2014.07.056

[20] Budryn G, Zaczynska D, Rachwal-Rosiak D, Oracz J. Changes in properties of food proteins after interaction with free and beta-cyclodextrin encapsulated hydroxycinnamic acids. European Food Research and Technology. 2015;**240**(6):1157-1166. DOI: 10.1007/s00217-015-2419-9

[21] Rawel HM, Rohn S. Nature of hydroxycinnamate-protein interactions. Phytochemistry Reviews. 2010;**9**(1):93-109. DOI: 10.1007/s11101-009-9154-4

[22] Shao P, Zhang JF, Fang ZX, Sun PL. Complexing of chlorogenic acid with beta-cyclodextrins: Inclusion effects, antioxidative properties and potential application in grape juice. Food Hydrocolloids. 2014;**41**:132-139. DOI: 10.1016/j.foodhyd.2014.04.003

[23] Mercader-Ros MT, Lucas-Abellan C, Fortea MI, Gabaldon JA, Nunez-Delicado E. Effect of HP-beta-cyclodextrins complexation on the antioxidant activity of flavonols. Food Chemistry. 2010;**118**(3):769-773. DOI: 10.1016/j.foodchem.2009.05.061

[24] Cutrignelli A, Lopedota A, Denora N, Iacobazzi RM, Fanizza E, Laquintana V, Perrone M, Maggi V, Franco M. A new complex of curcumin with sulfobutylether-beta-cyclodextrin: Characterization studies and *in vitro* evaluation of cytotoxic and antioxidant activity on HepG-2 cells. Journal of Pharmaceutical Sciences. 2014;**103**(12):3932-3940. DOI: 10.1002/jps.24200

[25] Davis ME, Brewster ME. Cyclodextrin-based pharmaceutics: Past, present and future. Nature Reviews. Drug Discovery. 2004;**3**(12):1023-1035. DOI: 10.1038/nrd1576

[26] Stella VJ, Rajewski RA. Cyclodextrins: Their future in drug formulation and delivery. Pharmaceutical Research. 1997;**14**(5):556-567. DOI: 10.1023/A:1012136608249

[27] Gong L, Li TT, Chen F, Duan XW, Yuan YF, Zhang DD, Jiang YM. An inclusion complex of eugenol into beta-cyclodextrin: Preparation, and physicochemical and antifungal characterization. Food Chemistry. 2016;**196**:324-330. DOI: 10.1016/j.foodchem.2015.09.052

[28] Aytac Z, Yildiz ZI, Kayaci-Senirmak F, Tekinay T, Uyar T. Electrospinning of cyclodextrin/linalool-inclusion complex nanofibers: Fast-dissolving nanofibrous web with prolonged release and antibacterial activity. Food Chemistry. 2017;**231**:192-201. DOI: 10.1016/j.foodchem.2017.03.113

[29] Ciobanu A, Mallard I, Landy D, Brabie G, Nistor D, Fourmentin S. Retention of aroma compounds from *Mentha piperita* essential oil by cyclodextrins and crosslinked cyclodextrin polymers. Food Chemistry. 2013;**138**(1):291-297. DOI: 10.1016/j.foodchem.2012.10.106

[30] Aytac Z, Ipek S, Durgun E, Tekinay T, Uyar T. Antibacterial electrospun zein nanofibrous web encapsulating thymol/cyclodextrin-inclusion complex for food packaging. Food Chemistry. 2017;**233**:117-124. DOI: 10.1016/j.foodchem.2017.04.095

[31] Kfoury M, Auezova L, Greige-Gerges H, Fourmentin S. Promising applications of cyclodextrins in food: Improvement of essential oils retention, controlled release and antiradical activity. Carbohydrate Polymers. 2015;**131**:264-272. DOI: 10.1016/j.carbpol.2015.06.014

[32] Bomfim LM, Menezes LRA, Rodrigues ACBC, Dias RB, Rocha CAG, Soares MBP, Neto AFS, Nascimento MP, Campos AF, Silva LCRCE, Costa EV, Bezerra DP. Antitumour activity of the microencapsulation of *Annona vepretorum* essential oil. Basic and Clinical Pharmacology. 2016;**118**(3):208-213. DOI: 10.1111/bcpt.12488

[33] Lima PSS, Lucchese AM, Araujo HG, Menezes PP, Araujo AAS, Quintans LJ, Quintans JSS. Inclusion of terpenes in cyclodextrins: Preparation, characterization and pharmacological approaches. Carbohydrate Polymers. 2016;**151**:965-987. DOI: 10.1016/j.carbpol.2016.06.040

[34] Menezes PD, Araujo AAD, Doria GAA, Quintans LJ, de Oliveira MGB, dos Santos MRV, de Oliveira JF, Matos JD, Carvalho FMD, Alves PB, de Matos IL, dos Santos DA, Marreto RN, da Silva GF, Serafini MR. Physicochemical characterization and analgesic effect of inclusion complexes of essential oil from *Hyptis pectinata* L. Poit leaves with beta-cyclodextrin. Current Pharmaceutical Biotechnology. 2015;**16**(5):440-450

[35] Falbe J. Surfactants in Consumer Products—Theory, Technology and Application. Berlin: Springer; 1986

[36] Holmberg K, Jonsson B, Kronberg B, Lindman B. Surfactants and Polymers in Aqueous Solution. Hoboken: Wiley; 2002

[37] von Rybinski W. Natural surfactants. Current Opinion in Colloid & Interface Science. 2001;**6**(2):146-147. DOI: 10.1016/S1359-0294(01)00081-4

[38] Hill K, Rhode O. Sugar-based surfactants for consumer products and technical applications. Fett/Lipid. 1999;**101**:25-33

[39] Dembitsky VM. Astonishing diversity of natural surfactants: 1. Glycosides of fatty acids and alcohols. Lipids. 2004;**39**(10):933-953. DOI: 10.1007/s11745-004-1316-1

[40] Queneau Y, Chambert S, Besset C, Cheaib R. Recent progress in the synthesis of carbohydrate-based amphiphilic materials: The examples of sucrose and isomaltulose. Carbohydrate Research. 2008;**343**(12):1999-2009. DOI: 10.1016/j.carres.2008.02.008

[41] Ruiz C. Sugar-based Surfactants Fundamentals and Applications. CRC Press; 2009

[42] Hill K, LeHen-Ferrenbach C. Sugar-based surfactants for consumer products and technical applications. In: Sugar-Based Surfactants Fundamentals and Applications. CRC Press; 2009. pp. 1-20

[43] Kitamoto D, Morita T, Fukuoka T, Konishi M, Imura T. Self-assembling properties of glycolipid biosurfactants and their potential applications. Current Opinion in Colloid & Interface Science. 2009;**14**(5):315-328. DOI: 10.1016/j.cocis.2009.05.009

[44] Razafindralambo H, Blecker C, Paquot M. Screening of basic properties of amphiphilic molecular structures for colloidal system formation and stability: The case of carbohydrate-based surfactants. In: Nagarajan R, editor. Amphiphiles: Molecular Assembly and Applications. Vol. 1070. Washignton: ACS; 2011

[45] Soderman O, Johansson I. Polyhydroxyl-based surfactants and their physico-chemical properties and applications. Current Opinion in Colloid & Interface Science. 1999;**4**(6): 391-401. DOI: 10.1016/S1359-0294(00)00019-4

[46] vonRybinski W. Alkyl glycosides and polyglycosides. Current Opinion in Colloid & Interface Science. 1996;**1**(5):587-597

[47] von Rybinski W, Hill K. Alkyl polyglycosides—Properties and applications of a new class of surfactants. Angewandte Chemie International Edition. 1998;**37**(10):1328-1345. DOI: 10.1002/(Sici)1521-3773(19980605)37:10<1328::Aid-Anie1328>3.0.Co;2-9

[48] Shinoda K, Carlsson A, Lindman B. On the importance of hydroxyl groups in the polar head-group of nonionic surfactants and membrane lipids. Advances in Colloid and Interface Science. 1996;**64**:253-271

[49] Lindman B, Medronho B, Karlström G. Clouding of nonionic surfactants. Current Opinion in Colloid & Interface Science. 2016;**22**:23-29. DOI: 10.1016/j.cocis.2016.01.005

[50] Fukuda K, Soderman O, Lindman B, Shinoda K. Microemulsions formed by alkyl polyglucosides and an alkyl glycerol ether. Langmuir. 1993;**9**(11):2921-2925. DOI: 10.1021/La00035a032

[51] Hill K, Von Rybinski W, Stoll G. Alkyl Polyglucosides: Technology, Properties and Applications. New York: VCH; 1997

[52] Costes F, Elghoul M, Bon M, Ricolattes I, Lattes A. Synthesis and structural-analysis of long-chain N-acetyl-N-alkyllactosylamines, a new series of surfactants derived from unprotected lactose. Langmuir. 1995;**11**(10):3644-3647. DOI: 10.1021/La00010a010

[53] Holmberg K. Natural surfactants. Current Opinion in Colloid & Interface Science. 2001; **6**(2):148-159. DOI: 10.1016/S1359-0294(01)00074-7

[54] von Rybinski W, Hill K. Alkyl polyglycosides. In: Holmberg K, editor. Novel Surfactants. New York: Marcel Dekker; 1998

[55] Saenger W, Mullerfahrnow A. Cyclodextrins increase surface-tension and critical micelle concentrations of detergent solutions. Angewandte Chemie-International Edition in English. 1988;**27**(3):393-394. DOI: 10.1002/anie.198803931

[56] Casu B, Grenni A, Naggi A, Torri G, Virtuani M, Focher B. Interaction of cyclodextrins (cyclomalto-oligosaccharides) with glycolipids-NMR-studies of aqueous systems of cyclomaltohexaose and alkyl glycosides. Carbohydrate Research. 1990;**200**:101-109. DOI: 10.1016/0008-6215(90)84185-W

[57] Carvalho RA, Correia HA, Valente AJM, Soderman O, Nilsson M. The effect of the head-group spacer length of 12-s-12 gemini surfactants in the host-guest association with beta-cyclodextrin. Journal of Colloid and Interface Science. 2011;**354**(2):725-732. DOI: 10.1016/j.jcis.2010.11.024

[58] AJM V, Nilsson M, Soderman O. Interactions between n-octyl and n-nonyl beta-D-glucosides and alpha- and beta-cyclodextrins as seen by self-diffusion NMR. Journal of Colloid and Interface Science. 2005;**281**(1):218-224. DOI: 10.1016/j.jcis.2004.08.018

[59] Reinsborough VC, Stephenson VC. Inclusion complexation involving sugar-containing species: Beta-cyclodextrin and sugar surfactants. Canadian Journal of Chemistry. 2004;**82**(1):45-49. DOI: 10.1139/V03-180

[60] Bernat V, Ringard-Lefebvre C, Le Bas G, Perly B, Djedaini-Pilard F, Lesieur S. Inclusion complex of n-octyl beta-D-glucopyranoside and alpha-cyclodextrin in aqueous solutions: Thermodynamic and structural characterization. Langmuir. 2008;**24**(7):3140-3149. DOI: 10.1021/la7034906

[61] Plot J. Formation and stability of inorganic complexes in solution. Annali di chimica. 1928;**9**:113-203

[62] Rymdén R, Carlfors J, Stilbs P. Substrate binding to cyclodextrins in aqueous solution: A multicomponent self-diffusion study. Journal of Inclusion Phenomena. 1983;**1**:159-167

[63] Castronuovo G, Elia V, Niccoli M, Velleca F, Viscardi G. Role of the functional group in the formation of the complexes between alpha-cyclodextrin and alkanols or monocarboxylic acids in aqueous solutions. A calorimetric study at 25 degrees C. Carbohydrate Research. 1998;**306**(1–2):147-155. DOI: 10.1016/S0008-6215(97)10046-5

[64] Castronuovo G, Elia V, Iannone A, Niccoli M, Velleca F. Factors determining the formation of complexes between alpha-cyclodextrin and alkylated substances in aqueous solutions: A calorimetric study at 25 degrees C. Carbohydrate Research. 2000;**325**(4):278-286. DOI: 10.1016/S0008-6215(99)00328-6

[65] Ribeiro ACF, Esteso MA, Lobo VMM, Valente AJM, Simoes SMN, Sobral AJFN, Burrows HD. Interactions of copper (II) chloride with sucrose, glucose, and fructose in aqueous solutions. Journal of Molecular Structure. 2007;**826**(2–3):113-119. DOI: 10.1016/j.molstruc.2006.04.035

[66] Marcus Y. Effect of ions on the structure of water: Structure making and breaking. Chemical Reviews. 2009;**109**(3):1346-1370. DOI: 10.1021/cr8003828

[67] Haller J, Kaatze U. Octyl glucopyranoside and cyclodextrin in water. Self-aggregation and complex formation. The Journal of Physical Chemistry. B. 2009;**113**(7):1940-1947. DOI: 10.1021/jp808733p

[68] Mwakibete H, Cristantino R, Bloor DM, Wynjones E, Holzwarth JF. Reliability of the experimental methods to determine equilibrium-constants for surfactant cyclodextrin inclusion complexes. Langmuir. 1995;**11**(1):57-60. DOI: 10.1021/La00001a013

[69] Brocos P, Banquy X, Diaz-Vergara N, Perez-Casas S, Pineiro A, Costas M. A critical approach to the thermodynamic characterization of inclusion complexes: Multiple-temperature isothermal titration calorimetric studies of native cyclodextrins with sodium dodecyl sulfate. The Journal of Physical Chemistry. B. 2011;**115**(49):14381-14396. DOI: 10.1021/jp208740b

[70] Lawtrakul L, Inthajak K, Toochinda P. Molecular calculations on beta-cyclodextrin inclusion complexes with five essential oil compounds from *Ocimum basilicum* (sweet basil). ScienceAsia. 2014;**40**(2):145-151. DOI: 10.2306/scienceasia1513-1874.2014.40.145

[71] Kfoury M, Landy D, Ruellan S, Auezova L, Greige-Gerges H, Fourmentin S. Nootkatone encapsulation by cyclodextrins: Effect on water solubility and photostability. Food Chemistry. 2017;**236**:41-48. DOI: 10.1016/j.foodchem.2016.12.086

[72] Menezes PD, Doria GAA, Araujo AAD, Sousa BMH, Quintans LJ, Lima RN, Alves PB, Carvalho FMS, Bezerra DP, Mendonca FJB, Scotti L, Scotti MT, da Silva GF, de Aquino TM, Sabino AR, do Egito EST, Serafini MR. Docking and physico-chemical properties of alpha- and beta-cyclodextrin complex containing isopulegol: A comparative study. Journal of Inclusion Phenomena and Macrocyclic Chemistry. 2016;**85**(3–4):341-354. DOI: 10.1007/s10847-016-0633-0

[73] Schneider HJ, Hacket F, Rudiger V, Ikeda H. NMR studies of cyclodextrins and cyclodextrin complexes. Chemical Reviews. 1998;**98**(5):1755-1785. DOI: 10.1021/Cr970019t

[74] Azzi J, Danjou PE, Landy D, Ruellan S, Auezova L, Greige-Gerges H, Fourmentin S. The effect of cyclodextrin complexation on the solubility and photostability of nerolidol as pure compound and as main constituent of cabreuva essential oil. Beilstein Journal of Organic Chemistry. 2017;**13**:835-844. DOI: 10.3762/bjoc.13.84

[75] Medronho B, Valente AJM, Costa P, Romano A. Inclusion complexes of rosmarinic acid and cyclodextrins: Stoichiometry, association constants, and antioxidant potential. Colloid & Polymer Science. 2014;**292**(4):885-894. DOI: 10.1007/s00396-013-3124-5

[76] Rakmai J, Cheirsilp B, Mejuto JC, Torrado-Agrasar A, Simal-Gandara J. Physico-chemical characterization and evaluation of bio-efficacies of black pepper essential oil encapsulated in hydroxypropyl-beta-cyclodextrin. Food Hydrocolloids. 2017;**65**:157-164. DOI: 10.1016/j.foodhyd.2016.11.014

[77] Kfoury M, Sahraoui ALH, Bourdon N, Laruelle F, Fontaine J, Auezova L, Greige-Gerges H, Fourmentin S. Solubility, photostability and antifungal activity of phenylpropanoids encapsulated in cyclodextrins. Food Chemistry. 2016;**196**:518-525. DOI: 10.1016/j.foodchem.2015.09.078

[78] Yang ZJ, Huang LY, Yao XD, Ji HB. Host-guest complexes of estragole with beta-cyclodextrin: An experimental and theoretical investigation. Flavour and Fragrance Journal. 2017;**32**(2):102-111. DOI: 10.1002/ffj.3358

[79] Cameron KS, Fielding L. NMR diffusion spectroscopy as a measure of host-guest complex association constants and as a probe of complex size. The Journal of Organic Chemistry. 2001;**66**(21):6891-6895. DOI: 10.1021/Jo010081x

[80] Kfoury M, Landy D, Ruellan S, Auezova L, Greige-Gerges H, Fourmentin S. Determination of formation constants and structural characterization of cyclodextrin inclusion complexes with two phenolic isomers: Carvacrol and thymol. Beilstein Journal of Organic Chemistry. 2016;**12**:29-42. DOI: 10.3762/bjoc.12.5

[81] Kato S, Nomura H, Miyahara Y. Ultrasonic relaxation study of aqueous-solutions of cyclodextrins. Journal of Physical Chemistry. 1985;**89**(25):5417-5421. DOI: 10.1021/J100271a021

[82] Hall D, Bloor D, Tawarah K, Wynjones E. Kinetic and equilibrium studies associated with the formation of inclusion-compounds involving normal-butanol and normal-pentanol in aqueous cyclodextrin solutions. Journal of the Chemical Society, Faraday Transactions. 1986;**1**(82):2111-2121. DOI: 10.1039/F19868202111

[83] Kaatze U. Acoustical spectroscopy of carbohydrate aqueous solutions: Saccharides; alkyl glycosides; Cyclodextrins. Part I. Conformer variations. Archives of Acoustics. 2010;**35**(4): 715-738. DOI: 10.2478/v10168-010-0054-9

[84] Tariq M, Freire MG, Saramago B, Coutinho JAP, Lopes JNC, Rebelo LPN. Surface tension of ionic liquids and ionic liquid solutions. Chemical Society Reviews. 2012;**41**(2):829-868. DOI: 10.1039/c1cs15146k

[85] Szejtli J. Utilization of cyclodextrins in industrial products and processes. Journal of Materials Chemistry. 1997;**7**(4):575-587. DOI: 10.1039/A605235e

[86] Pineiro A, Banquy X, Perez-Casas S, Tovar E, Garcia A, Villa A, Amigo A, Mark AE, Costas M. On the characterization of host-guest complexes: Surface tension, calorimetry, and molecular dynamics of cyclodextrins with a non-ionic surfactant. The Journal of Physical Chemistry. B. 2007;**111**(17):4383-4392. DOI: 10.1021/jp0688815

[87] Wadso L, Li YJ, Li X. Isothermal titration calorimetry in the student laboratory. Journal of Chemical Education. 2011;**88**(1):101-105. DOI: 10.1021/ed100649e

[88] Nilsson M, Valente AJM, Olofsson G, Soderman O, Bonini M. Thermodynamic and kinetic characterization of host-guest association between bolaform surfactants and alpha- and beta-cyclodextrins. The Journal of Physical Chemistry. B. 2008;**112**(36):11310-11316. DOI: 10.1021/jp802963x

[89] Russell DJ, Hansen LD. Calorimeters for biotechnology. Thermochimica Acta. 2006; **445**(2):151-159. DOI: 10.1016/j.tca.2005.08.023

[90] Wen P, Zhu DH, Feng K, Liu FJ, Lou WY, Li N, Zong MH, Wu H. Fabrication of electrospun polylactic acid nanofilm incorporating cinnamon essential oil/beta-cyclodextrin inclusion complex for antimicrobial packaging. Food Chemistry. 2016;**196**: 996-1004. DOI: 10.1016/j.foodchem.2015.10.043

[91] Khanna S, Chakraborty JN. Optimization of monochlorotriazine beta-cyclodextrin grafting on cotton and assessment of release behavior of essential oils from functionalized fabric. Fash Text. 2017;**4**:1-18. DOI: 10.1186/S40691-017-0089-X

[92] Yang ZJ, Yao XD, Xiao ZB, Chen HY, Ji HB. Preparation and release behaviour of the inclusion complexes of phenylethanol with beta-cyclodextrin. Flavour and Fragrance Journal. 2016;**31**(3):206-216. DOI: 10.1002/ffj.3302

[93] Russo M, Rigano F, Arigo A, Sciarrone D, Calabro ML, Farnetti S, Dugo P, Mondello L. Rapid isolation, reliable characterization, and water solubility improvement of polymethoxyflavones from cold-pressed mandarin essential oil. Journal of Separation Science. 2016;**39**(11):2018-2027. DOI: 10.1002/jssc.201501366

[94] Kfoury M, Pipkin JD, Antle V, Fourmentin S. Captisol (R): An efficient carrier and solubilizing agent for essential oils and their components. Flavour and Fragrance Journal. 2017;**32**(5):340-346. DOI: 10.1002/ffj.3395

[95] Kfoury M, Auezova L, Greige-Gerges H, Fourmentin S. Development of a total organic carbon method for the quantitative determination of solubility enhancement by cyclodextrins: Application to essential oils. Analytica Chimica Acta. 2016;**918**:21-25. DOI: 10.1016/j.aca.2016.03.013

[96] Loftsson T, Hreinsdottir D, Masson M. Evaluation of cyclodextrin solubilization of drugs. International Journal of Pharmaceutics. 2005;**302**(1–2):18-28. DOI: 10.1016/j.ijpharm.2005.05.042

[97] Benesi HA, Hildebrand JH. A spectrophotometric investigation of the interaction of iodine with aromatic hydrocarbons. Journal of the American Chemical Society. 1949; **71**(8):2703-2707. DOI: 10.1021/Ja01176a030

[98] Ulatowski F, Dabrowa K, Balakier T, Jurczak J. Recognizing the limited applicability of job plots in studying host-guest interactions in supramolecular chemistry. The Journal of Organic Chemistry. 2016;**81**(5):1746-1756. DOI: 10.1021/acs.joc.5b02909

[99] Draper NR, Smith H. Applied Regression Analysis. New York: Wiley-Interscience; 1998

[100] Cabrer PR, Azvarez-Parrilla E, Meijide F, Seijas JA, Nunez ER, Tato JV. Complexation of sodium cholate and sodium deoxycholate by beta-cyclodextrin and derivatives. Langmuir. 1999;**15**(17):5489-5495

[101] Calderon V, Schwarz G, Garcia F, Tapia MJ, Valente AJM, Burrows HD, Garcia JM. Synthesis and characterization of new aromatic polyamides bearing crown ethers and acyclic ethylene oxide units in the pendant structure. III. Benzo-18-crown-6 systems and their open-chain counterparts. Journal of Polymer Science, Part A: Polymer Chemistry. 2006;**44**(21):6252-6269. DOI: 10.1002/pola.21710

[102] Tapia MJ, Burrows HD, Garcia JM, Garcia F, Pais AACC. Lanthanide ion interaction with a crown ether methacrylic polymer, poly(1,4,7,10-tetraoxacyclododecan-2-ylmethyl methacrylate), as seen by spectroscopic, calorimetric, and theoretical studies. Macromolecules. 2004;**37**(3):856-862. DOI: 10.1021/ma0353888

[103] Valente AJM, Dinis CJS, Pereira RFP, Ribeiro ACF, Lobo VMM. Interactions between β-cyclodextrin and some sodium alkyl sulfates and sulfonates as seen by electrical conductivity measurements. Portugaliae Electrochimica Acta. 2006;**24**:129-136

Chapter 4

Use of 2-Hydroxypropyl-Beta-Cyclodextrin for Niemann-Pick Type C Disease

Juan Eduardo Megías-Vericat,
María José Company-Albir,
Ana Alejandra García-Robles and José Luis Poveda

Additional information is available at the end of the chapter

http://dx.doi.org/10.5772/intechopen.71970

Abstract

Niemann-Pick disease type C (NPD-C) is a rare neurodegenerative disorder characterized by a lysosomal storage disorder. Treatment has been supportive and symptomatic. In animal studies, 2-hydroxypropyl-β-cyclodextrin (HP-β-CD) showed a significant decrease in cerebellar damage, neurological progression, and higher lifespan. Based on these results, HP-β-CD has been tested in NPD-C patients for last 8 years. The first compassionate uses of intravenous HP-β-CD obtained a limited improvement in neurological symptoms, probably associated to the non-permeation of the blood-brain barrier. The change or combination with intrathecal administrations of HP-β-CD achieved higher benefits, especially improvement or stabilization of NPD-C progression. Biomarkers of neurological cholesterol homeostasis are being investigated in order to quantify the response of HP-β-CD treatment. The results of a clinical trial recently published have reproduced the slowing of NPD-C progression in 14 patients treated with a dose-escalation protocol of HP-β-CD intrathecal monthly infusions, with respect to a historical comparison cohort. The safety profile of this therapy is acceptable, being the loss of hearing as the most frequent adverse event. However, some severe toxicities have been reported in relation with HP-β-CD, including chemical meningitis and fever. The short experience with HP-β-CD suggested that it could be effective in the management of NPD-C.

Keywords: Niemann-pick type C, 2-hydroxypropyl-β-cyclodextrin, cholesterol, neurodegenerative diseases, blood-brain-barrier

IntechOpen

1. Introduction

1.1. Etiology

Niemann-Pick disease type C (NPD-C) is a rare autosomal recessive disorder characterized by lysosomal lipid storage in which faulty intracellular lipid transport leads to accumulation of unesterified cholesterol and glycosphingolipids in several neurovisceral tissues [1, 2].

NPD-C is caused by mutations in either the *NPC1* or *NPC2* genes. The *NPC1* gene, located on the long arm of chromosome 18 (18q11.2), encodes a large (142 kDa) membrane glycoprotein placed in endosomes and lysosomes. This protein mediates intracellular cholesterol trafficking via binding of cholesterol to its N-terminal domain. *NPC2* gene, located on the long arm of chromosome 14 (14q24.3), encodes a small (16 kDa) lysosomal protein that binds to cholesterol. NPC1 and NPC2 proteins seem to act in the cooperative transportation of molecules within cells. Recently, it has been shown, the N-terminal domain of NPC1 protein may interact with NPC2 protein to facilitate cholesterol efflux from the late endosome and lysosomes [3, 4].

The majority of patients show mutations in the *NPC1* gene (95%), whereas a much smaller number suffer mutations in the *NPC2* gene, but the resulting phenotypes are clinically indistinguishable. Loss of function of either of these proteins results in an accumulation of cholesterol and other lipids, including sphingomyelin, sphingosine, and gangliosides (GM2 and GM3), within the endosomes and lysosome [3, 4].

1.2. Diagnosis

The diagnosis of NPD-C requires a combination of clinical, cellular, and molecular criteria. NPD-C is suspected on the basis of the clinical features. Systemic manifestations such as hepatosplenomegaly neonatal, cholestatic jaundice, or splenomegaly can lead to diagnosis, but, due to the heterogeneous clinical phenotype, diagnosis is often delayed for many years or missed altogether [4]. A Suspicion Index tool might be useful as a screen for NPD-C, a risk prediction score ≥70 indicates a strong suspicion for NPD-C [5]. The diagnosis of NPD-C is confirmed by biochemical testing that demonstrates impaired cholesterol esterification and positive filipin staining in cultured fibroblasts obtained from a skin biopsy. Filipin staining demonstrates intense punctate pattern of fluorescence concentrated around the nucleus, consistent with unesterified cholesterol. Molecular genetic testing of *NPC1* and *NPC2* are commercially available and they detect pathogenic variants in approximately 94% of individuals with NPD-C [4].

The filipin assay is unable to provide a firm diagnosis in fibroblasts with "variant" phenotypes that represent one-third of NPD-C cases. Moreover, the assay is invasive and the results can delay for 3 months. Genetic analysis is an important diagnostic tool, though, due to cost considerations, generally applied as a confirmatory rather than screening test. In recent years have been identified promising NPD-C biomarkers. One of these markers, cholestane-3β,5α,6β-triol, which is a cholesterol oxidation product, has emerged as a sensitive diagnostic for NPD-C [6, 7].

1.3. Clinical presentation and symptoms of Niemann-Pick disease

Systemic involvement of liver, spleen, or lung, is present in ≥85% of patients, and precedes the development of neurologic symptoms. The age of onset and clinical presentation of NPD-C is highly variable.

Neonatal and infantile presentations: Occasionally, ultrasound examination in late pregnancy has detected fetal ascites; infants thus identified typically have a severe neonatal liver disease with jaundice and persistent ascites. Infiltration of the lungs with foam cells can be present. Many infants die at this stage. Of children who survive, many have hypotonia and psychomotor retardation whereas others may have complete resolution of symptoms, only to present with neurologic disease many years later [4].

Childhood presentations: These patients typically have cerebellar involvement characterized by clumsiness and gait problems progressing to frank ataxia and slow cognitive deterioration Vertical supranuclear ophthalmoplegia is another early manifestation. Progressive dystonia, dysarthria, and dysphagia occur, eventually impairing oral feeding, and approximately one-third of patients develop seizures [4].

Adolescent and adult presentations: The clinical presentation is similar to childhood onset with ataxia, supranuclear vertical gaze palsy, cognitive impairment, except that progression is generally much slower. Other adults present with cognitive dysfunction or psychiatric disturbances as major depression, schizophrenia, or bipolar disorder [4].

1.4. Previous treatments employed

There is no curative treatment for NPD-C. The disease management is individualized and it consists mainly of symptomatic treatment. Seizures, dystonia, and cataplexy can respond to drugs. Other symptomatic measures like physiotherapy in spasticity, gastrostomy tube placement to prevent aspiration and/or inadequate nutrition in patients with progressive dysphagia, and bronchoalveolar lavage to improve pulmonary function are useful in the disease management. Combination of drug regimens have been shown to lower hepatic and plasma cholesterol but there is no evidence that these results affect the progression of the disease in humans or murine models. Behavioral and speech problems or schooling difficulties should be referred to as psychiatric team and special schooling [8, 9].

To date, miglustat is the only disease-specific drug approved in Europe, Canada, and Japan. The drug works by inhibiting the glucosylceramide synthase enzyme that is responsible for the first step in the synthesis of most glycosphingolipids. Miglustat has shown to stabilize key parameters of neurological disease progression in patients of all ages, but it has no effect on the systemic manifestations or intracellular cholesterol accumulation associated with this disorder [10].

There is a persistent search for new treatments to prevent or slow down the progression of NPD-C. Investigational therapies in course are 2-hydroxypropyl-beta-cyclodextrin (HP-β-CD) or hematopoietic stem cell transplantation.

VTS-270 (Kleptose® HPB, Roquette Pharma, France) and Trappsol® Cyclo™ (CTD Holdings, Inc., Alachua, FL) are HP-β-CD products under investigation as novel treatments for NPD-C. Differences between these two products have been studied based on ion distribution and abundance profiles using mass spectrometry methodology as a means to assess key molecular distinctions between products. Trappsol® Cyclo™ was found to have a higher degree of substitution compared with VTS-270, with a greater number of hydroxypropyl groups and increased levels of dimeric ions. Additional differences in ion mobility profiles were found, there is a much greater level of non-specific chemical "noise" associated with Trappsol® Cyclo™.

These two products are not chemically equivalent and therefore may not be biochemically equivalent or lead to comparable formulations from a clinical development perspective. The data suggest that biological and potential therapeutic equivalence should not be assumed. Further studies are needed to examine potential differences in biological and therapeutic effects of Trappsol® Cyclo™ and VTS-270 [11].

2. Preclinical studies

2.1. *In vitro* experience

The efflux of cholesterol from cells in culture to cyclodextrin acceptors has been reported to be substantially more rapid than efflux induced by other known acceptors of cholesterol. A comparison of the time course of cellular [3H]cholesterol efflux mediated by HDL3 or by various concentrations of cyclodextrins showed the release of 50–90% of L cell [3H]cholesterol after 8 hours of incubation with HP-β-CD and methyl-β-cyclodextrin (M-β-CD) at 10 mM. The order of efficiency in accepting cholesterol was found to be M-β-CD > HP-β-CD > beta-cyclodextrin. The kinetics of the cholesterol efflux time course studies suggested that incubation of L cells with cyclodextrin resulted in the rapid equilibration of labeled cholesterol between cells and medium. [12].

Several studies have shown that cholesterol released from late endosomes/lysosomes of NPC-proteins deficient cells by HP-β-CD reaches the cytosolic compartment and is accessible to the endoplasmic reticulum (ER). In cultured cerebellar neurons, astrocytes, and microglia from NPC1-deficient mice, the sequestered cholesterol was mobilized to the ER by low concentrations (0.1–1.0 mM) of HP-β-CD [13].

In murine models of NPC1, cell culture studies have shown that M-β-CD is more potent in exchanging cholesterol than HP-β-CD. Efficacy comparison of M-β-CD and HP-β-CD in reducing cholesterol accumulation in late endosome/lysosome in human fibroblasts NPC1/NPC2 deficient after treatment with 300 μM M-β-CD and HP-β-CD for 1 day have shown to reduce the cholesterol accumulation detected by filipin labeling. After 2 and 3 days after treatment, cholesterol accumulation started to increase, but there was still a significant reduction. In concordance with animal studies, M-β-CD produces effects equivalent to those of HP-β-CD at lower concentrations [14].There are several theories about the mechanism by which cyclodextrins affect cholesterol homeostasis but none have been confirmed.

In vitro beta-cyclodextrins showed a high affinity for sterols as compared to other lipid and, because of the relatively high specificity of this substance for cholesterol, it was suggested that beta-cyclodextrins might be effective in modifying cholesterol metabolism in vivo.

2.2. Animal studies

The unexpected discovery of the utility of cyclodextrin in NPD-C was observed in a study in Npc1−/− mice treated with a combination therapy of two drugs, one of them (allopregnanolone) formulated in a cyclodextrin complex [15]. A later study showed the same cholesterol intraneuronal storage reduction and longevity increase with the combination of allopregnanolone and cyclodextrin than with the control arm, treated only with the cyclodextrin [16]. These results led researchers to perform studies to address the role of cyclodextrin as a possible treatment. The Npc1−/− mice receiving subcutaneous (SC) or intraperitoneal cyclodextrin every other day for 2 weeks revealed a slight decrease of intraneuronal accumulation of either cholesterol or gangliosides. Both routes showed similar outcome, but SC administration seemed to be a slightly more efficacious [16]. In another study, Npc1−/− mice treated with a single SC injection at 7 days of age of HP-β-CD (4000 mg/kg body weight) prolonged the average life (108 days).

Treatment with cyclodextrin improved hepatic dysfunction and decreased neurodegeneration, increasing the number of Purkinje cells surviving at 49 days of age nearly threefold respect to untreated mice [16]. In previous studies, no significant toxicity was observed following the administration of HP-β-CD except for increased macrophage infiltration of the lungs found at post mortem examination [16, 17]. Other studies in Npc1−/− mice showed that 1500 mg/kg HP-BC-D administered weekly caused a decrease in hepatic unesterified cholesterol concentrations without substantial effect on neurological signs The slight effects of the HP-β-CD on neurological symptoms at low doses may be partially due to their apparent non-permeation of the blood-brain barrier (BBB) [18].

The efficacy of HP-β-CD was also tested in a feline model of NPD-C. Cats affected with NPD-C were first administered the drug at 3 weeks of age, prior to the onset of clinical signs of disease, and continued to receive the drug weekly. Cats were placed into one of five groups: received no HP-β-CD; received a weekly dose of 1000 mg/kg; 4000 mg/kg; 8000 mg/kg body weight HP-β-CD subcutaneously; or 4000 mg/kg brain weight (120 mg for a 30 g brain weight) HP-β-CD intrathecally every 2 weeks. The preliminary data suggested a similar requirement for doses equal to or greater than 4000 mg/kg to positively affect neurological disease. Nevertheless, doses of 4000 mg/kg body weight resulted in an increase in hearing threshold only after repeated dosing and doses of 8000 mg/kg body weight resulted in significant increases in hearing threshold in both normal cats and cats with NPD-C following the administration of a single dose [19].

Studies in mouse models have shown that systemic administration of HP-β-CD, starting in early neonatal life, diminishes unesterified cholesterol accumulation in most organs, slows disease progression, and extends lifespan. Studies in adult Npc1−/− mice who received four weekly subcutaneous 4000 mg/kg body weight HP-β-CD at 49 days of age showed reduced whole-liver cholesterol content at 77 days. Comparable improvements were seen in other organs, such as spleen, and lifespan was extended [20]. On the whole, preclinical studies in animals showed that young animals respond more favorably, whereas the older ones may benefit less.

Brain uptake of 2-hydroxypropyl-[14 C]-propyl-β-cyclodextrin was determined in Npc+/+ and Npc1−/− mice using two methods: *in situ* brain perfusion and multi-time-point regression analysis flowing intraperitoneal administration. None of the data collected indicated that HP-β-CD enters the brain [21]. Other experiments examining cyclodextrins with regard to permeability using an *in vitro* model of the BBB have indicated that a small percentage of cyclodextrin may be transported across the barrier [22].

Intrathecal (IT) HP-β-CD (120 mg in 0.6 ml saline) every 2 weeks therapy of feline NPD-C delayed the clinical manifestations of neurological disease, but had no effect on hepatic or pulmonary disease. IT-treated cats showed amelioration of neuronal swelling and axonal spheroid formation in many but not all brain regions, and preservation of Purkinje cell numbers [23]. Research in mouse models of NPD-C also has shown that direct administration of HP-β-CD into the IT or intracerebroventricular (ICV) space at low concentrations has a similar or superior effect on delaying the onset of neurological symptoms as that observed following high systemic doses [24].

Recent researches in NPD-C cats showed that direct administration of HP-β-CD into the cisterna magna prevented the onset of cerebellar dysfunction for greater than a year and reduced in Purkinje cell loss and near normal concentrations of cholesterol and sphingolipids. Cats receiving 1000 mg/kg SC HP-β-CD had a similar occurrence of neurological dysfunction and survival than untreated cats. Nevertheless, cats that received 4000 mg/kg SC HP-β-CD showed modest amelioration of neurological disease and survived age than any untreated cats. Pulmonary toxicity limited the continued dosing of cats in the 8000 mg/kg group. Dose-dependent elevations in mean hearing threshold in cats receiving SC HP-β-CD were observed. In cats receiving intracisternal HP-β-CD, a significant elevation in the auditory threshold was also observed. [25].

3. Experience in humans: first experiences of compassionate use of cyclodextrin

Based on the promising results of preclinical studies in animal models of NPD-C, in November 2008, Dr Hastings applied for individual investigator new drug exemptions (INDs) to the Food and Drug Administration (FDA) for the use of with HP-β-CD in humans. In January 2009, the first NPD-C patient in the world, an Indian child, was treated with HP-β-CD by intravenous (IV) route. Few months later, the first two INDs of HP-β-CD in USA were approved by FDA for two identical twins aged 5-year-old [26]. The "Oakland protocol" of IV HP-β-CD employed in USA was also used in two Brazilian sisters with NPD-C in January 2010 [27].

The orphan designation of HP-β-CD was obtained in May 2010 by FDA [28] and 2 months later by the European Medicines Agency (EMA) [29]. The orphan drug brand Trappsol® Cyclo™ was approved specifically for treating NPD-C. New findings suggested that the passage of HP-β-CD across the BBB was limited [21, 24, 25, 30], and physicians who treated the first patients reported slight benefit with IV route [26]. The FDA approved the request for IT delivery of HP-β-CD in September 2010 [31]. Another orphan drug of HP-β-CD was designated by FDA in February 2013, Kleptose™ (brand VTS-270). Both drugs were being evaluated in clinical trials, moreover compassionate use outside clinical trials has been reported worldwide [32].

Five different approaches have been employed in HP-β-CD therapy:

- IV only.
- IT only.
- ICV only.
- IV at the beginning, and later changed to IT or ICV.
- IV and IT simultaneously.

The IV administration was the first route tested for HP-β-CD in humans. When different reports in animals demonstrated that HP-β-CD administered intravenously cross in very little proportion the BBB [21, 24, 25, 30], this route was changed or combined with IT/ICV administration of the same drug. Theoretically, IT and ICV routes deliver directly the drug in the central nervous system (CNS), the area mainly affected in NPD-C, especially in the neurological symptoms. There is still controversy over whether the IV route contributes in NPD-C therapy. Preclinical reports showed effectiveness in the peripheral manifestations of the disease in organs [20, 25], including the liver, the spleen, and, to a lesser extent, the lungs. However, the influence of HP-β-CD in CNS after IV infusion remains in humans unknown.

4. Experience in humans: efficacy of the intravenous administration of cyclodextrin

The first two patients who received IV HP-β-CD (Trappsol® Cyclo™) in USA were treated since April 2009 with 2500 mg/kg weekly over 8 hours, and later modified to administrations every 2 weeks for convenience. The twins were diagnosed at 3 years old with initial symptoms of hepatosplenomegaly, ataxia and seizure activity, and previously were treated with miglustat, without significant benefits. After 18 months of therapy, any objective improvement was reported, and the disease continued to progress as evidenced by positron emission tomography (PET) imaging and neurological assessments [26]. Both patients were described a transient appearance of slight scattered nodules in lungs during bronchoscopy, which were identified as xanthomas, a deposition of yellowish cholesterol-rich material typical of NPD-C, resolved without treatment changes in IV therapy [32].

In October 2010, the FDA allowed the addition of HP-β-CD IT administration at doses of 350 mg every 2 weeks (initial dose 175 mg/2 weeks) [32, 33]. During the first months of concurrent IV/IT therapy, a significant improvement was observed in the validated NIH NPC Clinical Severity Score (NNCSS), as well as in the brainstem auditory evoked response (BAER) [33]. After 18 months, ICV route was authorized by FDA in the twins through an Ommaya reservoir system. One of them tolerated perfectly the ICV and continues today with the treatment, whereas in her sister, ICV was discarded by insertion bleeding and IT route was recovered. Both patients have experienced improvements in alertness, swallow, head control, and ataxia [32, 34].

The second reported use of IV HP-β-CD (Trappsol® Cyclo™) was in two Brazilian sisters at age of 12 and 16 years old with concomitant miglustat treatment. Dr Vieira employed the same IV protocol with a dose escalation from 1200 to 2500 mg/kg/week. Both patients showed, after 1 year of treatment, objective benefits in PET and NNCSS, as well as improvements in motor dysfunction, cataplexy, psychiatric symptoms, behavior, cognitive functions, and memory [27]. Two later reports informed about the simultaneous treatment with IV weekly and IT twice monthly at least 3 years, maintaining the good results previously reported with IV infusions [35, 36]. No adverse events (AEs) were reported, although the published data were limited and only congress abstracts were reported [27, 35, 36].

Two reported cases of Japanese NPD-C patients diagnosed at 2 months and 13 years were initially treated at 4 and 14 years old with the IV "Oakland protocol", with doses from 80 mg to 2500 mg/kg/three or two times a week, respectively [37, 38]. These were the first published cases of VTS-270 use in NPD-C. The younger patient reported with IV administration, a mild decrease in hepatosplenomegaly and temporary improvements in EEG and stabilization of disability scores the first 6 months. However, rest of the 2 years with only IV infusions a rapid progression of neurological dysfunction and worsening of swallow, speech, rigidity, and seizures were reported [37]. Furthermore, repeated fever with transient diffuse pulmonary cloudiness episodes were observed, an AE probably related to HP-β-CD IV infusion. After 2 years of IV therapy, IT first and later ICV administrations were combined with IV infusions [38]. The combined therapy obtained objective benefits in EEG, PET and magnetic resonance spectroscopy (MRS), and CSF T-tau level reduction. Furthermore, improvements in eye movement, language, and speech were reported, as well as and stabilization of clinical progression for 2 years. No AEs were detected over the 2 years of IV/ICV combined therapy.

The other patient, a girl of 14 years old at the time of HP-β-CD therapy onset, was only treated with IV infusions for 3 years. At the beginning of the treatment the patient showed benefits in EEG, MRS and visual evoked potential (VEP), but after 3 months disappeared. A decrease in hepatosplenomegaly was observed in abdominal ultrasound and body computed tomography (CT). The general condition, neurological tests, and seizure control were stabilized with therapy, reporting only an improvement in alertness. Tolerance of IV therapy was excellent [37].

Drs Hrynkow and Hastings recently reported in a Congress the clinical experience with 11 patients initially treated with IV HP-β-CD more than 6 years [39]. Two of these patients were only treated with IV infusions, whereas in nine patients IT therapy was added later. In the two patients with only IV therapy, one showed a progression in NNCSS similar

AEs associated with administration	AEs recognized as features of NPD-C	Other AEs of interest
Rash	Seizures	Port-a-Cath infection
Generalized rash (trunk, elbow)	Pneumonia	Removal of Ommaya
Tremor/chills/vomiting/fever	Thrombocytopenia	Reservoir
Headache	Viral illnesses	Post-operative delayed
Nausea	Viral syndrome	Parenchymal hemorrhage
Stomach pain		Meningitis

Table 1. Adverse events related to IV with/without IT HP-β-CD administrations.

than previous scores before initiation, whereas the other patient reported a stabilization of NNCSS after IV therapy. In two patients with IV/IT therapy significant decrease in NNCSS were observed comparing pre and post-therapy scores. In the other seven patients was only available post-therapy information, showing in most of the cases an initial de decrease in NNCSS followed with a progressive increase. Other significant improvements were reported, as reduction in hepatomegaly, restoration of language skills, resolution of interstitial lung disease and improvements in fine motor control. The tolerability was favorable, although some AEs were reported during 6 years of therapy (**Table 1**).

Despite the positive results observed in the compassionate use of IV HP-β-CD [26, 27, 32–39], there is no clear evidence to date that IV route has a clinical benefit in CNS symptoms of NPD-C in humans. Most of the reports with neurological improvements combined the use of IV and IT/ICV administrations, making it impossible to demonstrate that these benefits were exclusively associated to IV and not IT/ICV. In addition, doses 1000 times larger than the IT dose should be required to achieve the same concentrations in the CSF reached by IT administration with IV infusions of HP-β-CD. This dosage would probably be unsafe in humans.

5. Experience in humans: efficacy of the intrathecal administration of cyclodextrin

The IT route was proposed when the efficacy of IV infusions of HP-β-CD was questioned. The first cases of IT administrations were combined with IV, as we previously explained [33–37, 39]. At this point, IT route was proposed as the main route for NPD-C therapy. The first published case report of IT HP-β-CD (Trappsol® Cyclo™) administration was a Spanish girl diagnosed and treated at 2 years old. The dosage varied from 200 to 450 mg every 2 weeks. At 38 months an IT Ommaya reservoir (IT-OR) was implanted in order to increase patient convenience and reduce number of punctures, modifying the dosage to 400 mg every 2 weeks. The patient showed improvements in BAER test, whereas neurological symptoms and NNCSS were stabilized. The patient suffered two episodes of seizures with a fever with IT therapy, resolved without any change in treatment and which could be related to HP-β-CD, the procedure or the disease [40].

Dr. Berry-Kravis reported the clinical outcomes of the use of IT HP-β-CD (Trappsol® Cyclo™) in three patients with NPD-C in several congresses [41–43]. Two of them were a couple siblings of 14 and 15 years old who were unable to be enrolled in the Phase 1 clinical trial (NCT01747135) and were treated as a compassionate use. The younger sibling with doses from 200 to 500 mg every 2 weeks obtained improvements in NNCSS, instrumented timed up and go (gait test) and Mullen Scales of Early Learning (cognitive scale), as well as benefits in cognitive functions, seizures, swallow, language and speech [40–43]. The other sibling received doses from 200 to 600 mg every 2 weeks but the benefit was poor, with stabilization of NNCSS, CFS reduction of lysozyme and improvement in cognitive functions [40–43]. After the first three IT infusions were reported post lumbar puncture headache and vomiting in both siblings, but it was not related to the drug and disappeared after a switch to use of Whitacre spinal needles. The clinical information regarding the third patient was insufficient to evaluate the efficacy, although an improvement in balance and gait was reported after 1 year treated with doses from 200 to 400 mg every 2 weeks [43].

The use of IT HP-β-CD with a fixed dose of 200 mg every 2 weeks was reported by Maarup et al. in a single case report. After 18 months of the therapy, the boy showed an improvement in vertical gaze (eye movement) and the consequent decrease in the NNCSS. This study also reported an increase in 24-OH cholesterol, a biomarker of cholesterol redistribution in CNS. On the other hand, an AE was associated to HP-β-CD administration, the well-known hearing loss to high frequency [44].

The use of IT-OR HP-β-CD in two Spanish boys diagnosed of NPD-C at 6 and 10 years old was briefly reported in congresses [45, 46]. The first patient was treated at 11 years with doses from 125 to 525 mg (Trappsol® Cyclo™) every 2 weeks for at least 37 months. Benefits in muscle tone and a decrease in seizures frequency were observed with IT-OR therapy. Initially was reported an improvement in BAER test, although it was alternated with auditory deteriorations [32, 45]. In the other case report, a 16 years patient received IT-OR fixed-doses of 350 mg every 2 weeks for 20 months. The objectives results included BAER tests and NNCSS. Furthermore, improvements in language and speech, ataxia and quality of life were obtained. The most relevant toxicities were intermittent fever and a suspicion of chemical meningitis [32, 46].

A recently published article described the IT therapy of a young NPD-C girl of 22 months [47]. The dosage of HP-β-CD was 175–325 mg every 2 weeks for 20 months. The treatment only achieved improvements in visual contact and motor function with the first doses, as well as slight retardation of disease progression and in the NPD-C disability scale during the first year. After the first year, MRI showed a progression of cerebral atrophy, which was consistent with a clinical disease progression (epilepsy, dysphagia, and worsening motor function). Despite the initial response and the absence of AEs, the IT HP-β-CD was discontinued after 20 months by lack of efficacy.

The employment of HP-β-CD in adult-onset NPD-C has been described with variable results in two publications [48, 49]. Sakiyama et al. reported the IT treatment (VTS-270) of two adult patients of 37 and 28 years with doses from 100 to 400 mg every month. The older patient showed better eye movement and neurological stabilization, whereas the younger patient reported improvement in NPD-C scales, balance and gait, language and speech, and swallow. In addition, reductions in oxysterol serum concentrations were observed in both patients, a sterol storage biomarker. No AEs were reported in these patients [32, 48]. The worse outcomes observed in two cases reported by García-Robles et al. could be related to the age (49 and 39 years old) and advanced disease at the HP-β-CD onset. The dosages of IT HP-β-CD (Trappsol® Cyclo™) ranged from 175 to 700 mg and 50 to 875 mg every 2 weeks, also using IT-OR route in the second patient. Any objective or subjective improvements were reported in both patients. The older patient received only four doses with optimal tolerance, but HP-β-CD therapy was discontinued when neuropsychiatric symptoms progressed. The other patient suffered two episodes of toxic meningitis as well as worsening respiratory symptoms and swallow. After second chemical meningitis and neurologic progression of the disease, HP-β-CD treatment was discontinued [49].

6. Experience in humans: ongoing clinical trials

Four clinical trials using cyclodextrin for the treatment of NPD-C have been found. One of them has been completed and their results have been published [50] and three are currently ongoing and no preliminary results have been yet published.

- Intrathecal 2-hydroxypropyl-β-cyclodextrin (VTS-270) for Niemann-Pick type C1 (NPC-1) disease. A non-randomized, open-label, Phase 1–2 trial. See ClinicalTrials.gov Identifier: NCT01747135 [50].

Phase 1–2, non-randomized, open-label, study, to assess the tolerability, safety, feasibility, and PK of HP-β-CD administered IT monthly via lumbar injection to drug naive cohorts of NPC-1 patients at doses of 50 mg escalated to a maximum of 1200 mg. The objective is to determine an active dose of HP-β-CD as measured by changes in plasma 24-(S)-hydroxycholesterol (24(S)-HC) concentration and to evaluate the use of biomarkers and potential clinical outcomes of NPC-1. NNCSS is used to assess clinical efficacy. The decision to dose-escalate is based on safety and biochemical data. Safety is assessed by the appearance of AEs with performance of clinical laboratory tests, physical examinations, and with special attention to audiological evaluation. Biochemical efficacy is measured by change from baseline in plasma 24(S)-HC. The PK analysis is assessed for plasma HP-β-CD concentrations. This is the only clinical trials of HP-β-CD with published results.

Eligible patients were aged 2–25 years and had NPC-1 with neurological manifestations. Fourteen patients were enrolled from National Institutes of Health (NIH-cohort). Cohort size was three participants for initial IT doses of 50, 200, 300, 400, and 900 mg (only two patients) administrated IT every month. Three participants were initially dosed with 50 mg ICV via an Ommaya reservoir approximately 6 months prior to initiation of the IT trial. Use of the Ommaya reservoirs was discontinued due to *P. acnes* infection/colonization in two subjects. Due to this problem, initial protocol was amended and ICV route was changed by IT. After initial dosing at the specified cohort dose, participants were dose-escalated based on tolerance and safety data.

As comparison control, a cohort of NPD-C subjects from Natural History study with longitudinal assessments was employed. These patients were not on HP-β-CD treatment. To explore a scheme every 2 weeks, three additional subjects were recruited with the same criteria mentioned above in Rush University Medical Center (RUMC-cohort).

The primary outcome was changed in 24[S]-HC area under the curve ($AUC_{8–72}$) response to drug administration compared with the response after saline administration. The $AUC_{8–72}$ of plasma 24(S)-HC concentrations were established after its determinations at pre-dose, 8, 24, 30, 48, and 72 hours post-dose after either HP-β-CD of saline infusion.

As a secondary objective, NNCSS was used to assess clinical efficacy. Audiological assessments were obtained monthly before each infusion. Also the concentrations of fatty acid binding protein 3 (FABP3) and calbindin D in cerebrospinal fluid (CSF) were assayed.

Finally, 14 NIH-patients and 3 RUMC-patients were enrolled. Twenty-one patients with similar characteristics were identified from the historical database for comparing. For primary outcome (change in 24(S)-HC $AUC_{8–72}$), 121 of 155 post-drug plasma value were greater than post-saline values. Despite the variability, the data suggest a dose-response relationship. All the patients of the study had either FABP3 or calbindin D, a significant negative linear regression slope (only one patient had a significant increase in calbindin D slope).

Regarding clinical efficacy, the total NNCSS for NIH-cohort increased at a slower rate than comparison cohort. These data show a significant reduction in disease progression in the cohort of HP-β-CD treated patients. In a secondary responder analysis, cohorts of treated and comparison subjects were classified as responders when their NNCSS minus hearing was stable or improved. Seven of 14 NIH-cohort subjects were classified as responders, 3 of 3 RUMC-cohort subjects were classified as responders, and none of 21 patients of comparison cohort were classified as responders. Safety will be discussed in Section 7.

The added value of this study was to provide a neurological disease progression comparison among a cohort of NPD-C subjects treated with IT HP-β-CD and a control cohort of NPD-C subjects from Natural History study, indicating a decrease rate of neurological disease progression in the treated cohort. Moreover, this study provides information on the safety of IT administered HP-β-CD and the measurement of biomarkers provided additional support for decreased neuronal damage and improved neuronal cholesterol homeostasis.

- A Phase 2b/3 prospective, randomized, double-blind, sham-controlled trial of VTS-270 (HP-β-CD) in subjects with neurologic manifestations of Niemann-Pick type C1 (NPC-1) disease. (See ClinicalTrials.gov identifier: NCT02534844 and EudraCT Number: 2015-002548-15).

Multicenter, multinational, prospective, randomized, double-blind, sham-controlled, three-part, efficacy and safety trial of HP-β-CD, administered by the lumbar IT route every 2 weeks, with a planned enrollment of approximately 51 subjects with NPC-1 disease. This study is ongoing, but not recruiting participants (male or female subjects, aged 4–21 years of age at time of screening with onset of neurological symptoms prior to 15 years of age).

This study has three parts with different objectives. The objective of Part A is to select the dose of HP-β-CD to be used in Part B and Part C. Three different HP-β-CD lumbar IT doses

(900, 1200, and 1800 mg) will be administered IT every 2 weeks for 8 weeks and 2 weeks for observation in 9 subjects; 3 subjects will receive sham treatment. The criteria for dose selection include safety and tolerability including a thorough audiological evaluation.

The objective of Part B is to evaluate, in a double-blind sham-controlled design, the progression of the neurologic manifestations of NPC-1 disease based on changes in the composite efficacy outcome (consisting of four components of the NNCSS: ambulation, fine motor skills, cognition, and swallowing), after 52 weeks of treatment in comparison to baseline. Part B will evaluate the safety and efficacy of the dose selected from Part A compared to sham control in 51 subjects (randomized 2:1), including the 12 subjects from Part A.

The objective of Part C is to evaluate the long-term safety, tolerability, and efficacy of the dose selected for Part B. This part is an open-label extension with IT treatment every 2 weeks to subjects who either complete Part B or are subjects in Part B who have met rescue therapy criteria. Additionally, subjects who are currently active in the NIH-sponsored Phase 1 protocol (NCT01747135, see above) will also be eligible to participate upon completion of their participation in the Phase 1 study. In this part, subjects will receive treatment until licensed product or end of the program.

The primary outcome measure of the study is NNCSS with a time Frame of 52 weeks. Data for NPD-C score rating will be provided to a centralized independent blinded rater, who will analyze all NPD-C information for all subjects and assign the NNCSS rate. As secondary outcome measures are: clinician and caregiver clinical global impression of change, time to get up and go test, 9-hole peg test, percentage of patients with clinical worsening, and European Quality of Life-5 dimensions quality of life rating (EQ-5D QoL). Moreover, CSF and plasma biomarkers will be measured.

The design enables a selection of best dose based on efficacy, safety, and tolerability according to an evaluation by a Committee. The dose chosen will be evaluated using changes in a composite efficacy outcome in order to assess the neurologic progression in participants. Highlight is that, it is the first study that uses quality of life rating. Moreover, assessment of biochemical markers of response and also due to a sufficient dosing duration will be performed to assess the effectiveness of HP-β-CD in NPC-1. This is a global, multi-site study with the largest planned number of participants, and this will allow a better knowledge about NPD-C and efficacy treatment.

- A Phase I/II study to evaluate the safety and pharmacokinetics of intravenous Trappsol Cyclo (HP-β-CD) in patients with Niemann-Pick disease type C (NPC-1) and the pharmacodynamic effects of treatment upon markers of cholesterol metabolism and clinical outcomes (see ClinicalTrials.gov Identifier: NCT02912793 and EudraCT Number: 2015-005761-23).

Phase 1/2, double-blind, randomized, multicentre, parallel group study based on data available and information from the administration via compassionate/named patient use in patients with NPC-1, and information of other cyclodextrin products in the literature. The study has two stages: the primary objective of Stage 1 is to compare the plasma pharmacokinetics (PK) of three different doses of IV HP-β-CD in the prevention/delay of NPC-1 progression whereas Stage 2 is to evaluate their efficacy and tolerability in the management of clinical manifestations. Secondary objectives include investigation of the effect of three different doses of HP-β-CD IV upon serum and lymphocytic markers of cholesterol metabolism (Stages 1 and 2) and evaluation of concentrations in the CSF following IV administration (Stage 1), evaluation of the impact of treatment upon behavioral aspects and the impact of treatment upon measures of neurological function including ataxia, aphasia, and saccadic eye movements of NPC-1 (Stage 2). The outcome measures are: plasma and CSF concentrations of HP-β-CD following IV administration, serum cholesterol markers, global impression of disease, quality of life scores, change in NNCSS, and changes in hepatic and splenic morphology.

In order to achieve these objectives, the primary endpoint of Stage 1 is plasma concentrations (at 0, 2, 4, 6, and 8 hours after the start of infusion and 30 minutes, 1, 2, 4, 8, and 12 hours after the end of the infusion) of HP-β-CD during and following infusion to evaluate time to maximum concentration (T max), maximum concentration (Cmax), volume of distribution and elimination half-life ($t_{1/2}$). The primary endpoint of Stage 2 is the change from baseline in global impression of disease severity at 48 weeks and the proportion of patients at 48 weeks with a reduction from baseline of at least one point in two or more domains of the NNCSS.

Patients taking miglustat are not excluded of the study because this drug is an approved treatment for NPC-1 in Europe and it would be unethical, but it is planned to balance randomization across groups for its use.

This clinical trial is already recruiting patients in United Kingdom and it is planned to recruit 12 patients (3 children of 2–11 years, 3 of 12–17, and 6 adults 18–64 years). Patients will be randomized 1:1:1 to one of the three dose levels (1500, 2000, or 2500 mg/kg; four patients per dose level). Treatment will be administered over 8 hours by slow IV infusion at a concentration of 250 mg/mL every 2 weeks. Patients completing Stage 1 of the study will continue into Stage 2 and receive treatment for 48 weeks.

The design enables early assessment of biochemical markers of response and also due to a sufficient dosing duration, to assess the effectiveness of HP-β-CD in NPC-1 and its pharmacokinetics.

- A Phase I study to evaluate the single and multiple-dose pharmacokinetics of intravenous Trappsol Cyclo (HP-Beta-CD) in patients with Niemann-Pick disease type C (NPC-1) and the effects of dosing upon biomarkers of NPC disease. (See ClinicalTrials.gov Identifier: NCT02939547).

Phase I, double-blind, randomized, single-center, parallel group study based on information and data available from the administration of HP-β-CD via compassionate/named patient use in patients with NPC-1, and data on other cyclodextrin products in the scientific literature.

The study has a first phase of screening (up to 4 weeks), a treatment phase of 12 weeks and a later phase of follow-up of 4 weeks. The primary objective is to compare the plasma pharmacokinetics of single and multiple doses of two different levels of IV HP-β-CD. Secondary objectives include investigation of the effect of different doses of IV HP-β-CD upon serum and lymphocytic markers of cholesterol metabolism and evaluation of HP-β-CD concentrations in the CSF following IV administration, evaluation of the impact of treatment upon measures of neurological function including aphasia, ataxia, and saccadic eye movements, and the impact of treatment upon behavioral aspects of NPC-1.

This study is currently recruiting participants. It is planned to recruit a total of 12 patients (all adults) which will be randomized 1:1 to one of the two dose levels (1500 mg/kg or 2500 mg/kg; 6 patients per dose level). Treatment will be administered every 2 weeks by slow IV infusion over 8 hours. Patients will receive treatment for a total of 12 weeks.

As primary outcome measures are pharmacokinetics parameters: Tmax, Cmax, volume of distribution, and $t_{1/2}$ of HP-β-CD in plasma from NPC-1 patients by measurement at pre-infusion then 2, 4, 6, 8, 8.5, 9, 10, 11, 12, 16, and 20 hours after the start of the infusions at weeks 1 and 12.

The design of the proposed study thus enables a better knowledge about pharmacokinetics of IV HP-β-CD administration, an early assessment of potential biochemical markers of response but allows for a sufficient dosing duration to enable the short-term effectiveness of HP-β-CD in NPD-C to be assessed.

In conclusion, there is a published clinical trial results, using IT administration of HP-β-CD and shows moderate response thought slowed disease progression with an acceptable safety profile. Another IT HP-β-CD clinical trial and two IV HP-β-CD clinical trials are ongoing but not results are been published yet. Regarding the route of administration, exist a debate, and treatment with HP-β-CD has used four different paradigms: IV only, IV followed by the addition of IT sequential, IV and IT initiated concurrently, and IT only. The main reason for IT route is that HP-β-CD does not cross the BBB, however, in animal models systemic HP-β-CD positively affects CNS disease thus CNS penetration may not be essential for neurologic efficacy. The ongoing clinical trials will lead to an improvement in knowledge of HP-β-CD for NPD-C treatment, setting the best route, dose and posology for NPD-C patients.

7. Experience in humans: potential toxicities of the different administration routes

There are three possible administration routes for HP-β-CD: IV, IT, or ICV via an Ommaya reservoir. Some adverse events observed could be related to the administration route, the disease progression, or the cyclodextrin itself.

The review performed by the EMA regarding the use of HP-β-CD as an excipient indicated that the IV administrations had low toxicity, being the most prevalent issue of the renal toxicity, especially with high doses [51]. However, any patients showed renal toxicity with HP-β-CD therapy [32]. This was the first route employed, but later a change to IT route was requested due to the discovery that HP-β-CD not cross the BBB and only a little quantity is able to enter to the brain [21, 24, 25, 30].

Based on animal models, ICV route was proposed as an alternative with potential benefits and this route was chosen in the Phase I clinical trial (NCT01747135) using an Ommaya reservoir to facilitate the administration. Unfortunately, complications due to colonization by *P. acnes* led to change ICV route to IT [52]. Currently, IV route is being tested in two ongoing clinical trials, whereas IT route is analyzed in another ongoing trial and in the recently published trial [50].

Megías-Vericat et al. reviewed the initial published cases of HP-β-CD treatment [32]. Regarding safety, 11 of 17 NPD-C patients included suffered AEs. Of the 17 AEs reported, 6 were related to the route of administration, specifically with the IT and ICV routes and 10 of them could be attributed by HP-β-CD. Six AEs associated to the route were: CNS bleed related to insertion of the Ommaya reservoir system in a patient treated with ICV HP-β-CD led to ICV administration suspended [34]; post lumbar puncture headache and vomiting resolved with switching a Whitacre spinal needles [41–43]; post lumbar puncture pain, headache, nausea, and vomiting resolved with symptomatic and postural treatment [44]; seizures with fever resolved without changes in treatment [40] and aspiration pneumonia; febrile syndrome; and candidiasis resolved with antibiotic and antifungal treatment [49].

Among AEs related to HP-β-CD itself, loss of hearing was reported in four reported cases of IT infusions [41–45], although it proved reversible in two of them. These patients were administered with reduced doses of HP-β-CD after hearing recovery, two times in one of them. In other two patients, despite delaying the next dose, hearing loss was not reversed. This AE was also observed in animal studies with IT [18], ICV [53], and intracisternal [25] routes of administration. At the dosage employed, hearing loss is an expected AE, as well as a well-known NPD-C symptom. At the clinical trial recently published, loss of hearing was reported in all the participants [50].

Four patients suffered from fever two or more times after HP-β-CD administrations [37, 40, 46, 49]. In one case, fever was accompanied by seizures [40], whereas another patient showed an infusion reaction with fever and transient diffuse pulmonary cloudiness [37]. In one patient, after intermittent fever episodes, a diagnosis of chemical meningitis was made after bacterial meningitis was discarded. The IT-OR was withdrawn, and HP-β-CD was reintroduced 2 months later because the patient's condition worsened [46]. Other patient showed two episodes of chemical meningitis (bacterial meningitis was discarded in both cases) after IT and IT-OR HP-β-CD administrations, although both were resolved quickly and without consequences [49]. This AE could be related to the method of administration or disease symptoms and not to HP-β-CD. However, the chemical meningitis observed seems to be associated to the drug after analyzed with the Naranjo algorithm [49]. Some acute neurological effects after the infusion of high doses that resolve after few days were described by some investigators in conferences (but unpublished yet).

Regarding the results of the first clinical trial published, no serious AEs were observed [50]. Marked expected AE included: ototoxicity (14 of 14 patients) and post lumbar puncture headache (9 of 14 patients). Among unexpected AE included, post-administration unsteadiness and fatigue occurs at doses above 600 mg. The degree of impairment varied between subjects but usually was transient and occurred 24–72 hours after administration.

Sensorineural hearing loss was present in all subjects of NIH-cohort (14 patients) and 2 of 3 RUMC-cohort, and according to the study results, it was associated with the administration of HP-β-CD. Moreover, the data obtained suggest that there is greater HP-β-CD ototoxicity in subjects who have not yet lost hearing due to NPD-C itself. Also, tinnitus was present in 6 of 14 patients in NIH-cohort and 1 of 3 in RUMC-cohort [50].

8. Conclusions

Until now, NPD-C treatment has been supportive and symptomatic with miglustat as the only disease-specific drug approved in some countries. HP-β-CD is a new option under investigation with two different products, Kleptose® and Trappsol Cyclo®, which are not chemically equivalent and therefore may not be biochemically equivalent or lead to comparable formulations from a clinical development perspective.

Based on preclinical animal studies, HP-β-CD has been tested in humans affected by NPD-C during the last years. In compassionate use outside clinical trials, HP-β-CD has been administrated with either IV, IT, or ICV. Despite the positive results observed with IV HP-β-CD, there is no clear evidence to date that IV route has a clinical benefit in CNS symptoms of NPD-C as most of the reports combined IV and IT/ICV administrations. The reported cases of IT infusions obtained higher improvements reducing the disease progression.

The results of a recently published clinical trial reproduced the findings observed with IT route. The trial has shown slowing of NPD-C progression in 14 patients with a dose escalation of IT HP-β-CD administrated monthly as well as in 3 patients with administration every 2 weeks regarding a historical comparison cohort.

Some adverse events observed could be related to the administration route, the disease progression, or the cyclodextrin itself. The safety profile of HP-β-CD seems acceptable, being the loss of hearing (related to HP-β-CD) the most frequent adverse reported in the clinical trial and published cases. However, some severe toxicities have been reported including chemical meningitis and fever although not in published clinical trial.

Furthermore, there are currently two IV HP-β-CD and one IT HP-β-CD ongoing clinical trials without published results. The findings of these trials will lead to an improvement in knowledge of HP-β-CD for NPD-C, setting the best route, dose, and posology. Currently, the short experience with HP-β-CD suggested that it could be effective in the management of NPD-C but the results of ongoing clinical trials will be definitive.

Acknowledgements

We would like to thank Asociación para la Investigación, Desarrollo e Innovación en Farmacia Hospitalaria (AIDIFH) for its economic support regarding the publication fees.

Author details

Juan Eduardo Megías-Vericat*, María José Company-Albir, Ana Alejandra García-Robles and José Luis Poveda

*Address all correspondence to: megias_jua@gva.es

Servicio de Farmacia, Área del Medicamento, Hospital Universitari i Politècnic La Fe, Valencia, Spain

References

[1] Vanier MT. Niemann-Pick disease type C. Orphanet Journal of Rare Diseases. 2010;**5**:16. DOI: 10.1186/1750-1172-5-16

[2] Mengel E, Klünemann H-H, Lourenço CM, Hendriksz CJ, Sedel F, Walterfang M, et al. Niemann-Pick disease type C symptomatology: An expert-based clinical description. Orphanet Journal of Rare Diseases. 2013;**8**:166. DOI: 10.1186/1750-1172-8-166

[3] Peake KB, Vance JE. Defective cholesterol trafficking in Niemann-Pick C-deficient cells. FEBS Letters. 2010;**584**:2731-2739. DOI: 10.1016/j.febslet.2010.04.047

[4] Patterson M. Niemann-Pick disease type C. GeneReviews. University of Washington; 2013. Available from: http://www.ncbi.nlm.nih.gov/books/NBK1296/ [Accessed: August 24, 2017]

[5] Wijburg FA, Sedel F, Pineda M, Hendriksz CJ, Fahey M, Walterfang M, et al. Development of a suspicion index to aid diagnosis of Niemann-Pick disease type C. Neurology 2012; 78:1560. doi: 10.1212/WNL.0b013e3182563b82

[6] Jiang X, Ory DS. Towards a new diagnostic standard for Niemann-Pick C disease. eBio-Medicine. 2016;**4**:18-19. DOI: 10.1016/j.ebiom.2016.01.004

[7] Reunert J, Fobker M, Kannenberg F, Du Chesne I, Plate M, Wellhausen J, Rust S, Marquardt T. Rapid diagnosis of 83 patients with Niemann Pick type C disease and related cholesterol transport disorders by Cholestantriol screening. eBioMedicine. 2016;**4**:170-175. DOI: 10.1016/j.ebiom.2015.12.018

[8] Santos-Lozano A, Villamandos D, Sanchis-Gomar F, Fiuza-Luces C, Pareja-Galeano H, Garatachea N, et al. Niemann-Pick disease treatment: A systematic review of clinical trials. Annals of Translational Medicine. 2015;**3**:360. DOI: 10.3978/j.issn.2305-5839.2015.12.04

[9] Alobaidy H. Recent advances in the diagnosis and treatment of Niemann-Pick disease type C in children: A guide to early diagnosis for the general pediatrician. International Journal of Pediatrics. 2015;**2015**:816593. DOI: 10.1155/2015/816593

[10] Lyseng-Williamson KA. Miglustat: A review of its use in Niemann-Pick disease type C. Drugs. 2014;**74**(1):61-74. DOI: 10.1007/s40265-013-0164-6

[11] Yergey AL, Blank PS, Cologna SM, Backlund PS, Porter FD, Darling AJ. Characterization of hydroxypropyl-beta-cyclodextrins used in the treatment of Niemann-Pick disease type C1. PLoS One. 2017;**12**:e0175478. DOI: doi.org/10.1371/journal.pone.0175478

[12] Kilsdonk EPC, Yancey PG, Stoudt GW, Bangerter FW, Johnson WJ, Phillips MC, et al. Cellular cholesterol efflux mediated by cyclodextrins. The Journal of Biological Chemistry. 1995;**270**:17250-17256. DOI: 10.1074/jbc.270.29.17250

[13] Abi-Mosleh L, Infante RE, Radhakrishnan A, Goldstein JL, Brown MS. Cyclodextrin overcomes deficient lysosome-to-endoplasmic reticulum transport of cholesterol in Niemann-Pick type C cells. PNAS. 2009;**106**:19316-19321. DOI: 10.1073/pnas.0910916106

[14] Rosenbaum AI, Zhang G, Warren JD, Maxfield FR. Endocytosis of betacyclodextrins is responsible for cholesterol reduction in Niemann-Pick type C mutant cells. Proceedings of the National Academy of Sciences of the United States of America. 2010;**107**:5477-5482. DOI: 10.1073/pnas.0914309107

[15] Griffin LD, Gong W, Verot L, Mellon SH. Niemann-Pick type C disease involves disrupted neurosteroidogenesis and responds to allopregnanolone. Nature Medicine. 2004;**10**:704-711. DOI: 10.1038/nm1073

[16] Davidson CD, Ali NF, Micsenyi MC, Stephney G, Renault S, Dobrenis K, et al. Chronic cyclodextrin treatment of murine Niemann-Pick C disease ameliorates neuronal cholesterol and glycosphingolipid storage and disease progression. PLoS One. 2009;**4**:e6951. DOI: 10.1371/journal.pone.0006951

[17] Liu B, Li H, Repa JJ, Turley SD, Dietschy JM. Genetic variations and treatments that affect the lifespan of the NPC1 mouse. Journal of Lipid Research. 2008;**49**:663-669. DOI: 10.1194/jlr.M700525-JLR200

[18] Camargo F, Erickson RP, Garver WS, Hossain GS, Carbone PN, Heidenreich RA, et al. Cyclodextrins in the treatment of a mouse model of Niemann-Pick C disease. Life Sciences. 2001;**70**:131-142. DOI: 10.1016/S0024-3205(01)01384-4

[19] Ward S, O'Donnell P, Fernandez S. Vite CH2-hydroxypropyl-beta-cyclodextrin raises hearing threshold in normal cats and in cats with Niemann-Pick type C disease. Pediatric Research. 2010;**68**:52-56. DOI: 10.1203/00006450-201011001-00099

[20] Lopez AM, Terpack SJ, Posey KS, Liu B, Ramirez CM, Turley SD. Systemic administration of 2-hydroxypropyl-β-cyclodextrin to symptomatic Npc1-deficient mice slows cholesterol sequestration in the major organs and improves liver function. Clinical and Experimental Pharmacology & Physiology. 2014;**41**:780-787. DOI: 10.1111/1440-1681.12285

[21] Pontikis CC, Davidson CD, Walkley SU, Platt FM, Begley DJ. Cyclodextrin alleviates neuronal storage of cholesterol in Niemann-Pick C disease without evidence of detectable blood-brain barrier permeability. Journal of Inherited Metabolic Disease. 2013;**36**:491-498. DOI: 10.1007/s10545-012-9583-x

[22] Monnaert V, Tilloy S, Bricout H, Fenart L, Cecchelli R, Monflier E. Behavior of α-, β-, and γ-cyclodextrins and their derivatives on an in vitro model of blood-brain barrier. The Journal of Pharmacology and Experimental Therapeutics. 2004;**310**:745-751. DOI: 10.1124/jpet.104.067512

[23] Vite CH, Mauldin E, Ward S, Stein V, Prociuk M, Haskins ME, et al. Intrathecal cyclodextrin therapy of feline Niemann-Pick type C disease. Molecular Genetics and Metabolism. 2011;**2**:44. DOI: doi.org/10.1016/j.ymgme.2010.11.148

[24] Aqul A, Liu B, Ramirez CM, Pieper AA, Estill SJ, Burns DK, et al. Unesterified cholesterol accumulation in late endosomes/lysosomes causes neurodegeneration and is prevented by driving cholesterol export from this compartment. The Journal of Neuroscience. 2011;**31**:9404-9413. DOI: 10.1523/JNEUROSCI.1317-11.2011

[25] Vite CH, Bagel JH, Swain GP, Prociuk M, Sikora TU, Stein VM, et al. Intracisternal cyclodextrin prevents cerebellar dysfunction and Purkinje cell death in feline Niemann-Pick type C1 disease. Science Translational Medicine. 2015;**7**:276ra26. DOI: 10.1126/scitranslmed.3010101

[26] Hastings C, Hempel C, Browne R. Clinical Experience with Intravenous Infusions of Hydroxy-Propyl-Beta-Cyclodextrin in Identical Twin Patients with Niemann-Pick Type C Disease under Individual Investigator New Drug Exemptions. Abstract unpublished from Michael, Marcia, and Christa Parseghian Scientific Conference for Niemann-Pick Type C Research, June 2010. Abstract. Available from: http://www.parseghian.org/documents/abstractsall.pdf

[27] Vieira C. The Use of Cyclodextrin in Niemann-Pick Type C Disease in Two Girls – Report After One Year of Treatment. Abstract unpublished from Michael, Marcia, and Christa Parseghian Scientific Conference for Niemann-Pick Type C Research June 9-11, 2011. Abstract 28:28. Available form: https://niemannpick.nd.edu/assets/52702/all_abstracts.pdf

[28] FDAs Office of Orphan Product Development. Orphan designation of HP-β-CD. Designation Request #10-3039. 2010

[29] European Medicines Agency. Public Summary of Opinion on Orphan Designation. 2015. EMA/COMP/546608/2011.2010. Available form: http://www.ema.europa.eu/docs/en_GB/document_library/Orphan_designation/2011/09/WC500112733.pdf

[30] Tortelli B, Fujiwara H, Bagel JH, Zhang J, Sidhu R, Jiang X, et al. Cholesterol homeostatic responses provide biomarkers for monitoring treatment for the neurodegenerative disease Niemann-Pick C1 (NPC1). Human Molecular Genetics. 2014;**23**:6022-6033

[31] Hastings C. Request for intrathecal delivery of HPBCD for Niemann Pick Type C patients, Caroline Hastings, M.D. Principal Investigator Department of Pediatric Hematology Oncology Children's Hospital & Research Center Oakland Submission Date to FDA: August 13, 2010. Available form: http://addiandcassi.com/wordpress/wp-content/uploads/Hempel-Cyclodextrin-Intrathecal-FDA-Filing-2010-Aug.pdf

[32] Megías-Vericat JE, García-Robles A, Company-Albir MJ, Fernández MJ, Pérez-Miralles FC, López-Briz E, et al. Early experience with compassionate use of 2-hydroxypropyl-beta-cyclodextrin for Niemann-Pick type C disease: Review of initial published cases. Neurological Sciences. 2017;**38**:727-743

[33] Hastings C, Torkildson J, Raphael R. Intrathecal hydroxy-propyl-beta-cyclodextrin reverses hearing loss in identical twin girls with Niemann-Pick type C disease. Molecular Genetics and Metabolism. 2012;**105**(2):S34. DOI: 10.1016/j.ymgme.2011.11.072

[34] Hastings C. Clinical update on 2-hydroxy-propyl-beta-cyclodextrin in twin girls. Presented at the 21st Annual NNPDF Medical Conference, Baltimore, MD, August 2013

[35] Vieira C, Lucena R. Use of Cyclodextrin in Two Brazilian Girls with Niemann-Pick Type C – Intrathecal Report. Abstract unpublished from Michael, Marcia, and Christa Parseghian Scientific Conference for Niemann-Pick Type C Research June 7-9, 2012. Abstract 10:18. Available form: https://niemannpick.nd.edu/assets/69394/abstracts_in_speaking_order.pdf

[36] Vieira C, Jagersbacher JB, Dórea I, Costa F, Lucena R. Use of cyclodextrin in two Brazilian girls with Niemann-Pick disease type C. ICNC, 2. Neurogenetics (P338-P372). Journal of the International Child Neurology. 2014;**1**(Suppl I):124. DOI: http://dx.doi.org/10.17724/jicna.2014.s1.pos12

[37] Matsuo M, Togawa M, Hirabaru K, Mochinaga S, Narita A, Adachi M, et al. Effects of cyclodextrin in two patients with Niemann-Pick type C disease. Molecular Genetics and Metabolism. 2013;**108**:76-81. DOI: 10.1016/j.ymgme.2012.11.005

[38] Matsuo M, Shraishic K, Wadad K, Ishitsukac Y, Doia H, Maedab M, et al. Effects of intra-cerebroventricular administration of 2-hydroxypropyl-β-cyclodextrin in a patient with Niemann-Pick type C disease. Molecular Genetics and Metabolism Reports. 2014;**1**:391-400. DOI: 10.1016/j.ymgmr.2014.08.004

[39] Hrynkow S, Hastings C. Niemann-Pick disease type C: Clinical experience in 11 patients with intravenous hydroxypropyl-β-cyclodextrin. CTD Holdings. Listening Circle Pres-entation. Oral communication from a scientific meeting (unpublished), June 7, 2016. Available form: http://content.stockpr.com/ctdholdings/media/828b54eb3981cfa8ab1c643cde8a1431.pdf

[40] Cruz-Pardos S, García-Poza P, Sánchez-García FJ. Treatment with cyclodextrin for Niemann Pick's disease. (Article in Spanish) Farm Hosp. 2013;**37**:271-272. doi: http://dx.doi.org/10.7399/FH.2013.37.3.555

[41] Berry-Kravis E, O'Keefe J, Pounardjian CJ, Porter FD, Ory D, et al. Intermediate size patient population IND for 2-hydroxypropyl-cyclodextrin (HP-CD) outpatient treatment of siblings with Niemann-Pick type C1 (NP-C1) and disparate rates of progression: Exploration of outcome measures and biomarkers. Abstract unpublished from Michael, Marcia, and Christa Parseghian Scientific Conference for Niemann-Pick type C Research. June 12-14, 2014. Abstract 6:17. Available form: https://niemannpick.nd.edu/assets/135126/2014abstractbooklet.pdf

[42] Berry-Kravis E, O'Keefe J, Pounardjian T, Chin J, Porter FD, Ory D, et al. Intermediate size patient population IND for 2-hydroxypropyl-ß-cyclodextrin (HP-ß-CD) outpatient treatment of siblings with Neimann-Pick type C1 (NP-C1) and disparate rates of progression: Exploration of outcome measures and biomarkers. Annals of Neurology. 2014;**76**(s18):S215-S216. DOI: 10.1002/ana.24226

[43] Berry-Kravis E, O'Keefe J, Hoffmann A, Winston A, LaGorio L, Robertson E, et al. Objective clinical efficacy outcome measures for Cyclodextrin treatment in Niemann-Pick type C (NP-C): A five-domain approach. Abstract unpublished from Michael, Marcia, and Christa Parseghian Scientific Conference for Niemann-Pick Type C Research. June 11-13, 2015. Abstract 13:23. Available from: https://niemannpick.nd.edu/assets/168424/2015abstractbooklet.pdf

[44] Maarup TJ, Chen AH, Porter FD, Farhat NY, Ory DS, Sidhu R, et al. Intrathecal 2-hydroxypropyl-beta-cyclodextrin in a single patient with Niemann-Pick C1. Molecular Genetics and Metabolism. 2015;**116**:75-79. DOI: 10.1016/j.ymgme.2015.07.001

[45] Valle AM, González MC, Morón R, Ruiz S, Cabeza J. Hydroxypropyl-beta cyclodextrin in Niemann-Pick type C disease. International Journal of Clinical Pharmacy. 2015;**37**:282. DOI: 10.1007/s11096-014-0039-2

[46] Rodríguez R, De Santos MT, Natera D, Fernández MI, Grasa CD, Taboas A, et al. Nuevas expectativas de tratamiento de Niemann Pick tipo C. (Article in Spanish) XXXVIII Anual Meeting SENEP. 2015. Abstract PO 147. Available from: http://www.senep2015.com/images/site/comunicaciones_aceptadas/Posters_Orales.pdf

[47] Roch Santed M, Cabañas Poy MJ, Del Toro Riera M, Cañete Ramírez C, Fernández Polo A, Clemente Bautista S. Intrathecal cyclodextrin in the treatment of Niemann-Pick disease type C. European Journal of Hospital Pharmacy. 2017;**24**:185-188. DOI: 10.1136/ejhpharm-2016-001067

[48] Sakiyama Y, Shibata S, Sanayama H, Ono S, Maekawa M, Matsuo M, et al. Intrathecal 2-hydroxypropyl-beta-cyclodextrin (HPBCD) therapy in adult-onset Niemann-Pick disease type C (NPC). Journal of the Neurological Sciences. 2015;**357**(S1):e20 http://dx.doi.org/10.1016/j.jns.2015.08.719

[49] García-Robles AA, Company-Albir MJ, Megías-Vericat JE, Fernández-Megía MJ, Pérez-Miralles FC, López-Briz E, et al. Use of 2-hydroxypropyl-beta-cyclodextrin therapy in two adult Niemann-Pick type C patients. Journal of the Neurological Sciences. 2016; **366**:65-67. DOI: 10.1016/j.jns.2016.04.048

[50] Ory DS, Ottinger EA, Farhat NY, King KA, Jiang X, Weissfeld L. Intrathecal 2-hydroxypropyl-β-cyclodextrin decreases neurological disease progression in Niemann-Pick disease, type C1: A non-randomised, open-label, phase 1-2 trial. Lancet. 2017;**390**(10104):1758-1768. DOI: 10.1016/S0140-6736(17)31465-4

[51] Committee for Human Medicinal Products (CHMP), Official document unpublished from European Medicines Agency (EMA). Background Review for Cyclodextrins Used as Excipients. In the Context of the Revision of the Guideline on 'Excipients in the Label and Package Leaflet of Medicinal Products for Human Use'. 2014. (CPMP/463/00 Rev. 1). Available from: http://www.ema.europa.eu/docs/en_GB/document_library/Report/2014/12/WC500177936.pdf

[52] Porter FD, Yanjanin N, Liu A, Shankar R, Ory D, Pavan B, et al. Therapeutic trials for Niemann-Pick disease, type C1: 2-Hydroxypropyl-β-cyclodextrin. Abstract unpublished from Michael, Marcia, and Christa Parseghian Scientific Conference for Niemann-Pick type C Research. June 12-14, 2014. Abstract 7:18. Available from: https://niemannpick.nd.edu/assets/135126/2014abstractbooklet.pdf

[53] Cronin S, Lin A, Thompson K, Hoenerhoff M, Duncan RK. Hearing loss and otopathology following systemic and intracerebroventricular delivery of 2-hydroxypropyl-beta-cyclodextrin. Journal of the Association for Research in Otolaryngology. 2015;**16**(5):599-611. DOI: 10.1007/s10162-015-0528-6

Chapter 5

Room at the Top as well as at the Bottom: Structure of Functional Food Inclusion Compounds

Aida Moreira da Silva

Additional information is available at the end of the chapter

http://dx.doi.org/10.5772/intechopen.74162

Abstract

Cyclodextrins (CDs) have been hanging around research laboratories, since their discovery by Villiers. When Schardinger identified the three naturally occurring forms—α, alpha; β, beta; and γ, gamma—they were called "Schardinger sugars." Then it was found that CDs have the ability to bind with many different types of molecules in their cavities. Cyclodextrins are oligosaccharides obtained by enzymatic means from starch-containing raw materials such as corn. The characteristic feature of cyclodextrins is their ring-shaped, three-dimensional structure, with a hydrophobic cavity in the center, which is capable of receiving a lipophilic "guest" molecule. The hydrophilic outer surface ensures compatibility with aqueous systems. The specific properties of CDs opened up a wide range of application in food fields. Molecular encapsulation of food ingredients through cyclodextrins is intended to improve the stability of the ingredients, by extending the shelf life of the products. The results of accelerated and long-term stability tests have demonstrated that the stability of food ingredients encapsulated by cyclodextrins has outpaced those of traditionally formulated ones. The technological advantages of using cyclodextrins in food systems and food processing technologies are also manifested in the improvement of sensory and nutritional properties. Examples of food products are presented to demonstrate the importance of cyclodextrin-based molecular inclusion technology in the food industry.

Keywords: cyclodextrins, Global Paradox, functional foods, novel food ingredients, inclusion compounds, supramolecular chemistry

1. Introduction

The Neolithic Revolution marked the end of nomadic peoples and the beginning of sedentarization of *Homo sapiens*, having originated 10,000 years of food inventions. Nowadays, it produced sufficient food to feed each of the 7 billion human beings. However, malnutrition and food waste related to production and consumption are big issues that pose major challenges for the future. The big question then arises: why exists malnutrition, if there is enough food quantity and quality for all people around the world?

The Global Paradox of Malnutrition (**Figure 1**), i.e., the conditions of malnutrition and overweight/obesity, have causes and consequences strongly related with the existence of inadequate food systems. This Global Paradox of obesity *versus* undernutrition is portrayed through the numbers: around 33% of the world population suffers from malnutrition. While one part of the world population has no access to food in quantity and quality necessary for a healthy life, another large group of people chooses to excessively high-calorie foods low in nutrients and fiber. Metabolic disorders as obesity and related diseases can be prevented but recently became pandemic. Even more serious is the fact that recently the number of children suffering from these metabolic disorders (155 million) exceeded the number of children suffering from malnutrition (148 million). If this process is not stopped, through appropriate measures at the community level and at the individual level, it is expected that the economic and social impact will be catastrophic.

One approach to reduce malnutrition is the development of functional foods or novel food ingredients to reduce the risk of disease, providing longevity and a healthy lifestyle. In this context, glucose cyclic oligosaccharides, the cyclodextrins, play an important role.

Cyclodextrins (CDs) are compounds derived from starch, modified industrially produced enzymatically. These starch derivatives are nontoxic ingredients, are not absorbed at the level of the upper gastrointestinal tract, and are completely metabolized by colonic microflora.

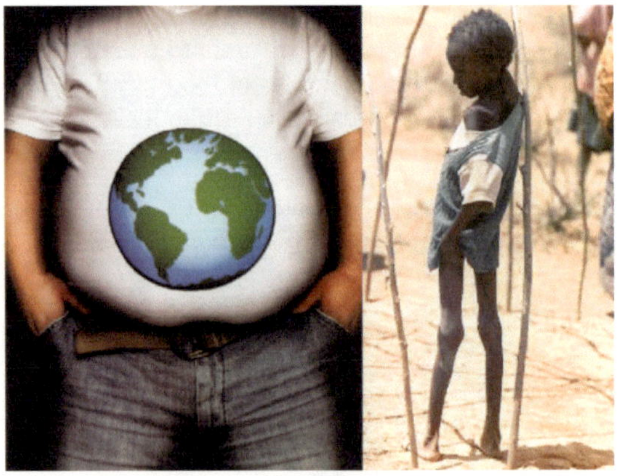

Figure 1. The Global Paradox: Obesity and Malnutrition. Obesity has doubled since 1980; In 2014 there were more than 1.9 billion adults overweight, of which 600 million people are obese; most people live in countries where overweight causes more mortality than malnutrition (WHO Fact sheet n°311 2015).

http://dx.doi.org/10.5772/intechopen.74162

CDs can be used in food as supports for molecular encapsulation of flavors and other ingredients, and there is a huge scope that goes beyond food applications.

2. Cyclodextrins discovery

Over 100 years have elapsed, since Villers first described the isolation of a crystalline substance from a medium of *Bacillus amylobacter* starch culture, corresponding to what is now recognized as cyclomalto-oligosaccharides or cyclodextrins. This crystalline substance was then called "cellusine" because of its similarity with the cellulose. Over the next 60 years, progress in purification products, elucidation of structures, and identification of unusual properties were remarkable and surprised the pioneering researchers in the field. Contributions from Schardinger, Pringshein, Freudenburg, and Cramer, in Germany, and from French in the United States of America are presented in a review article that has become a true classic, published in Advances in Carbohydrate Chemistry [1]. In this article, French anticipated that the cyclodextrins would serve, teach, delight, and intrigue the scientific community. It was from this time that the scientific community recognized one of the most characteristic properties of cyclodextrins: the ability to form inclusion compounds with a wide variety of substrates or guest molecules.

However, the path taken by the pioneers in the area was not easy. Cramer felt strong opposition when mentioned that cyclodextrins in solution could include other molecules. M.L. Bender recognized the synthesis of inclusion compounds based on cyclodextrins and the ability to discriminate enantiomers during the inclusion process and the catalytic capacity in reactions on bound substrates for the very first time. The result of these discoveries fascinated a large number of researchers: Bender and Breslow in the United States, Saenger in Germany, and Tabushi, Komiyama, and Hirai in Japan. They felt captivated by scientific research involving cyclodextrins [2].

During the decades of 60 and 70 of the last century, cyclodextrins have been widely studied as simulants enzyme systems (cyclophanes) or very similar to the behavior of the various enzymes The advent of high-pressure liquid chromatography techniques (HPLC), fast atom bombardment mass spectrometry atoms (FAMS), and nuclear magnetic resonance spectroscopy (NMR) has made possible the characterization of chemically modified cyclodextrins (e.g., methylated cyclodextrins). A complete and unambiguous characterization of the structures was made by X-ray diffraction and neutron diffraction techniques.

On the other hand, the commercial interest in cyclodextrins has grown and is growing at a phenomenal rate, particularly in Japan and Hungary, where Professor Joseph Szejtli's contribution was enormous. The pharmaceuticals, agrochemicals, food, and cosmetics industries have been influenced by these outstanding molecules to a lesser or greater extent [2].

Scientific and technological impact of cyclodextrins are associated, on the one hand, the diversity of situations in which they operate and on the other hand, the enormous challenge that its use has caused in the design of new molecular systems reminiscent functions of biological, chemical, or physical nature.

3. Chemistry and structure and cyclodextrins

3.1. Biological synthesis of cyclodextrins

The CDs are produced by degradation of the prehydrolyzed starch and their subsequent cyclization-mediated cyclodextrin glucosyltransferase enzyme (CGTase, EC 2.4.1.19) produced by bacteria that belong to the genus *Bacillus*. Due to the helical structure of the starch molecules, the primary cleavage product undergoes an intermolecular reaction forming cyclic products joined by α-1,4 linkages, generally designated by cyclodextrins. To distinguish them, Greek letters are used to specify the number of D-glucose units (in brackets): α (6) β (7) γ (8) δ (9) ε (10) ξ (12) η (13).

The shapes α, β, and γ are the natural cyclodextrins and most commonly used (**Figure 2 (c)**). Higher numbers of counterparts of glucose units also exist but are difficult to purify, with weaker inclusion properties. Cyclodextrins with a number of glucose units less than 6 do not exist, probably due to steric hindrance.

The preparation of cyclodextrins can be subdivided into the following main stages:

- Culture of the producing microorganism CGTase enzyme
- Separation of the enzyme from the medium, their concentration, and purification
- Enzymatic conversion of prehydrolyzed starch a mixture of cyclic and noncyclic dextrins
- Separation of the CDs are from the conversion mixture, purification and crystallization

In industrial production of cyclodextrins, the most frequently used source of enzyme is *Bacillus macerans*, renamed as *Paenobacillus macerans*. Other enzymatic sources used are *Klebsiella pneumonia* and *Alkalophilic bacterium* 38–2. The forms α, β, and γ are dependent from the source of CTGase enzyme. The *Bacillus macerans* and *Klebsiella pneumonia* CTGase mainly produce the α form. *Alkalophilic bacterium* 38-2 mainly produces β-cyclodextrin. However, the relationship between the CD formed also depends on the incubation time of the enzyme in starch medium culture because most CTGases initially produce the α form, while the synthesis of other forms is slower [3].

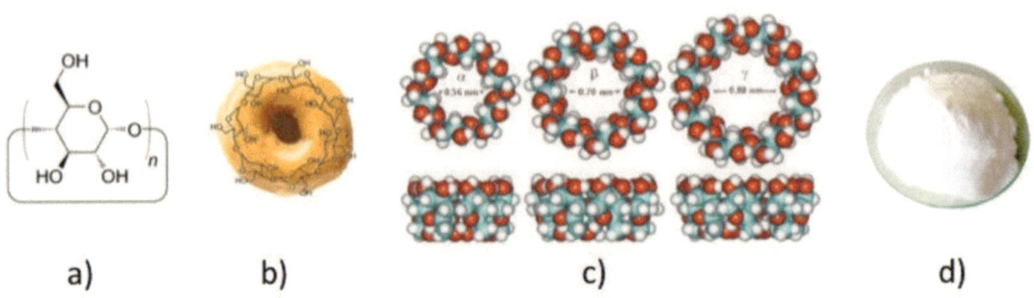

Figure 2. Structure of a cyclic oligosaccharide, cyclodextrin, CD (a); The "donut" Molecular (b); Paragraph equal to six, seven or eight rings of D-glucopyranose, joined by glycosidic linkages of the type α-1.4, representing α-CD, β-CD and γ-CD, respectively (c); white crystalline powder β-CD (d).

3.2. The structure of cyclodextrins

The native cyclodextrin molecules (α-CD, β-CD, and γ-CD) have the shape of a short truncated cone with a cavity inside, i.e., a toroidal shape. The length is determined by the height of the glucose unit (7.9 Å = 0.79 nm), and the diameter of the cavity is determined by the number of glucose units (**Figure 2 (a)** and (**c**)).

The glucose rings linked together by α-1.4 linkages as in amylose. They are oriented in the same direction, and thus, the narrow end of the torus is formed by the primary hydroxyl groups (O (6) H), while the wider edge of the truncated cone is occupied by the secondary hydroxyl (O (2) H, O (3) H) groups. These peripheral hydroxyl groups confer hydrophilic properties to the CD surface. Moreover, the internal cavity has mainly hydrophobic characteristics due to the methine group (CH) and the oxygen atoms of the ether type (O (4) and (5)).

The CDs may crystallize in the form of hydrate or inclusion compound, and the crystal structure was mainly determined by the following factors:

1. The nature and size of the cavity included in the molecule;
2. Hydrogen bonding between the included molecules and between CD and CD drives.

The interstices between the CD units are occupied by water molecules incorporated in the overall structure (see **Figure 3**) [3, 4].

The CDs cavity in the center, with predominantly hydrophobic character, is large enough to hold, accommodate, or include other molecules. When this occurs, there is the formation of an inclusion compound. These compounds, or complexes, may be described as a molecular-level nanoencapsulation. Food ingredients formulated with cyclodextrins become stable to heat and oxidation processes and are not affected by dispersion forces and are readily dispersed for use in liquid products [5].

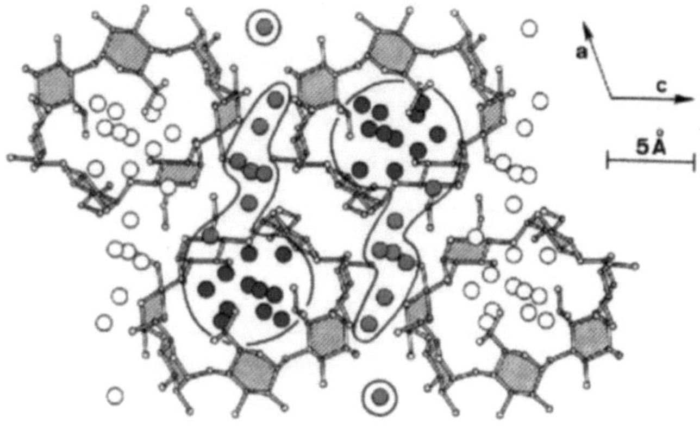

Figure 3. The crystalline hydrate of β-CD. The blue are represented statistical locations of the interstices water molecules; red are represented statistical locations of cavity water molecules [3, 4].

The food industry uses the native cyclodextrins in different ways owing to the above-described properties, being used in various applications due to their ability to form inclusion compounds. The α-cyclodextrin acts even as prebiotic. Thus, formulations with CDs are used in food and also in the designated functional food markets in order to circumvent the problems of stability, taste, and flavors of special ingredients. In this context, natural functional foods are food systems enriched, e.g., with bacterial cultures, omega 3 fatty acids, anthocyanins, dietary fibers, etc., which can contribute to the maintenance of health and reduction of disease risk. [6]

4. α-CD as a novel food ingredient

α-cyclodextrin was approved in Europe as a novel food ingredient for use as a prebiotic in 2008 [6, 7]. The term prebiotic is used to describe nondigestible carbohydrates necessary to maintain healthy intestinal flora [7, 8].

Currently, there is considerable interest in these food ingredients, since that the common diet only consumes half the indigestible portion of carbohydrate necessary to maintain the intestinal flora. The properties of the smaller native cyclodextrin make it particularly suitable for this function, since it is soluble in water, does not affect the solution viscosity or change its taste, and can be added to beverages. Thus, the daily intake of dietary fiber essential to the health can be increased. Soluble fibers also have a beneficial effect on the levels of fat and sugar in blood. Apart from adding the liquor, the α-cyclodextrin may also be added to pasta to baking or other finished products, since it is stable even at high temperatures. The Maillard reactions (a kind of browning reaction) are not supported, since the α-cyclodextrin is not a

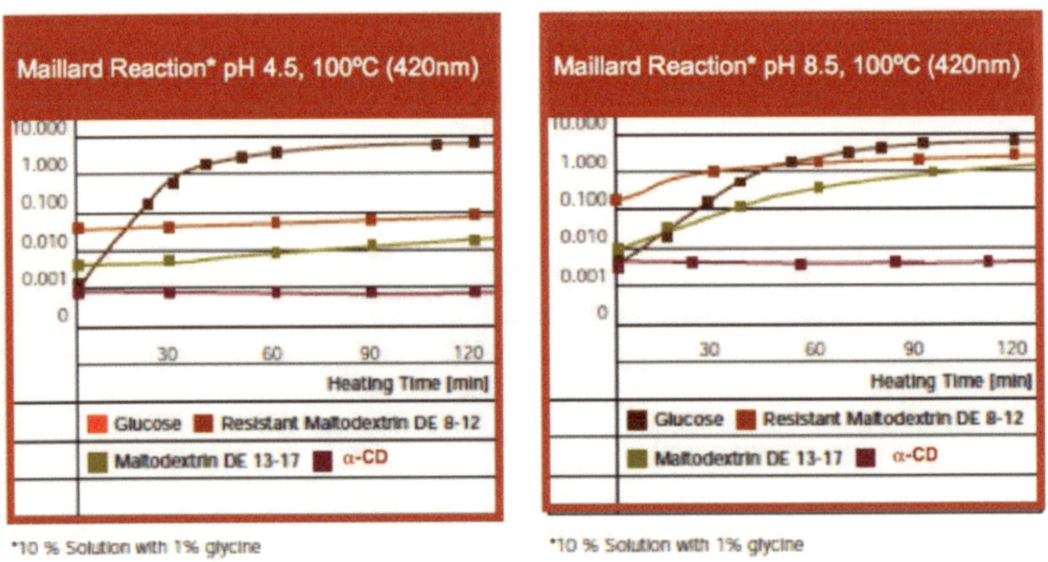

Figure 4. Speed of the Maillard reaction as measured by the formation of HMF ([HMF]/(mg/kg)); λmax = 420 nm, pH = 4.5 and 100°C. The α-cyclodextrin does not take part in Maillard reactions [10].

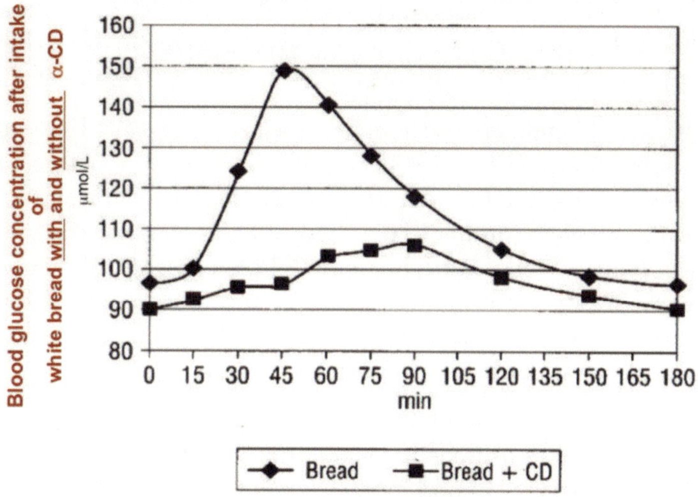

Figure 5. Effect of addition of α-cyclodextrin white bread glycemic index (GI ratio between the area calculated in bread curve (100 g) and the area calculated in bread curve (100 g) + *α-CD (10% (m/m))* [10].

reducing oligosaccharide. The same is not true with glucose and maltodextrin that react with glycine in the formation of undesirable pigments, in addition to nutritional loss (**Figure 4**) [9].

For bakery products remain fresh and crispy during the storage period, the ingredients should not have hygroscopic characteristics. In this respect, and contrary to inulin, the α-cyclodextrin has a low hygroscopicity.

Beyond the benefits of not participating in browning reactions and low hygroscopicity, this soluble fiber also has beneficial effects on the levels of fat and sugar in the blood. Two months clinic studies performed by researchers of the University of California, Davis, CA (UC Davis), was used controlled placebo in 28 volunteers who were overweight (25 > Body Mass Index >30) but not obese, there was the following: after ingesting 6-cyclodextrin daily for a period of 2 months, these people lost weight without changing diet or lifestyle. During the study, blood parameters as total cholesterol and low density lipoprotein (LDL) decreased along with insulin level. The positive effect of α-cyclodextrin on the glycemic index (GI) was also confirmed by other studies [9].

Figure 5 shows the effect on glycemic index after adding α-cyclodextrin to the white bread.

5. Inclusion compounds based on cyclodextrins: applications in food and food processing

The three-dimensional nature of native CDs makes these molecules very important, since these starch derivatives are ingredients that do not have toxic characteristics; they are not absorbed in the digestive tract and are completely metabolized by the intestinal microflora. These CDs have the necessary requirements to be neutral in terms of aroma and flavor, although they are made from glucose units; the α-CD and β-CD have no sweet taste, while the γform only has a slight sweet taste. Since the CDs are occurring in the form of a colorless powder, it makes them easier to process [4, 11].

The molecular inclusion phenomenon is one chemistry field also called supramolecular chemistry. Jean-Marie Lehn, Nobel Prize in 1997, is one of the creators of this area of chemistry that deals with complex entities resulting from the association of two or more chemical species held together by non-covalent intramolecular bonds. Lehn, paraphrasing Richard Feynman (and his well-known speech on nanotechnology "There is plenty of room on the bottom") with the expression "There's even more room at the top", has indicated that chemistry not only has to look toward the extremely small but can also go beyond the molecular size, studying the supramolecular complexity.

The main advantage of using CDs in food systems lies summarized in **Table 1**. The prolongation of the shelf-life of the compositions and standardization and ease in dosing and transport of inclusion compounds are very important features of this inclusion nanotechnology.

Cyclodextrins can encapsulate biocides that can be applied to food packaging materials. By changing humidity conditions, there is controlled release of the biocide thus preventing, for example, the proliferation of microorganisms (bacteria, fungi, and yeasts). The CDs are good carriers for flavors and fragrances. In bakery products, it can be reduced to one-third the amount of aroma needed, if they are CD encapsulated. The CDs can also improve the bread dough and the crispy effect of rice crackers. The aroma of fresh vegetables can also be preserved by reduction of the degradation rate and preventing discoloration. Mixed

Benefit	Food system
Stabilization effect	
Protection against deterioration of sensitive substances	• Fats, • Terpenos • Flavors
Elimination or reduction	
Unpleasant flavors/aromas	• Removal of bitter taste in grapefruit juices
Microbial contamination; Hygroscopicity	• Smell of Vitamin B1
Selective cholesterol removal	• Cheese • Butter
Improvement of the solubility of lipophilic ingredients	• Carotenes • Curcumin
Improvement of emulsion stability	• Mayonnaise • Dairy products
Improvement in the uniformity of content	• Standard flavors

Table 1. Summary of some beneficial effects of the use of cyclodextrins in food systems.

with spices, the CDs can help stabilize emulsions (mayonnaises and salad dressings). Cyclodextrins may act as sweetener by aspartame encapsulation to make it more water soluble (**Table 2**).

In beverages, the CDs allow the use of aromas, control the dissolution of aspartame preventing its breakage, maintain the color of fruit juices, and also allow the encapsulating of carbon dioxide.

In Japan, CDs are considered a natural product, used to deodorize meats and fish, improving the defrosting properties through water and red pigments retention and reduction of undesirable aromas.

Some of the foods that contain plant extracts have undesirable bitter flavors. The CD molecules are suitable to encapsulate these components. For example, grapefruit juice can be treated during preparation CD are to remove the bitter taste caused by naringin and the limonene (**Figure 6**). The other grapefruit aromas are encapsulated in a small extent, and the treatment does not alter the contents acid or vitamin C [11].

The polymers of cyclodextrins are used to prevent the juice to precipitate. The polymer molecules are synthesized using a bifunctional crosslinking agent (*e.g.* glutaraldehyde). The CD loaded polymer naringin, adding sodium hydroxide and subsequent washing regenerate limonene.

Brand	Type of food	CD'S function
Natural[a]	Cheese	Cholesterol removal
Cyroma-line[b]	Flavored sugar for baking	Preserve the flavor after heating
Balade [c]	Butter	Cholesterol removal
Simply eggs[d]	Eggs	Cholesterol removal
FlavorAktiv Standard Kit [e]	Patterns of beer aromas	Preserve the flavor
Flavono [f]	Chewing gum	Estabilize the flavor
Choco bar [f]	Chocolate	Emulsifier
Poder tea [f]	Instant tea	Preserve color
Gymet [f]	Dietary fiber drink	Mask flavor
Stick lemon [f]	Instant tea	Preserve the flavor

[a]France.
[b]Hungary.
[c]Bélgium.
[d]USA.
[e]Great Britain.
[f]Japan.

Table 2. Foods that include cyclodextrins in their formulation.

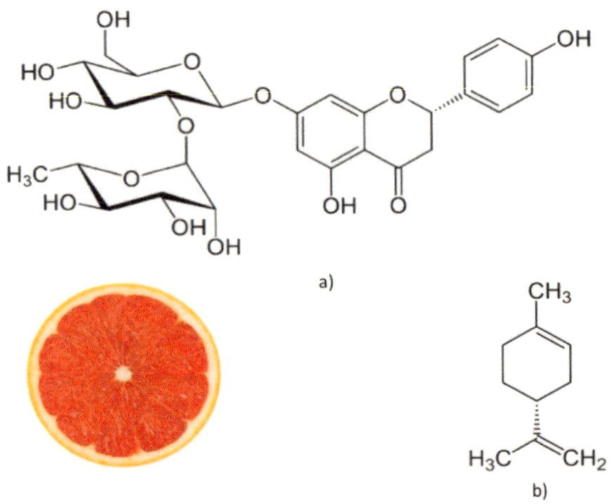

Figure 6. Naringin (a) and Limonene (b) give a bitter taste to grapefruit juice.

The ability to form inclusion complexes is also used for the production of cholesterol-free food due to the cholesterol molecule is retained in the cavity of seven glucose units from β-cyclodextrin (**Figure** 7).

The results of the preliminary study on the effect of β-cyclodextrin in removing red wine unpleasant smells have also been promising. This study was conducted triangle sensory analysis by 14 untrained panelists using a red wine enriched with 4-ethylphenol (5000 g/L) or a mixture of 4-ethylphenol (750 g/L) + 4-ethylguaiacol (75 g/L) and different levels of β-CD concentration. According to the sensory panel, the concentration of 11.52 g/l CD-β (ratio 2:1) was sufficient to reduce the perception of red wine unpleasant odors. Additionally, an increase of colour intensity and the total polyphenol content of red wine treated with β-CD was measure and statistically significant [12].

β-cyclodextrin and γ-cyclodextrin are the most commonly used forms, because its size is usually more favorable. The complex formation facilitates controlled release of the encapsulated molecules, since the dissociation of the complex requires low humidity. For example, the flavors may be stored and released in a chewing gum, when it is chewed. An additional advantage of

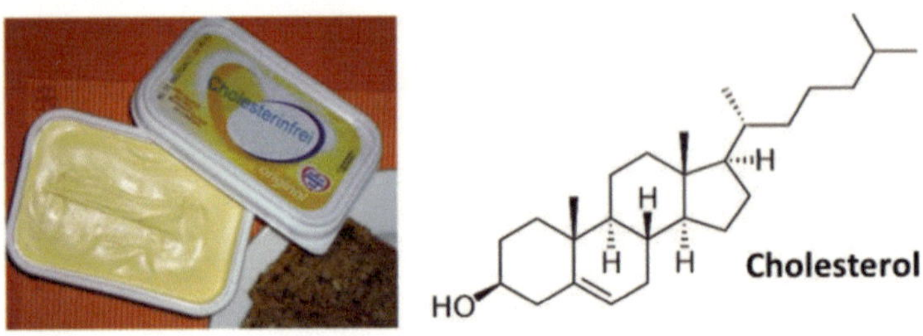

Figure 7. Cholesterol free butter is obtained by treatment with cyclodextrin to sequester (encapsulate) the cholesterol molecule.

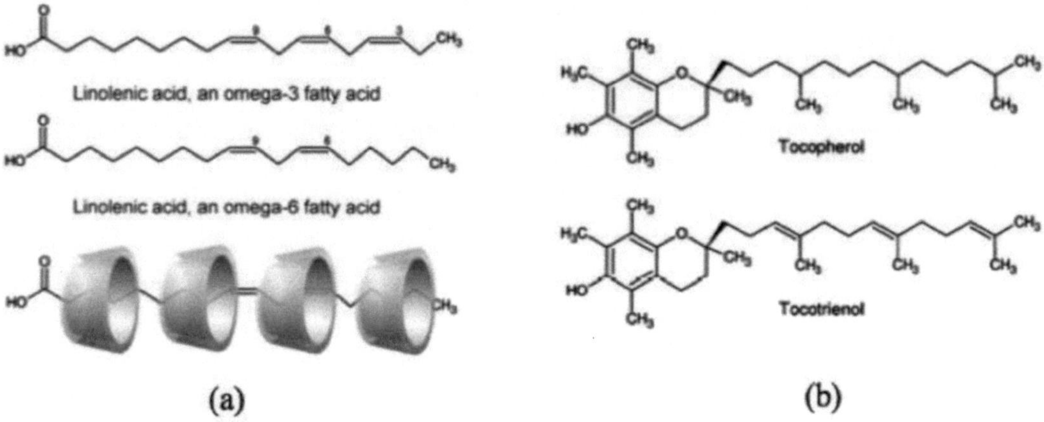

Figure 8. (a) Fatty acids ω3 and ω6 complexed with several cyclodextrin molecules; (b) The product contains OmegaDry® Cranberry inclusion compounds with components obtained from cranberry essential oil [13].

the inclusion of these molecules is the protection against oxidation, sublimation, and evaporation. The complexation may also be used to mask unpleasant flavors and odors. One of the advantages of cyclodextrins nanoencapsulation is exploited in the case of omega 3 (ω^3) and omega 6 (ω^6) fatty acids because they have proven positive effects in reducing fat levels in the blood plasma, with a consequent reduction in the risk of cardiovascular diseases. Although the ω^3 and ω^6 fatty acids are mostly derived from fish oils and algae extracts that in the native state have an unpleasant taste and aroma. Therefore, the encapsulation via CD is necessary to obtain a white odorless powder, which is easily processable, preventing also the oxidation reactions.

Currently, there is a product on the market, registered under the name OmegaDry® *Cranberry*, which contains several cranberry oil components encapsulated in γ-CD (**Figure 8 (a)**) [13]. There are also formulations with isomeric vitamin E with functional tocotrienols (**Figure 8 (b)**) and myricetin (3,5,7-trihydroxy-2- (3,4,5-trihidroxifenil) - 4-chromenona) and quercetin [14, 15]. Considerable interest in these formulations results of whether they are lipophilic substances with antioxidant properties, enabling the cell protection from oxidative degradation of lipid membrane structures of the molecules, preventing premature aging of the skin, caused by damage from ultraviolet radiation. Thus, the complexes with tocopherol and tocotrienol can be used in food systems and also in cosmetics.

6. Legal status and patents

From very early stages, cyclodextrins were used by the food industry. In Japan, cyclodextrins were considered natural products, and its use occurred in the late 70s of last century after the development of industrial manufacturing processes. In 1987, there were already 88 Japanese patents, making Japan the first country to functionalize food with CDs. Since 2000, the three native cyclodextrins can be used as food additives in the United States of America (USA). In

Australia and New Zealand, the α and γ forms were classified as new ingredient since 2003 and 2004, respectively.

However, there was no information on the legal situation of β-cyclodextrin in this region of the globe. In Europe, since 1998, β-cyclodextrin was approved under the designation E-459 in the list of food additives. More recently, α-cyclodextrin and γ-cyclodextrin were approved as new ingredient in 2008 and 2012, respectively. In **Table 3**, the information is gathered about the legal status of native cyclodextrins worldwide [16–21].

In 2014, Deorsola et al. presented a research paper on patents involving cyclodextrins. These authors used free databases: Espacenet, USPTO, PATENT SCOPE, INPI, and DERWENT. The

Cyclodextrin	α-CD	β-CD	γ-CD
WHO/FAO[1]	ADI = Not specified June 2001 to June 2004	ADI = 5 mg kg/day 4/1/1995	ADI = Not specified 1999 e 2000
USA	GRAS[2] 2004	GRAS[3] 2001	GRAS[2] 2000
CANADA	Filed for Novel food July 2006		
UE	Novel food Approved 2008	Carrier for food additives (< 1 g/kg) E-459	Novel food Approved 2012
JAPAN	Natural product	Natural product	Natural product
MEXICO	Approval of FDA Import licence	Follow FDA approvals with an import license	Approval of FDA Import licence
MERCOSUR[4]	Approved		
FSANZ[5]	Novel food 12/1/2004		Novel food 2003
COREIA	Approved for dietary supplement	Approved for dietary supplement	Approved for dietary supplement
PHILLIPINES	Approved		Approved
THAILAND			Approved for dietary supplement
TAIWAN	Approved for dietary supplement		

[1]WHO/FAO – World Health Organization /Food & Agriculture Organization of the United Nations.
[2]GRAS in a wide range of intended use in food.
[3]GRAS as flavor protectant.
[4]State Mercosur – Argentina, Brazil, Paraguai, Uraguai & Venezuela.
[5]FSANZ – Food Standards Australia and New Zealand.

Table 3. Cyclodextrins approval status as food ingredients/additives [15–20].

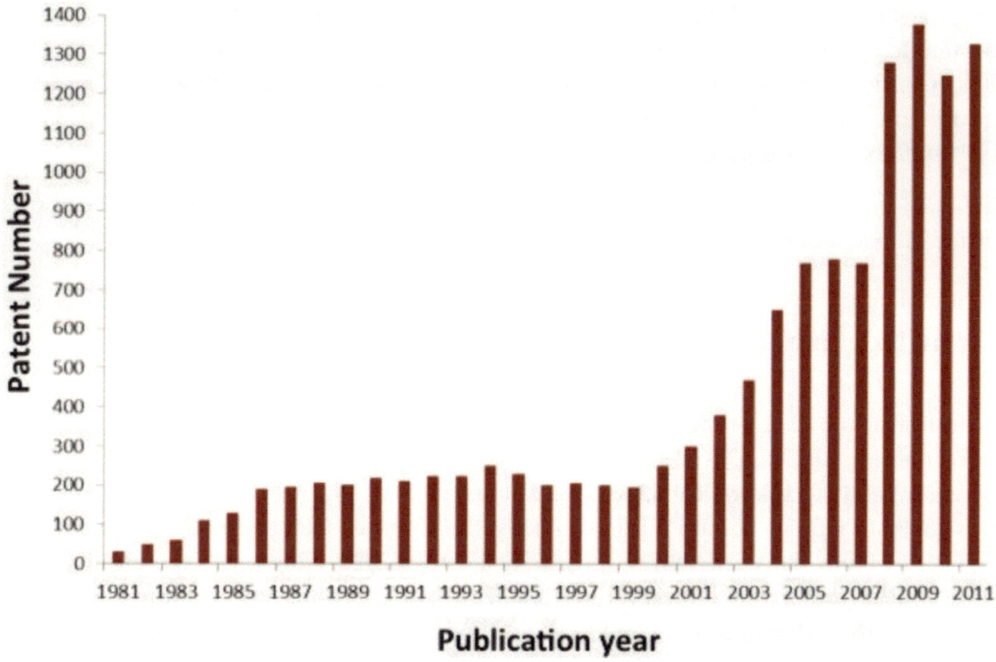

Figure 9. Number of cyclodextrin patents [22].

30-year evaluation period was from 1981 to 2011, having found 14,572 records. The increased number of patents in three stages: (I) to 1990 followed by a slight decline; (II) a faster growth 2000–2005 and remained for 2 years; and (III) from 2008 to 2011 the number nearly doubled patent (**Figure 9**).

Companies that recorded the highest number of cyclodextrin patents are Procter & Gamble Co. (>300), Kao Corp. (99), Schering AG (92), the Ensuiko Sugar Refining Co. (83), and Ono Pharm. Co. (79). The classification of patents allows an assessment of the topic of invention, and this analysis shows that the most selected classification relates to medicinal and/or food preparations [23].

7. Final notes

Cyclodextrins passed the benches of research and development laboratories (R&D) for industrial-scale production, since they were discovered by Villers in 1891. This is due to the fact that CDs had a very special feature: the ability to include different types of molecules in the cavity of the molecular "donut."

Currently, the global market for cyclodextrins is estimated above 10 tons per year, mainly applied to food systems, or related to them (*e.g.*, retardant product of fruit maturation, SmartFresh® product from Agrofresh company).

Therefore, it is expected a huge increase in applications of CDs in this next century of research dedicated to cyclodextrins in food systems. This observation is supported on the exponential patent records, scientific publications, inclusion of the topic in school programs, and pedagogical manuals [21, 24], and also the growing organization of conferences devoted to the subject is globally widespread.

In conclusion, there is "*Room at the Top*" as well as "*Room at the Bottom*" for the use of cyclodextrins in food systems.

Acknowledgements

The author acknowledges financial support from the Portuguese Foundation for Science and Technology: PTDC/QEQ-MED/1890/2014, within Project 3599 – to Promote Scientific Production and Technological Development as well as the formation of thematic networks (3599-PPCDT) – jointly co-financed by the European Regional Development Fund (FEDER), through the partnership agreement Portugal2020—Regional Operation Program CENTRO2020, under the project CENTRO- 01-0145-FEDER-023631 SoSValor: Sustainable Solutions for the Valorization of Plant Natural Products and Industrial Waste.

Author details

Aida Moreira da Silva[1,2*]

*Address all correspondence to: aidams@esac.pt

1 Molecular Physical-Chemistry, R&D Unit, Department of Chemistry, University of Coimbra, Coimbra, Portugal

2 Coimbra College of Agriculture, Polytechnic Institute of Coimbra, Coimbra, Portugal

References

[1] French D, The Schardinger dextrins, Advances in Carbohydrate Chemistry, 1957;**12**:187; in Moreira da Silva A. Behavior of ß-cyclodextrin in hydration/dehydration processes, Water Diffusion and aromas of molecules Inclusion, PhD Thesis, Instituto Superior Técnico, Technical University of Lisbon; 1997

[2] Szejtli J. Past, present, and future research of cyclodextrin. Pure and Applied Chemistry. 2004;**76**(10):1825-1845

[3] Moreira da Silva, Aida MG. Behavior of ß-Cyclodextrin in Hydration Processes/ Dehydration, Diffusion of Water and aromas of molecules Inclusion, PhD Thesis, Instituto Superior Técnico, Technical University of Lisbon, 1997

[4] Steiner T, Moreira da Silva A, Tan-day J, Muller J, Saenger, W, Rapid water diffusion in cage-type crystal lattice: Beta-cyclodextrin dodecahydrate. Angewandte Chemie International Edition in English. 1995;**34**(13/14):1452-1453

[5] Szejtli J, Osa T editors. Comprehensive supramolecular chemistry. Cyclodextrins. Vol.3. Oxford: Elsevier; 1996

[6] Moreira da Silva A. Food products que help to promote human health. Chapter 4, Food, Diet and Health: Past, Present and Future Tendencies. New York: New Science Publishers, Inc. 2010

[7] Official Journal of the European Union (L146/12) (05.06.2008). Commission decision on the placing on the market of alpha-cyclodextrin as a novel food ingredient under Regulation (EC) No.^{O} ? 258/97 of the European Parliament and of the Council, available at http://eur-lex.europa.eu/legal- content / EN / TXT / uri = CELEX: 32008D0413

[8] Roberfroid M. Prebiotics: The concept revisited. Journal of Nutrition. 2007;**137**:830S

[9] Chemie W. Cavamax®Alpha Cyclodextrin: The new soluble dietary fiber. [Internet]. Available from http://www.wacker.com/cms/media/publications/downloads/6752_EN.pdf. [Accessed: 18-05-2017]

[10] Szente L, Szejtli J. Cyclodextrins the food ingredients. Trends in Food Science & Technology. 2004;**15**(I3-4):137-142

[11] Mourtzinos I, Kalogeropoulos N, Papadakis R, Konstantinou K, Karathanos VT. Encapsulation of monoterpenes in nutraceutical and beta-cyclodextrin modified starch. Journal of Food Science. 2008;**73**(1):S89-S94

[12] Botelho G, Valiau C, Moreira da Silva A. Effect of cyclodextrins on off-odors removal of red wine: an innovative approach. Ciência e Tecnologia Vitivinícola. 2011;**26**:63-68

[13] Chemie W. Cavamax® OmegaDry® Cyclodextrins in Food and Nutraceutical Applications. [Internet]. Available from http://www.wacker.com/cms/media/publications/downloads/6088_EN.pdf. [Accessed: 20-05-2017]

[14] Lucas-Abellan C, Fortea I, Gabaldon J, Nunez A. Soft and encapsulation of cyclodextrins in myricetin quercetin and at acidic pH. Journal of Agricultural and Food Chemistry. 2008;**56**(1):255-259

[15] Pereira A B, Moreira da Silva A, Barroca MJ, Marques MPM, Braga SSM. Physicochemical properties, antioxidant action and practical application in fresh cheese of the solid inclusion compound γ-cyclodextrin·quercetin, in comparison with β-cyclodextrin·quercetin. Arabian Journal of Chemistry. 2017. in press. DOI: 10.1016/j.arabjc.2017.04.001

[16] US Food and Drug Administration. GRAS notice for Alpha-Cyclodextrin. 2004 June 25. GRAS Notice No. GRN 000155 [Internet]. Available from http://www.fda.gov/Food/IngredientsPackagingLabeling/GRAS/NoticeInventory/ucm154385.htm [Accessed: 20-05-2017]

[17] US Food and Drug Administration. GRAS notice for Gamma-Cyclodextrin. 2000 May 22. In the NRM 46. [Internet]. Available from http://www.fda.gov/downloads/Food/IngredientsPackagingLabeling/GRAS/NoticeInventory/UCM261675 [Accessed: 20-05-2017]

[18] US Food and Drug Administration. GRAS notice for Beta-Cyclodextrin. 2001 Mar 23. In GRN 74. [Internet]. Available from: http://www.fda.gov/downloads/Food/Ingredients PackagingLabeling/GRAS/NoticeInventory/UCM261320 [Accessed: 29-07-2017]

[19] Official Journal of the European Union (L144/41) (05.06.2012), the Commission Implementing Decision authorizing the placing on gamma-cyclodextrin market as novel food ingredient under Regulation (EC) No.º 258/97 of the European Parliament and of the Council [Internet]. Available in http://eur-lex.europa.eu/legal- content/EN/TXT/PDF/uri =CELEX: 32012D0288 & from = EN [Accessed: 20-05-2017]

[20] Official Journal of the European Union (L160/4) (12.06.2013), Regulation (EU) No 536/2013 of the Commissionñthe June 11, 2013 [Internet]. [Consulted in May 2014]. Available at http://EUR-lex.europa.eu/legal-content/PT/TXT/PDF/?uri=CELEX:32013R0 536&from=EN [Accessed: 20-05-2017]

[21] Official Journal of the European Union (L83 / 1) (22.3.2012), Regulation (EU) No 231/2012 of the Commissionñthe 9 marçO 2012 establishing specificationOes for food additives listed in Annexes II and III of Regulation (EC) No 1333/2008 of the European Parliament and of the Council [Internet]. [Consulted in May 2013]. Available in http://eur-lex.europa.eu/legal- content/EN/TXT/PDF/uri=CELEX: 32012R0231 & from = EN [Accessed: 23-07-2017]

[22] Deorsola AC Mothe CG, de Oliviera LG, Deorsola AB; Technological monitoring of cyclodextrin - World panorama. World Patent Information, 2014;**39**:41-49. [Accessed: 20-08-2017]

[23] Astray G, C-Barreiro G, Mejuto JC, Rial-Otero R, Simal-Gandara J. A review on the use of cyclodextrins in foods. Food Hydrocolloids. 2009;**23**(7):1631-1640

[24] Belitz HD, Grosch W, Schieberle P. Food Chemistry. 4th revised and extended ed. Berlin: Springer; 2009

Cyclodextrin-Based Nanofibers and Membranes: Fabrication, Properties and Applications

Mandla B. Chabalala, Bonisiwe C. Seshabela, Stijn W.H. Van Hulle, Bhekie B. Mamba, Sabelo D. Mhlanga and Edward N. Nxumalo

Additional information is available at the end of the chapter

http://dx.doi.org/10.5772/intechopen.74737

Abstract

Cyclodextrin (CD)-based electrospun nanofibers have become critical role players in the water treatment arena due to their high porosities, small diameters, high surface area-to-volume ratio and other unique properties they exhibit. Investigations demonstrate that nanofibers containing CD molecules can be facially blended with other polymeric species and/or photocatalytic and magnetic nanoparticles to enhance their rates of adsorption, inclusion complexation and selective photodegradation. These properties make them excellent candidates for the removal of water pollutants. On the other hand, the electrospinning process has become the method of choice in the fabrication of various types of CD nanofibrous mats due to its versatility, cost-effectiveness and its potential for the mass production of uniform nanofibers. CDs and CD-derivatives have also found application in membrane technologies, particularly in mixed matrix and thin film composite membranes. CD-blended membranes display improved performances in terms of selectivity, rejection, permeation and flux with reduced fouling propensities and can be used for drinking water purification and removal of emerging micropollutants. This chapter critically reviews CD-based electrospun nanofibers looking at their production, characterization methods and various applications. The use of CDs as membrane materials and how they can be fully explored in water treatment are also investigated.

Keywords: cyclodextrins, electrospun nanofibers, membranes, adsorption, inclusion complexation, water treatment

1. Introduction

Common water pollutants such as toxic micro-organisms and inorganic and organic waste are a serious threat to the health of humans, animals and the aquatic biota. Diverse methods and materials are currently used to treat wastewater in order to overcome the high level of water pollution. However, emerging micropollutants (EMPs) such as hormones, pharmaceuticals, detergents, phenols, fragrances, illicit drugs, endocrine disruptors, steroids and personal care products have proven to be persistent and difficult to remove from aqueous systems [1, 2]. In particular, EMPs are continuously found in trace amounts in waste and treated water. In addition, trace organic pollutants find their way into wastewater after being excreted from human bodies and when disposed to the sewage system [1–3]. Pharmaceutical waste and other trace organic pollutants leave the wastewater treatment plant without being treated because they are present in small amounts, biologically active or thermally and chemically stable [2]. EMPs and their derivatives are of great concern since their fate and behavior is not well understood. Moreover, current treatment methods are not effective in the removal of these EMPs. This has therefore led to the development of diverse technologies including nanotechnology-based technologies to remove EMPs in water systems.

Nanotechnology-based techniques frequently applied in water treatment make use of various nanomaterials including, among others, nanofibers, nanowires, nanotubes, nanorods and nanospheres. These distinct nanomaterials are endowed with several advantageous properties that render them suitable in the removal of micropollutants from aqueous systems. These properties include high porosity, small diameters and high surface area per unit volume. In particular, nanofibers are easy to handle, reusable and recyclable making them ideal candidates for use in water treatment applications [4, 5]. Workers have also applied photocatalytic nanomaterials such as TiO_2 and ZnO and found them to be excellent candidates for use in the photodegradation of most micropollutants using their inherent quantum and surface properties [6]. TiO_2 and ZnO nanomaterials have been applied in various areas, which include textile, wastewater treatment, particulate separation, health care, desalination, energy, liquid filtration and sensors [7, 8]. Hollow-structured nanomaterials such as nanotubes and nanofibers (**Figure 1**) can encapsulate active additives such as photocatalysts, antioxidants and antibacterial agents and then applied in water treatment, food packaging and biotechnology [8, 9]. Photocatalysts have been incorporated with other materials such as the natural polymer cyclodextrins (CDs) for use in water treatment applications [10]. CDs are gaining extensive popularity as adsorbents and as membrane materials due to their toroidal structure and distinct characteristics.

CDs and their derivatives such as methyl-βCD (m-βCD) (**Figure 2**) are known to significantly remove organic pollutants from water systems *via* adsorption and inclusion complexation. They can also be polymerized to form supramolecular structures with high surface areas [12]. The geometry of CDs demonstrates a hydrophobic cavity capable of inclusion complexation with a wide range of pollutants [7, 13]. The ability of CDs to capture and form inclusion complexes with other molecules in solution and their ability to be electrospun with ease as well as other physicochemical properties render them ideal candidates for water treatment applications [7, 12, 14]. The release of water molecules from the hydrophobic cavity, coupled with

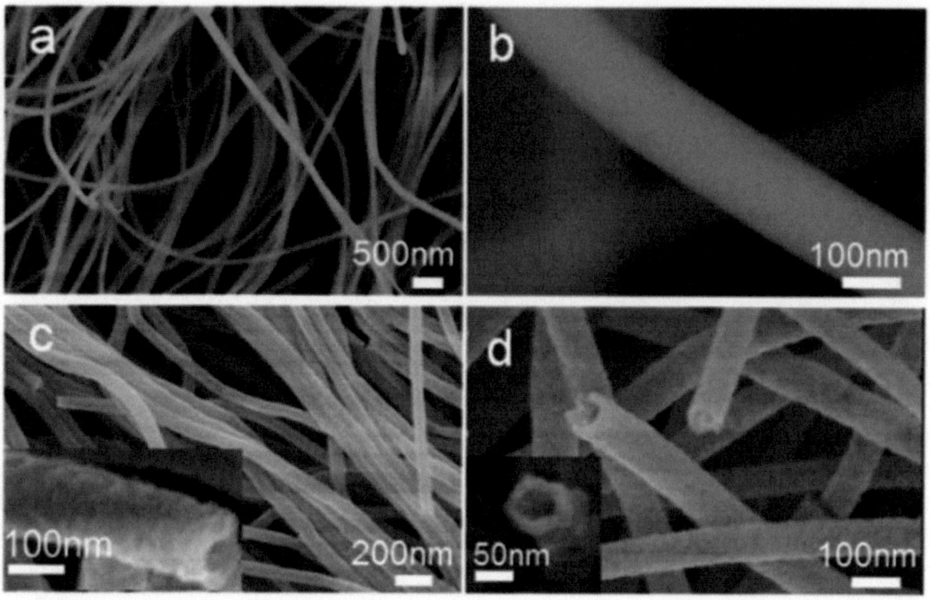

Figure 1. Microscopic view of hollow-structured nanofibers (a and b) the as-electrospun nanofibers. (c and d) The images of nanotubes at low and high magnification. The inset shows the surface and cross profile of the nanotube. Reproduced with permission from [11].

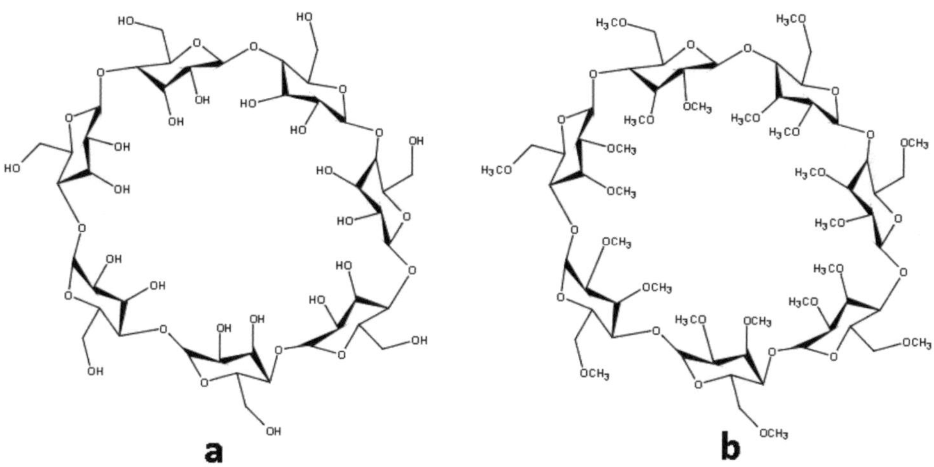

Figure 2. Schematic illustration of (a) βCD and (b) m-βCD.

hydrogen bonding, charge transfer interactions, hydrophobic interactions, van der Waals interactions, release of conformational strain and electrostatic interactions are the driving forces for inclusion complexation through apolar-apolar interaction of CDs and the guest compounds [12, 15]. It is for this reason that CDs are used in water treatment to remove EMPs and various other pollutants as well as in medicine for drug delivery applications. Besides their use in nanofiber production, CDs and their derivatives have become popular in membrane technology.

Recently, workers prepared electrospun nanofibers using chitosan and incorporated silver and iron nanoparticles for water disinfection processes. These nanofibers were later effectively modified using CDs and cellulose to increase their thermal and chemical stability [16]. Somewhere else, thermally and mechanically stable βCD/cellulose acetate nanofibers were synthesized using an environmentally benign procedure and used for enhanced antimicrobial treatment of water [17]. In membrane technology, Adams et al. utilized CD molecules as modifying agents for the preparation of polysulfone-polyurethane (PSF-PU) composite nanofiltration membranes, which were used for the removal of undesirable salts [18].

In this chapter, electrospun CD-based materials are discussed in view of water treatment, their properties and advantages toward improving current water treatment methods by removing EMPs in waste and treated water. We also critically investigate CD-based membrane techniques in terms of their production and characterization methods with focus placed on their application in water treatment. Other applications of these CD-based nanocomposites such as drug delivery, antimicrobial uses, biomedical uses, filtration, photocatalysis and environmental protection are covered.

2. Fabrication methods of nanofibers

Various methods are used in the production of nanofibers today. These include, among others, polymer blending, sea/island cross-section conjugation and electrospinning techniques [19, 20]. These technologies have several disadvantages that include sizes in the microscale instead of nanoscale and low tensile strength. They also form nonwoven sheets that need further treatment using organic solvents [21]. However, several researchers have described electrospinning as the best nanofiber fabrication method compared to the other methods. In the next sections, we explore the various methods of nanofiber production.

2.1. Polymer blend method

The polymer blend method is a method that uses two or more polymers to produce materials with superior properties [22]. This method is divided into three main categories, which are: miscible, immiscible and compatible polymer blending. To produce fibers with nanoscale diameters and uniform continuous length in large scale, this method is often coupled with the electrospinning method. In this way, blended polymer solutions can be electrospun to produce fibers with desired properties [23, 24].

Miscible polymer blends are characterized by a homogeneous morphology/mixture on the segment level; however, the local chain dynamic may exhibit different dependences on temperature and blend composition [25]. The presence of nanoheterogeneities has been observed in miscible polymer blends where Lodge and McLeish have described this as "self-concentration" [26]. This illustrates that high glass transition temperature components often have segmental dynamics much closer to the bulk blend, while the low glass transition temperature is closer to the pure component [25]. However, miscible polymers often have one glass transition temperature that is dependent on the composition. Polymers can be miscible in melt state and immiscible in solid state due to fast crystallization of one component compared to the other [22]. When blended polymers do not crystalize at the same rate, it results in phase separation, which will affect the final product and their envisaged properties [22]. However, due to the miscibility of the components, they can each reside in the interlamellar and/or interspherulitic regions of each other during crystallization, thus reducing separation rates [27].

Immiscible solutions are often referred to as emulsions where one component is dispersed on top of the other as small-sized droplets depending on the quantity of each solution as shown in **Figure 3** [28]. Most polymer blends are immiscible because of the weak interfacial interactions between components and different molecular weight of each component [22]. Immiscible polymer blends also have enhanced properties compared to their separate components [29]. Immiscible polymer blends limit full access of each component properties and application due to their incompatibility. Producing nanofibers through this method requires the use of stabilizing agents such as fillers and metal organic frameworks. Produced fibers are in microscale and requires subsequent polymer matrix extraction [30, 31].

Compatible polymer blends are immiscible polymer blends that have uniform macroscopically physical properties. Compatible polymer blends are often used to enhance the properties of components such as elastic modulus, crystallinity and glass transition temperature [32, 33]. Polymers often require the use of fillers/compatibilizer to induce compatibility between the components (**Figure 4**). To be effective enough, fillers must have a particle radius of the same order of magnitude as the gyration radius of the polymers used. Examples of fillers include ethylene-acrylic acid and ethylene-vinyl alcohol [33, 34].

2.2. Sea/island cross-section conjugation

Sea or island cross-section conjugation is a type of a conjugate spinning method used to fabricate fibers with diameters of less than 1 μm with a predetermined component arrangement in its cross-section. Two polymer components of a conjugate type are elongated and extruded together from a spinneret. These polymers then combine in the back of a spinning nozzle. The produced conjugate fibers with two components are then split into filaments. This technique involves spinning a bicomponent filament consisting of polyester, polyethylene, nylon or polypropylene used as an island component and a polymer like polystyrene is used as a sea component. The fabric is then exposed to a solvent, thermal or mechanical treatment whereby the immiscible components separate as the polystyrene sea component dissolves in a solvent,

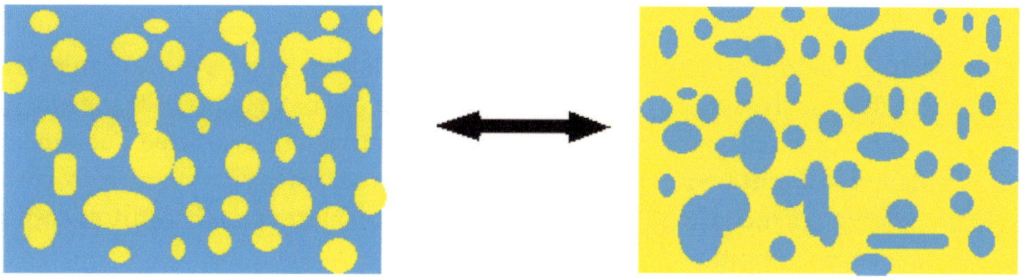

Figure 3. Schematic representation of immiscible polymer solution with varied concentrations of each polymer.

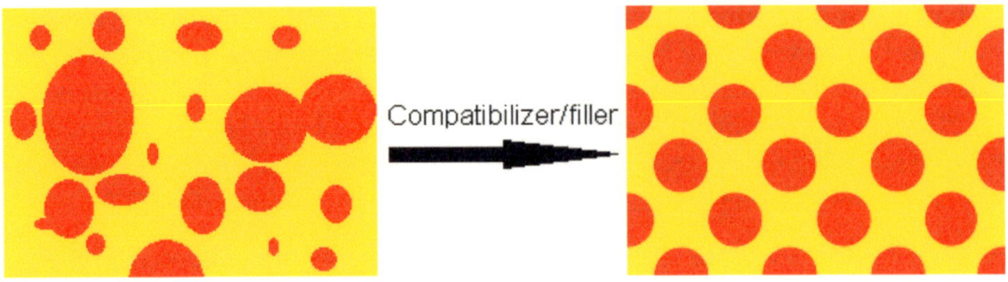

Figure 4. Illustration of the effects of compatibilizer or filler on compatible polymer blends.

which is a nonsolvent for the polyester island component after conventional processing into fibers. This results in individual polyester island filaments. The ratio of the two components in the ultrafine filament yarn, the shape and the number of the resulting individual segments can be varied depending on the design of the spinneret [20, 21].

2.3. Electrospinning technique

The electrospinning technique is a versatile, flexible and cost-effective method for producing nanofibers. It has become very attractive and common in the synthesis of nanofibers for various applications. Electrospinning is often preferred over other methods since it readily produces nanofibers from a number of materials, which include polymers, ceramics, composites and semiconductors [35, 36]. Electrospun nanofibers can be easily modified to improve certain properties. This can be achieved during electrospinning or by posttreatment methods [36–38].

Electrospinning can produce nanofibers of long length, diversified composition, high surface area-to-volume ratio, uniform diameter, flexible surface functionalities and superior mechanical properties [39]. In electrospinning, nanofibers are formed as the polymer solution is stretched between two surface charges and as the solvent evaporates due to electrostatic repulsion forces [40]. During electrospinning, a polymer solution in a syringe is stretched to the collector in a cone shape (Taylor cone) under high voltage. The collector screen used can be either a stationery flat screen or a rotating drum collector. The type of collector can greatly influence the morphology of the nanofibers [41]. The diameter, morphology and distribution of electrospun nanofibers depend on the applied voltage, solution viscosity, tip to collector distance, temperature and flow rate. **Figure 5** shows a setup of an electrospinning instrument with a flat stationary collector. The setup consists of a high-voltage supply, polymer solution in a syringe and the collector screen [42].

When the electrospinning parameters and polymer solution properties are not properly optimized, beaded fibers such as those depicted in **Figure 6(a)** can be obtained. On the other hand, when all the properties and parameters are precisely optimized, bead-free nanofibers such as those displayed in **Figure 6(b)** can be obtained [43]. In electrospinning, it is important to optimize all parameters including the polymer solution properties such as the concentration before spinning large quantities.

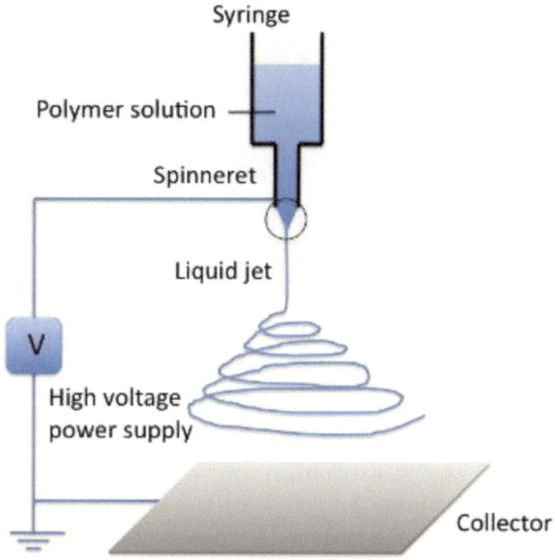

Figure 5. Schematic illustration of electrospinning setup with flat stationery collector. Reproduced with permission from [42].

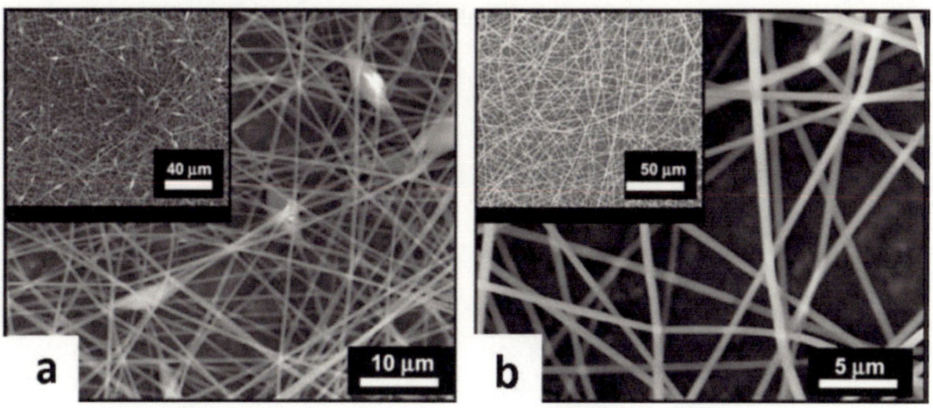

Figure 6. Microscopic representation of electrospun nanofibers without proper optimization (a) and with proper optimization (b) of all parameters. Reproduced with permission from [43].

2.3.1. Mechanism for the formation of electrospun CD nanofibers and influencing factors

Electrospinning CDs into nanofibers is still a challenging task because of their small cyclic structure. However, electrospinning these glucose derivatives would result in nanofibrous mats with excellent properties such as high surface area-to-volume ratio and high possibilities of specific surface functionalization [44, 45]. A number of factors can affect the electrospinning mechanism of CDs. Besides the fact that these cyclic oligosaccharides cannot be easily stretched into nanofibers, factors such as solvent type, concentration, copolymers and compatibility play a critical role in the successful electrospinning of CD nanofibers. Fortunately, CDs are soluble in most organic solvents such as water, dimethylacetamide, dimethyl sulfoxide and dimethylformamide, with water and DMF being the most used solvents [46, 47]. **Figure 7** demonstrates that indeed the type of solvent used to prepare the solution has a pivotal effect on the formation and surface roughness of nanofibers. The most important factors to look at when choosing a solvent include the conductivity, density, solubility with other solvents and viscosity.

When electrospinning CDs, concentration also plays a critical role. At high concentrations, cyclic molecules like CDs and phospholipids can form aggregates and have sufficient electrostatic and intermolecular interactions [44, 48, 49]. The aggregation and molecular interaction act as chain entanglements making the molecules overlap and entangle like polymers in dilute solutions. **Figure 8** clearly depicts the low- and high-concentration effects on electrospun phospholipids and CDs. When using copolymers with CDs, the compatibility and dissolution of the two should be checked since this will affect the intramolecular interactions of the two polymers as well as the formation and morphology of the ultimate nanofibers [36, 47, 50, 51].

2.3.2. Polymer-free cyclodextrin nanofibers

As highlighted earlier, electrospinning cyclic polymers such as CDs is very challenging. However, CDs can form aggregates in their concentrated solution via intermolecular hydrogen bonding and interactions resulting in chain entanglements making it possible to electrospin CDs [44, 52].

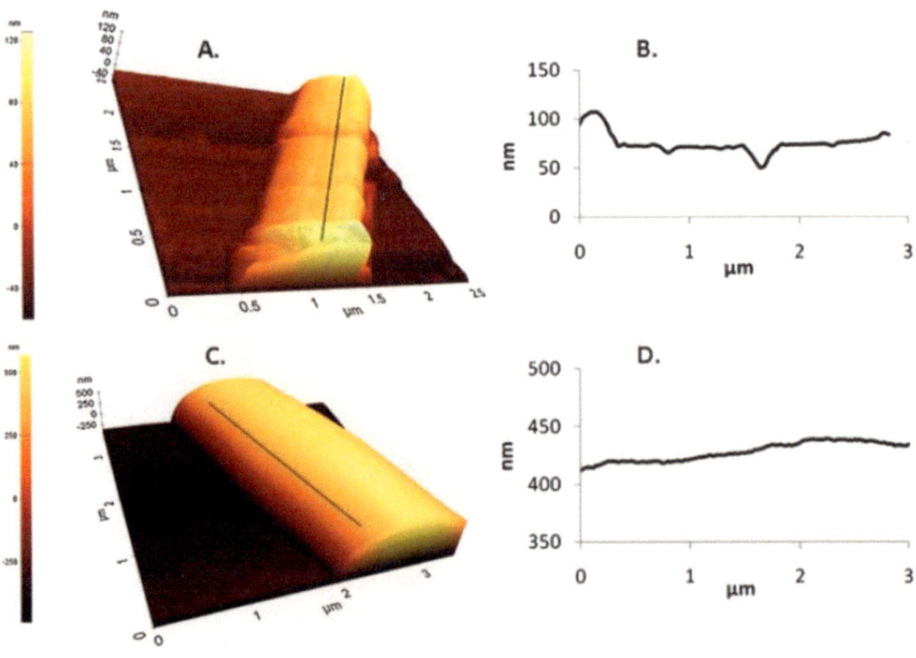

Figure 7. (A) AFM image and (B) fiber axis cross-section profile of the MβCD nanofiber prepared in water. (C) AFM image and (D) fiber axis cross-section profile of the MβCD nanofiber dissolved DMF. Reproduced with permission from [44].

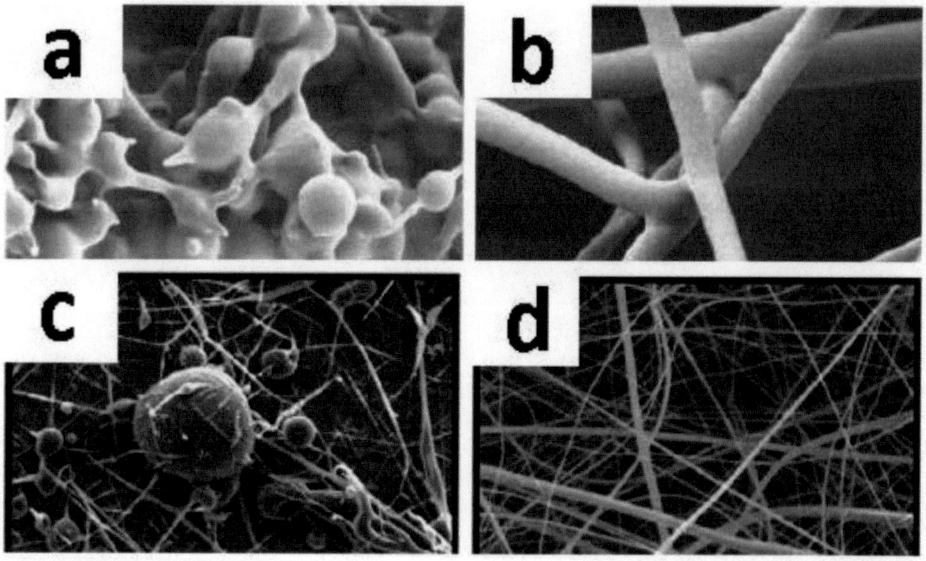

Figure 8. SEM micrographs revealing the concentration effect on electrospun phospholipids (a 35% and b 50%) and CD nanofibers (c and d) at low and high concentrations. Reproduced with permission from [49].

Celebioglu and Uyar reported the first successful electrospinning of carrier polymer-free CD nanofibers. In their report, highly concentrated solutions of methyl-βCD (140 and 160% w/v) were prepared in water and DMF. Electrospinning these solutions yielded nanofibers with diameter ranges of 20–100 and 100–1200 nm using water and DMF, respectively [44, 53]. At concentrations lower than 140% w/v, beaded nanofibers were obtained. They also reported another successful electrospinning of hydroxypropyl-βCD (HP-βCD) and HP-βCD inclusion complex with triclosan. Bead-free HP-βCD and HP-βCD/triclosan were obtained at higher concentrations of 160% w/v. **Figure 9** shows the SEM images of polymer-free HP-βCD and HP-γCD nanofibers dissolved in DMF and water [52]. Celebioglu used HP-βCD and HP-γCD for the entrapment of volatile organic compounds (VOCs), aniline and benzene [45]. The results indicated that CD nanofibers had better performances compared to powdered CDs, while βCD nanofibers performed better than γCD nanofibers. The performance was dependent on the type of CDs, solvent and VOC type. Therefore, electrospinning of CDs and CD derivative nanofibers strongly depends on the type of CDs, solution concentration, solvent used and intermolecular interactions [46]. In **Table 1**, we show electrospun CD nanofibers prepared using different types of CD derivatives in different solvents without an additional polymer being used and the effect on the fiber sizes thereof.

2.3.3. Copolymerized cyclodextrin nanofibers

In order to improve the propensity of electrospinning CDs into excellent nanofibrous mats and take advantage of the CD properties, CDs can be blended with other polymers such as polyethylene terephthalate (PET), polyvinyl alcohol (PVA), polycaprolactone (PCL), poly(methyl methacrylate) (PMMA), polyvinylpyrrolidone (PVP), polylactic acid (PLA) and cellulose. This

Figure 9. SEM images of (a-i) HP-βCD/powder, (a-ii) HP-βCD/water-nanofiber, (a-iii) HP-βCD/DMF-nanofiber, (b-i) HP-γCD/powder, (b-ii) HP-γCD/water-nanofiber and (b-iii) HP-γCD/DMF-nanofiber. Reproduced with permission from [45].

CD-type	Solvent	Conc. (% w/v)	Method	Size (nm)	Ref
α and βCDs	Water, DMF, DMAc and DMSO	120–160	Electrospinning	80–940	[53]
HP-βCDs	Water	50–70	Electrospinning	933–990	[54]
M-βCDs, HP-βCDs and HP-γCDs	Water, DMF and DMAc	100–160	Electrospinning	250–1860	[55]
HP-βCDs	Water	100–160	Electrospinning	200–1600	[52]
M-βCDs	Water and DMF	100–160	Electrospinning	20–1200	[44]

Table 1. Electrospun polymer-free CD nanofibers, conditions of preparation and size of nanofibers.

greatly improves the general properties of the nanofibers and expands their possible areas of application [19, 20, 50, 51, 56]. For example, **Figure 10** depicts electrospun CD-modified PS nanofibrous mats. Nanofibrous mats or membranes with high permeability are excellent candidates as microporous support substrates for thin film composite membranes with high flux and for application in microfiltration processes [57–60]. Copolymerization of cross-linked or modified CDs with other polymers reduces their solubility in water and makes them excellent candidates for water treatment applications.

Uyar and coworkers electrospun poly(methyl methacrylate) functionalized with CDs (PMMA/βCDs) for the treatment of organic vapors. It was found that PMMA/βCD nanofibers

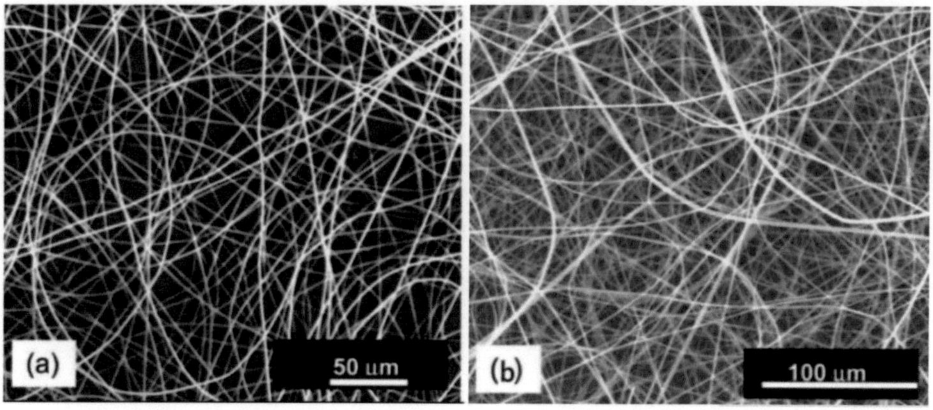

Figure 10. SEM micrographs for electrospun (a) PS and (b) PS-βCD nanofibers. Reproduced with permission from [46].

successfully encapsulated organic vapors such as aniline and styrene. The organics were inclusion complexed by βCD cones in solution [46]. CD/polymer nanofibers have also been used as reducing and stabilizing agents for other nanoparticles. Celebioglu et al. demonstrated this when they used polyvinyl alcohol/HP-βCD (PVA/HP-βCD) nanofibers as reducing and stabilizing agents for Ag nanoparticles. In this case, PVA was used as a primary agent, while HP-βCDs were used as secondary agents [56].

In water treatment, materials with high adsorption capacity are useful when it comes to the removal of pollutants such as dyes. Teng et al. used mesoporous PVA/SiO_2/βCD nanofibers as adsorbents for the removal of indigo carmine dye in wastewater. The nanofibers were found to have adsorption capacities of up to 495 mg/g and equilibrium was reached in less than 40 min, due to the presence of CDs [61]. Zhang et al. prepared composite nanofibrous membranes from PVA/βCDs using electrospinning for molecular entrapment of organics. It was found that these nanofibrous membranes could effectively and readily capture organic molecules. This was attributed to the inclusion complexation of organic molecules by the CDs. It was further suggested that these kinds of membranes can also be applied in areas such as drug delivery, separation/purification and electrochemical sensors, among others [62].

Research shows that copolymerized CD nanofibers and their derivatives find application in various areas due to the properties and advantages induced by the incorporation of CDs. **Table 2** shows various copolymerized CD nanofibrous mats and their applications.

2.3.4. *CD nanofibers incorporated with nanoparticles*

Blending electrospun CD nanofibers with nanomaterials results in unique properties from large surface area of nanofibers and excellent properties of the nanomaterials to specific structural and functional properties [67]. Nanomaterials supported on other materials have the disadvantage of low efficiency because of small interface surface available compared to powder photocatalysts. However, powdered nanomaterials cause secondary contamination with low recovery and require further treatment after usage. The high efficiency of powdered

CD type	Copolymer	Solvent	Method	Size (nm)	Application	Ref.
HP-βCDs	PVA	Deionized water	Electrospinning	290–485	Reducing and stabilizing agent for Ag antibacterial nanoparticles	[56]
βCDs	PCL	DMF and DCM	Electrospinning	336–389	Drug delivery of naproxen	[50]
βCDs	PMMA	DMF and toluene	Electrospinning	625–977	Organic vapor waste treatment	[51]
βCDs	PVC	DMF and THF	Electrospinning	420–450	Membrane modification	[19]
βCDs	Carbonaceous nanofiber membrane	Water	Electrospinning	120–144	Membrane filtration of phenolphthalein	[63]
βCDs	Cellulose	DMF	Electrospinning	100–800	Antibacterial activity (*E. coli* and *S. aureus*)	[20]
βCDs	PVP	Water and ethanol	Electrospinning	450	Stabilizing and reducing agent for Au nanoparticles	[47]
βCDs	Chitosan and PVA	Water and chloroacetic acid	Electrospinning	84–285	Drug delivery	[64]
α, β and γCDs	PET	TFA and DMF	Electrospinning	12.4–15.3	Phenanthrene removal in aqueous solutions	[7]
α, β and γCD-menthol IC	PS	DMF	Electrospinning	300–4250	Enhancement of durability and stability of fragrances	[65]
α, β and γCDs	PLA	DMF, DMSO and chloroform	Coprecipitation and electrospinning	140–1580	Antibacterial growth (*E. coli* and *S. aureus*)	[66]
α, β and γCDs	Zein	DMF	Electrospinning	90–185	—	[5]
α, β and γCDs	PMMA	DMF	Electrospinning	625–1024	Molecular filters and water treatment	[9]

Table 2. Electrospun copolymerized CD nanofibers, solvent type, sizes and their applications.

nanomaterials is outweighed by the recovery, recyclability and reusability of supported nanomaterials [68]. **Figure 11** shows SEM and TEM images of electrospun HP-βCD containing Au nanoparticles which also shows the d-spacing of 0.235 nm between Au {111} planes (**Figure 11**) [69].

CD nanofibers incorporated with nanoparticles have been prepared via electrospinning and self-assembly. These two techniques produced hierarchical structures of nano- to macroscale catalysts such as CD-TiO_2 and CD-ZnO [70, 71]. CD-TiO_2 nanofibers prepared by Yoon and Nichols displayed a hierarchical structure of nanoparticles interconnected into a 3D network as demonstrated in **Figure 12**. The 3D structure was found to have high surface area and large porosity [72, 73]. CD-TiO_2 composites have been found to be stable over a wide range of pH. Somewhere else, it was found that CDs guide the assembly of TiO_2 nanomaterials into nanowires hosted in a CD nanotube structure. In this process, CD molecules coat TiO_2 nanomaterials and produce long

Figure 11. (a) SEM and (b) TEM images of electrospun HP-βCD nanofibers incorporated with Au nanoparticles. Inset HR-TEM image of a single Au-nanoparticle indicating the d-spacing between the planes [69].

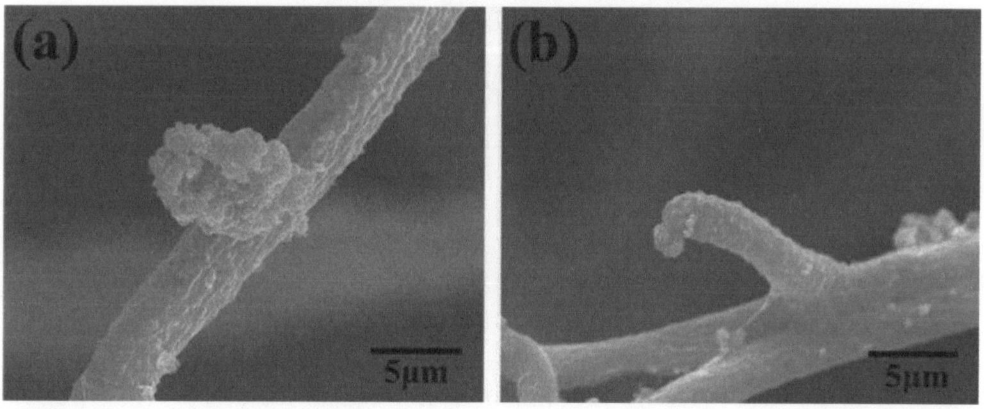

Figure 12. SEM images illustrating the 3D nature of CD-directed TiO_2 nanoparticles forming nanofibers (a) A small aggregate attached on the side of the nanofiber. (b) The aggregate smoothes and continues to grow into a new fiber. Reproduced with permission from [73].

CD-TiO_2 nanotubes. These CD-TiO_2 nanofiber catalysts have shown enhanced photodegradation of organic materials [74–76].

Using CDs, Zhao and Chen have also demonstrated the preparation of ZnO nanofibers and multipetals (**Figure 13**). When analyzed, the ZnO nanomaterials were decorated with CD molecules [77]. In another study, CD-ZnO nanofibers were synthesized under mild conditions of thermal decomposition [78, 79]. In their study, zinc acetate was coated with CD molecules and ZnO synthesis took place within the CD molecules resulting in CD-ZnO nanofibers. Rakshit and Vasudevan prepared CD-ZnO fibers with high degradation performance of Nile red [80]. In all these studies, CD-ZnO nanofibers were prepared using self-assembly processes.

CDs can also act as electron donors and molecular recognition agents when incorporated with photocatalytic nanoparticles [76]. CD chemisorption onto photocatalytic nanomaterials improve their stability against aggregation and enhance their charge transfer reactions.

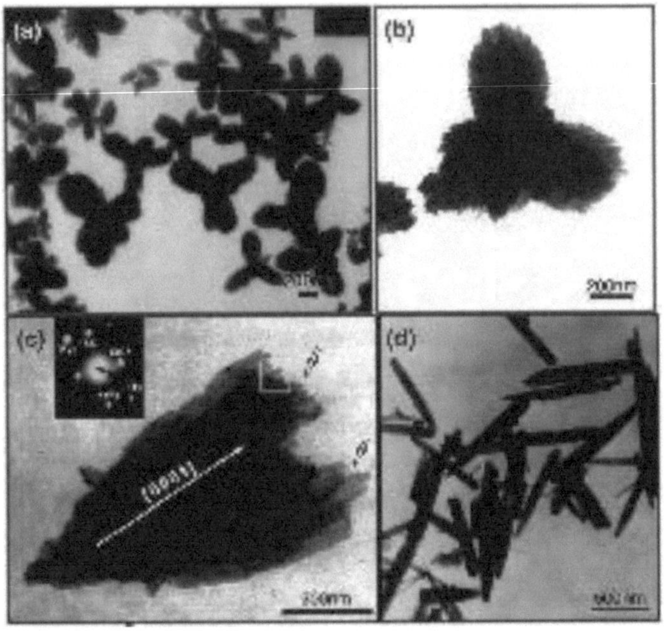

Figure 13. Micrographs showing multipetals of (a–c) ZnO nanomaterials prepared with βCD and (d) ZnO nanofibers prepared without βCD. Reproduced with permission from [77].

3. Cyclodextrins in membrane technology

Membrane technology is one of the methods that have been used in water treatment for decades. Its merits have made its use continuous throughout the decades; however, membranes have several demerits, which cannot be ignored. Membranes suffer from fouling, selectivity and low flux among other problems. This results in poor performance, which implies high operational cost for poor water output. Modifying membranes with materials such as nanomaterials and polymers via surface functionalization greatly improves the overall performance of membranes. Currently, polymeric membranes are the most used in water treatment because of their advantages, which include higher flexibility, easy pore formation mechanism, smaller footprint for installation and low cost compared to other membrane types [81–83]. When it comes to energy efficiency and cost-effectiveness, it is required that membranes should have high permeability, high rejection and good fouling properties [81].

The use of CDs and their derivatives has been found to enhance membrane performance in terms of improved porosity, flux, rejection and efficiency, which makes the use of CDs considerably effective. Porous CD-based membranes have interconnected pores with high permeability and find application in various filtration processes [84, 85]. Even though the use of CDs in membrane technology is not as advanced as the use of other polymers, several researchers have dedicated their time into studying and exploring new properties and advantages brought by CDs in various membrane types. The use of CDs in water treatment has been motivated by the ability of CDs to allow water to pass through their cavities and their surface functionalities, which greatly improve the hydrophilicity and permeability of membranes [86]. Mixed matrix membranes (MMMs) and thin film composite (TFC) membranes are two types of membranes where CDs and their derivatives have been used as modifying agents to improve their total performance.

3.1. Mixed matrix membranes

Mixed matrix membranes (MMMs) are known for their high flux and low pressure drops. MMM high flux capacity and selectivity are often a result of functionalized modifying agents [87]. MMMs are used mostly for the removal of heavy metals, natural organic matter (NOM), EMPs and disinfection by products such as trihalomethanes, haloacetic acids, trihaloacetaldehydes, haloacetones and trihalonitromethanes in water [88]. Adams et al. prepared MMMs using polysulfone/βCD-polyurethane (PSf/βCD/PU) for the selective removal of Cd^{2+} ions and improved structural properties of PSf MMMs. Upon studying their characteristics, it was found that βCD-polyurethane enhanced the water sorption and hydrophilicity and achieved 70% removal of Cd^{2+} ions [18]. Adams et al. used the same material (PSf/βCD/PU) in 2014 to study the effect of βCD/PU on the rejection of NOM and fouling resistance of PSf MMMs. It was concluded that βCD/PU improved the effective pore sizes and molecular-weight cut-off of PSf membranes due to their conical structure and larger pore sizes, which allows water molecules to pass easily [89]. Other workers used ceramic membranes modified with cross-linked silylated dendritic polymers and CDs for the removal of organic pollutants in water. The modified membranes removed pollutants such as monocyclic aromatic hydrocarbons (93%), pesticide (43%), polycyclic aromatic hydrocarbons (99%) and trihalogen methanes (81%). The high removal percentage was attributed to the dendritic polymers and CDs [90]. **Figure 14** shows

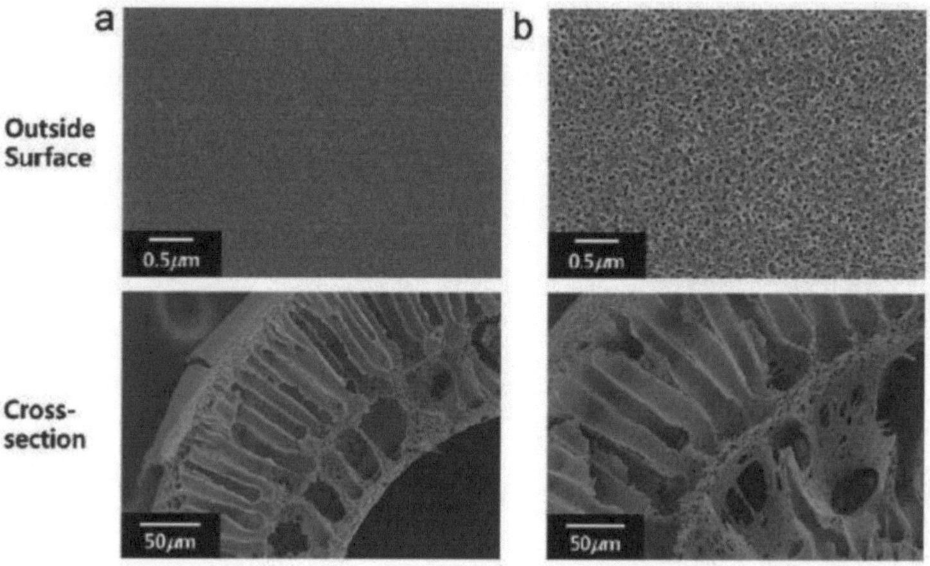

Figure 14. SEM images comparing morphologies for (a) PSf and (b) PSf/βCD outside surface and cross-section. Reproduced with permission from [91].

comparison between PSf and PSf/βCD membranes and it is shown that βCDs improved the surface and cross-sectional morphology in terms of pore size dimensions and size distribution.

3.2. Thin film composite membranes

Nanofiltration (NF) membranes are pressure-driven membranes and are mostly thin film composite (TFC) membranes [86]. TFC membranes include reverse osmosis (RO) and ultrafiltration (UF) membranes prepared by interfacial polymerization. The unique structure of TFC membranes that consists of a UF support, a nonwoven support and a polymer membrane brings advantages such as low operation pressure, high retention of multivalent ions or salts, low maintenance cost and high permeation flux [86, 92, 93]. This type of membranes is mostly used in water treatment for the production of drinking water from wastewater, seawater and brackish water [94]. Other areas of application include organic solvent nanofiltration and pharmaceuticals and biochemical industries [86, 93]. The layers of TFC membranes can be modified independently for maximum preferred properties such as water uptake, fouling resistance, chemical resistance, thermal stability, hydrophilicity and mechanical strength [92]. The excellent properties of TFC membranes are due to modifying agents such as CDs and their derivatives.

Wu et al. prepared NF TFC membranes using polyester/βCD as a polymer material [81]. The addition of βCDs was found to improve the membrane performance as shown by double flux and high rejection of Na_2SO_4 compared to bare membranes. When sulfated-βCDs were used, the membrane had improved negative charge density and salt rejection. Both membranes were reported to have enhanced antifouling properties [86]. On the subject of TFC membranes, Mbuli et al. used amino-CDs and diethylamino-CDs to modify polyamide TFC membranes. The addition of modified CDs enhanced the membrane's permeability because of improved hydrophilicity and additional water channels. In a separate study, modified TFC membranes containing CDs also demonstrated high flux and good NaCl rejection [94]. Mao et al. prepared CD-modified PEI membranes for organic solvent nanofiltration. In the study, they

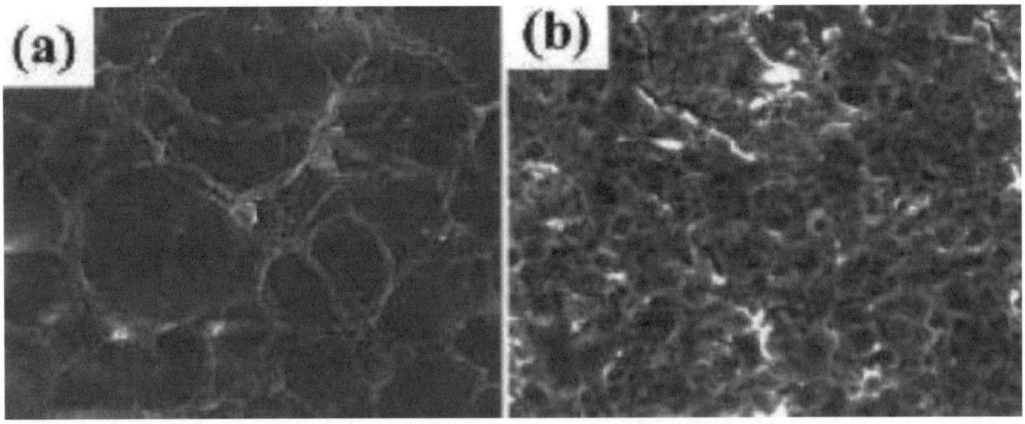

Figure 15. SEM images of (a) unmodified and (b) amine f-CD m-phenyldiamine TFC membranes. Reproduced with permission from [95].

Material	Type	Method	Application	Ref.
PSf-βCD-polyurethane	MMM	Phase inversion	Nanofiltration for removal of Cd^{2+} ions	[18]
βCD-polyurethane	MMM	Phase inversion	Rejection of NOM (humic acid)	[89]
Azo dye-modified βCD-epichlorohydrin	MMM	Polymerization and phase inversion	Detection of chloroform, 1,3-dichloropropane, 1,2-dichloroethane, 1,2-dichloropropane, 1,1,2-trichloroethane and dichloromethane	[88]
Fe-Ni/f-CNT/ βCD-polyurethane	MMM	Polymerization and precipitation	Degradation of trichloroethylene	[96]
PSf-βCD	MMM	Phase inversion	Removal of endocrine disruptive chemicals	[91]
CD-polymers	—	—	Adsorption and separation of pesticides	[97]
PES/m-phenyldiamine and PES/m-phenyldiamine/ amine-f-CDs	TFC	Interfacial polymerization	Rejection of NaCl and Na_2SO_4	[95]
PE/βCDs	TFC	Interfacial polymerization	Rejection of Na_2SO_4 and antifouling properties	[86]
PA/amino-βCDs PA/diethylamino-βCDs	TFC	Interfacial polymerization	Rejection of $MgSO_4$ and fouling-resistant studies	[92]
PEI/α, β and γCDs	TFC	Interfacial polymerization	Organic solvent nanofiltration	[93]
PA/amino-α and βCDs PA/diethylamino-α and βCDs	TFC	Interfacial polymerization	Rejection of NaCl	[94]

Table 3. CDs and CD derivatives used as additives in traditional membranes.

prepared a membrane with dual pathway nanostructures from CDs (hydrophobic pathway) and the fractional free volume of PEI (hydrophilic pathway). Toluene permeation was improved from 0.13 to 2.25 L/mhbar when CD loadings were increased [93]. **Figure 15** shows m-phenylenediamine (a) and amine (b) f-CD-modified PES membranes. The presence of uniform pores is observed on membranes (b) due to the addition of amine f-CDs, while the addition of m-phenylenediamine produced membranes with layered structures on top (a). The incorporation of amine f-CDs improved the general performance of the membrane in terms of hydrophilicity, flux and salt rejection [95]. In **Table 3**, we show recent works on the use of CDs and their derivatives in the production of MMMs and TFC membranes.

4. Other applications of CD-based materials

The ability of CDs to form inclusion complexes with other materials and alter their properties has enabled electrospun CD-based materials and membranes to be used in many applications such as drug delivery, filtration, templates, biomedical, catalysis, water treatment, reinforcement, electronics, pharmaceuticals and optical devices [39, 98, 99]. Celebioglu and coworkers formed inclusion complexation using the antibacterial agent, triclosan, with two types of CD derivatives (HP-βCDs and HP-γCDs). The electrospun inclusion complexes were tested against Gram-negative (*Escherichia coli*) and Gram-positive (*Staphylococcus aureus*) bacteria. The antibacterial activity against the two bacteria strains was found to be higher for the inclusion complexes compared to the bare triclosan. The interactions of triclosan with the CD derivatives improved its antibacterial activity [100]. Li and coworkers used βCDs with maleic anhydride (MAH) and 3-(4-vinylbenzyl)-5,5-dimethylhydantoin (VBDMH) for antibacterial studies. The composite βCD-MAH-VBDMH was electrospun with cellulose acetate and the antibacterial activity was tested against *E. coli* and *S. aureus* bacteria. The nanofibers achieved 99.7 and 80.3% activity against *E. coli* and *S. aureus*, respectively, within 10–30 min contact time [20]. In another study, Dong and coworkers used ciprofloxacin hydrochloride (CipHCl) as the antibacterial agent with electrospun citric acid cross-linked cellulose and βCDs. The CipHCl loaded on the electrospun nanofibers demonstrated high antibacterial activity against *E. coli* and *S. aureus* [101].

In drug delivery systems, CDs and their derivatives have also been used for targeted delivery and control of release rate as well as solubility control. Bazhban and coworkers electrospun a drug delivery system from carboxymethyl-βCDs and chitosan blended with PVA in the presence of 1-ethyl-3-(3-dimethylaminopropyl) carbodiimide as the condensing agent and *N*-hydroxysuccinimide as a hydrolyzing agent. The electrospun nanofibrous mats were observed to have slower release rates of the entrapped salicylic acid compared to the nanofibrous mats without βCDs [64]. Canbolat and coworkers complexed naproxen (NAP) with βCDs and electrospun the inclusion complex with poly(ε-caprolactone) (PCL/NAP-βCDs). The electrospun PCL/NAP-βCDs had high release rates of NAP compared to the electrospun PCL/NAP [50]. Electrospun CD nanofibers have also been used in the syntheses of metal nanoparticles as reducing agents and size-controlling agents. Celebioglu and coworkers synthesized Ag nanoparticles in the presence of PVA/CD electrospun nanofibers. They obtained Ag nanoparticles of 2 nm in size without aggregation compared to the 8 nm aggregated nanoparticles obtained with the use of bare nanofibers [56]. Bai and coworkers used electrospun PVP/βCDs as stabilizing and reducing agents for the synthesis of Au nanoparticles. The Au nanoparticles were found to be evenly distributed and well dispersed in the nanofibers and induced antibacterial behavior on the nanofibers [47].

By forming inclusion complexes with other materials, CDs can improve their stability and shelf life. Kayaci and coworkers enhanced the thermal stability of eugenol (EG) by means of inclusion complexation with β and γCDs. The inclusion complex EG-CD was incorporated and electrospun together with PVA. The complexed EG demonstrated thermal evaporation at high temperature and slowed release at temperatures as high as 100°C compared to poor thermal stability of pure EG [66]. Uyar and coworkers prepared inclusion complexes between menthol with α, β and γCDs and electrospun the complexes with polystyrene in order to enhance the thermal stability of menthol. The thermal stability of menthol was improved up to 350°C by the electrospun nanofibers [65].

This is an indication that electrospun CD nanofibrous mats have a wide range of applications. Whenever CDs are used, they enhance certain properties of the materials incorporated with to achieve excellent outcomes.

5. Mechanism for the interaction of CD nanofibers/membranes with various species

As discussed earlier, CDs and their derivatives have the ability to form inclusion complexes with a number of liquid, solid or gaseous compounds [12], which can alter the physicochemical properties of the guest molecule [102]. This happens by taking up a whole or part of a guest molecule into the hydrophobic cavity, which is lined by skeletal carbons as well as the ethereal oxygen of the glucose units. During the formation, no covalent bonds are made or broken, and as a result the guest molecule is not permanently hosted, it is rather in a dynamic equilibrium with the host [103–105]. **Figure 16** shows an example of an inclusion complex between a CD molecule and an organic compound. The hydrophobic cavity provides an environment for appropriate guests to settle in and form a complexation with the CD molecule [12]. In solid-state inclusion complexation, guest molecules can be enclosed within the cavity or can aggregate outside the CD, while solution state inclusion complexation is controlled by equilibrium between the complexed and noncomplexed molecules [102, 106]. For successful inclusion complexation to occur the guest or part of the guest must have the size, polarity and shape that are compatible with those of the host [107]. Physicochemical properties of

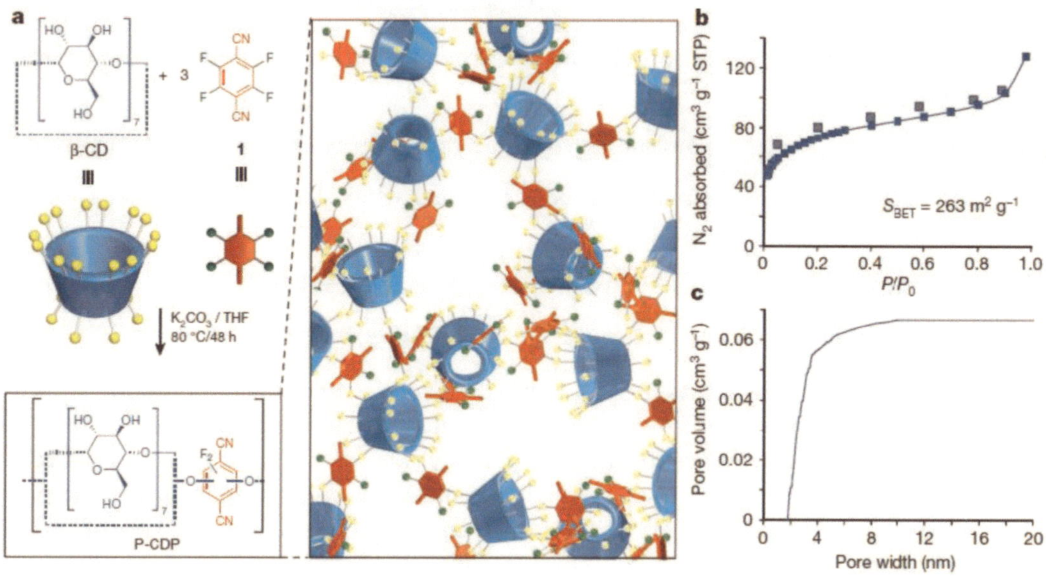

Figure 16. Illustration of (a) interaction of βCD with an organic molecule forming a polymeric network, (b) N_2 adsorption and desorption and (c) pore volume measurements of the polymeric structure. Reproduced with permission from [109].

guests that can be altered during inclusion complexation include taste modification, solubility enhancement, physical isolation and stabilization of labile guests [108]. As shown by **Figure 16**, CDs and their derivatives can interact with other molecules to form supramolecular polymeric structures [109]. The high surface area and pore volume and permanent porosity of the porous CD polymer enable the rapid removal of the organic contaminants [109].

Marques et al. used CDs for the encapsulation of essential oils such as chamomile oil via inclusion complexation. Inclusion complexation of essential oils is used for various reasons such as avoiding degradation induced by oxygen, light or heat, improving water solubility and stabilizing fragrances [110]. βCDs were also used by others to enhance the solubility of Gliclazide by employing the coprecipitation and kneading methods. It was found that the complex prepared by kneading method was more suitable for the improvement of Gliclazide solubility compared to the one prepared by coprecipitation [111].

By inclusion complexation, the CD moiety does not only bind the guest molecules but also brings them close to the functional groups for other reactions such as photocatalytic degradation to take place [112]. CDs can also alter the physicochemical properties of the guest molecules by making it easy to modify during that period. For example, it can promote fast dissolution rates, efficient absorption and short drug release times. As a result, CDs find application in cosmetics, food, drug delivery, bioconversion and environmental protection [103, 105]. It should be noted that complexation between the host and a guest depends mostly on several properties of the host and guest, dosage, thermodynamics and equilibrium kinetics.

The index of the change in physicochemical properties of guest can be shown by the stability, equilibrium constant (K_c) and dissociation constant (K_d) measurements. The formation of inclusion complexation toward equilibrium is assisted by four energetically favorable interactions, which are:

i. The displacement of water molecules from the hydrophobic cavity by nonpolar molecules.

ii. The increasing number of hydrogen bonds formed when the displaced water is driven to the outer pool.

iii. Reduced repulsive interactions between the nonpolar guest and the aqueous environment.

iv. Increasing hydrophobic interactions as the guest inserts itself into the cavity [12].

Clearly, the common interaction of CDs with other species in aqueous solutions is inclusion complexation. Kayaci and coworkers reported that the filtration efficiency of PET was improved after surface modification with α, β and γCD. The filtration process was tested against phenanthrene compounds and the improvement was credited to inclusion complexation between the CDs and phenanthrene [7]. Uyar and coworkers reported the use of PMMA nanofibers modified with βCD for molecular entrapment of organic vapors such as styrene, aniline and toluene. The reported interaction between the vapors and nanofibers was inclusion complexation (CD) and adsorption (PMMA and CDs). The interactions between the two were analyzed by direct pyrolysis mass spectrometry and thermogravimetric analysis [51].

Chen and coworkers prepared a molecular filtration membrane using carbonaceous nanofiber membranes (CNFs) modified with βCDs for the filtration of phenolphthalein in aqueous solutions. Again, the removal of this compound was credited to complexation with CDs and absorption by both CDs and CNFs [63]. Workers reported the reduction of Ag and Fe supported on βCD/cellulose acetate nanofiber membranes for antibacterial studies. The CD molecules facilitated the charge transfer that occurred between ionized water molecules and Ag^{+} and Fe^{3+} since they were able to alter physicochemical properties of guest molecules (and in this case metal nanoparticles) [17].

Several publications reported the use of CDs in various applications such as drug delivery, catalysis, water and air treatment, sensors and energy storage devices. The outstanding performance of all the materials is credited to inclusion complexation and absorption ability that are caused by hydrophobic and intermolecular interactions between the compounds of interest and CDs [16, 64, 113, 114].

6. Characterization tools for CD materials

There are several methods that can be used to study and understand the properties and characteristics of CDs, CD-derivatives and CD-guest inclusion complexes. To study, understand and confirm changes on CD physicochemical properties during applications, analysis has to be conducted using a series of conventional techniques that can complement each other and give conclusive data. Some of the techniques are discussed in the subsections below.

6.1. X-ray diffraction

X-ray diffraction (XRD) is very useful for the analysis of CD materials in powder or microcrystalline states. This is simply because the XRD pattern of the parent CD will be different from that of the derivative or the inclusion complex, which will confirm successful modification, functionalization and/or inclusion complexation [110, 111]. This technique can be used on ground and homogenized samples even on unknown samples. The intensity of peaks helps understand the interactions between CDs and other materials as well as their degree of crystallinity [115].

6.2. Nuclear magnetic resonance

Nuclear magnetic resonance (NMR) is mostly used to study inclusion complexation in solution and has been very useful in understanding the bonding configuration of functionalities present. This is mostly because when a guest is hosted the interior hydrogens are shielded by the guest resulting in a shift on the NMR spectroscopy [110]. NMR can also be used to determine the atoms that stabilize host-guest complexes using ^{13}C-NMR by monitoring the shifts of the carbon atoms involved [110, 115]. Furthermore, NMR can also provide information on the orientation of the guest within the host's cavity [115, 116].

6.3. Infrared spectroscopy

Infrared spectroscopic techniques such as the Fourier transform infrared (FTIR) spectroscopy can be used to reveal CD functionalization and inclusion complexation. This can be validated by the appearance of new peaks, shift in peak position or change in peak intensity as a result of changes on pure CD molecules. Noticeable changes may include disappearance of the —OH peak for functionalization with the appearance of new peaks depending on the type of functional groups introduced. For inclusion complexation, —CO stretching peaks may be observed [110]. Vibrational modes of the host and guest can be studied using this technique to understand the process of complexation and/or functionalization. Vibrational modes can be restricted to a certain level during complexation and this can result in weak interatomic bonds due to the altered environment around the bonds [116, 117].

6.4. Ultraviolet/Visible spectroscopy

The absorption properties of the host and guest molecules (such as dyes) can be easily altered by functionalization or inclusion complexation. When that happens, ultraviolet-visible spectroscopy confirms the successful complexation or functionalization by monitoring the band broadening or narrowing and/or bathochromic shift [110]. In fact, inclusion complexation can result in hypsochromic or bathochromic shift and/or increase or decrease in the intensity of the absorption maxima. However, this technique does not provide conclusive results on complexation or functionalization [116, 118].

6.5. Fluorescence spectroscopy

The environment of molecules can greatly influence their fluorescence properties; hence, fluorescence spectroscopy can be used to determine the geometry of complexation. Fluorescence quantum yield is high in complex formation and the maxima emission is often shifted to shorter wavelengths [110]. The enhancement of fluorescence in complexation is a result of shielding caused by quenching and nonradioactive decay processes [116]. This technique can only be used for fluorescent molecules.

6.6. Differential scanning calorimetry

Differential scanning calorimetry (DSC) is an analytical technique based on thermal analysis of compounds. For physical and energetic properties, DSC is one of the most used techniques for CD complexation especially in CD-drug complexes. Endothermal dehydration peaks and decomposition peak are the main characteristics of CDs and are found at 90–130° C and 300°C, respectively. The appearance of a sharp enhanced endothermal peak indicates the formation of a host-guest complex, which is a sum of the individual compound peaks. Since physicochemical properties of guests can be changed during complexation, DSC can show the loss of guest crystallinity by broadening, size reduction and lower temperature shift of guest-melting peaks [116, 119]. However, guest-melting peaks may also indicate the presence of free guest molecules meaning that equilibrium has been reached [120]. In this case, chromatographic techniques can be used to separate the complex and free molecules.

http://dx.doi.org/10.5772/intechopen.74737

6.7. Chromatography

Chromatographic analysis such as thin layer chromatography (TLC) can be very useful for the verification of complexation and modification by monitoring the alterations of the retardation factor R_f values. The complex of modified CDs is found between the R_f values of the CDs and that of the functional group or guest [110, 121]. Another way to study CDs and their complexes using chromatography is by monitoring their volatility using head-space chromatography. This chromatographic technique is specifically for volatile compounds. The increase or decrease in volatility can be observed as influenced by the host-guest interaction and the stability of the complex can be determined [110, 122].

6.8. Microscopic techniques

Microscopic techniques such as SEM and TEM are used as complementary techniques to analyze the surface morphology, topography and composition of various samples including nanofibers and membranes containing CD species. These two techniques give critical details on the size, size distribution and alignment of fibers as well as the nature of the nanofiber or membrane surfaces. These microscopic techniques are mostly used in the analytic investigation of nanofibers and membranes because of their capability of imaging at high resolutions [11, 43, 49, 69, 73, 77].

6.9. Other characterization techniques

Other popular techniques often used to study CDs, their derivatives and nanocomposites are thermogravimetric analysis to probe their thermal stability, circular dichroism spectroscopy to study inclusion complexation of ideally sized molecules in the CD cavity, contact angle analysis to understand the hydrophilicity of surfaces, nanosizer instruments for surface charge and Brunauer-Emmett-Teller (BET) to measure the surface area and pore volume.

7. Toxicology and safety of CDs and their derivatives

Even though CDs have several advantages for applications in areas such as water treatment, tissue engineering and drug delivery, their toxicity, biological fates and safety issues need to be evaluated since they eventually find their way into animal and human bodies. The three most common natural CDs and their hydrophilic derivatives are known to only permeate lipophilic biological membranes, which include the gastrointestinal mucosa, skin and cornea of the eye with certain difficulty. CDs have been reported to be nontoxic to a certain level due to low absorption from the gastrointestinal track [12, 123, 124]. αCDs were found to bind with some lipids resulting in eye irritation in rats when they were orally administered, whereby 60% of the dose was excreted as CO_2, 26–33% as metabolite incorporation and 7–14% as metabolites in urine and feces [12, 125]. Oral administration in rats showed that βCDs have less irritation compared to αCDs; however, small amounts were absorbed in the upper intestinal track [12, 126]. Even though they are nontoxic when administered orally, βCDs cannot be administered parenterally due to their low solubility in aqueous solutions and their nephrotoxicity [127].

Oral administration of γCDs has insignificant irritation followed by rapid and complete degradation to glucose by intestinal enzymes. They are therefore deemed the least toxic [12, 128]. α and βCDs are also known for their renal toxicity [127]. βCD is not used in parenteral formulations and the use of αCD is seriously limited due to toxicological consideration [123]. Parent CDs (α and β) and lipophilic CD derivatives such as m-βCD are also not suitable for parenteral formulation due to their rapid absorption by the gastrointestinal track; however, they are suitable for oral formulations [123, 127].

The newly discovered CD derivatives with better safety profiles have sparked a renewed interest in the use of CDs, especially for those that will find way into human and animal bodies. For example, HP-βCDs and sulfobutylether-βCDs are used in parental formulations in very high concentrations [123, 129]. The concentration, type of administration and time of exposure play a critical role in determining the level of toxicity and safety of CDs and their derivatives. It is therefore generally thought that CDs and their derivatives can be safely used in membrane technology and other applications without toxicological problems in case they leach out and be ingested by humans or animals, especially in areas such as wound dressing, water treatment and air purification.

8. Challenges and perspectives

One of the major challenges in the production of CD nanofibers is the inability to electrospin CDs directly, i.e., without the need to blend them with other copolymers or modifying their structure. The poor solubility of CDs in water and organic solvents makes it impossible to electrospin CDs directly. Hence, the most feasible way is to blend the CDs with other flexible polymers. Indeed, many studies have reported the fabrication of CD-based nanofibers using copolymers. In membrane technology, the main challenge is their high water solubility, which results in the loss of structure and functionality of membranes once the CDs are dissolved and washed away.

The application of these materials varies significantly. Workers are currently exploring the commercial viability of CD-based fibers especially in areas of water treatment. Their use in various applications is possible due to their ability to form inclusion complexes with various compounds. The chemistry responsible for this complexation is now well understood. The performance of CD-based materials can thus be ascertained or monitored using simple phase-solubility diagram.

Further, the high surface area-to-volume ratio of nanofibers and modification possibilities of CD molecules have motivated the use of CD nanofibers in areas such as tissue engineering and water treatment. The nanofiber morphology is preferred over other morphologies of the CD polymers, which tend to possess lower surface areas. The electrospinning technique can now produce large quantities of nanofiber membranes and makes it viable to use the materials in large-scale quantities. However, the cost is yet to be ascertained.

9. Conclusion

In general, CDs are less toxic and environment-friendly materials due to their biodegradable nature. CDs and their derivatives have unique properties such as high porosities,

http://dx.doi.org/10.5772/intechopen.74737

small diameters and high surface area-to-volume ratio rendering them practically usable in a wide range of applications including water treatment. Indeed, electrospinning these glycosidic sugars into nanofibrous mats improves their surface area-to-volume ratio further and enhances their adsorption and inclusion complexing properties. Undesirable species of specific sizes can be encapsulated *via* inclusion complexation and filtered by CD-based nanofibers and membranes. Electrospun CD nanofiber mats and membranes find further applications in various areas such as drug delivery, filtration, catalysis, water treatment, reinforcement, electronics, pharmaceuticals and optical devices. Considering that they are natural, nontoxic, cost-effective and readily available, their properties can be explored for further applications. Scaling-up and subsequent commercialization of these superior nanofibers and membranes can be explored once their cost has been determined.

Acknowledgements

We gratefully acknowledge the University of South Africa (UNISA) and Ghent University.

Author details

Mandla B. Chabalala[1], Bonisiwe C. Seshabela[1], Stijn W.H. Van Hulle[2], Bhekie B. Mamba[1], Sabelo D. Mhlanga[1] and Edward N. Nxumalo[1*]

*Address all correspondence to: nxumaen@unisa.ac.za

1 Nanotechnology and Water Sustainability Research Unit, College of Science, Engineering and Technology, University of South Africa, Johannesburg, South Africa

2 Department of Green Chemistry and Technology, Ghent University Campus Kortrijk, Belgium

References

[1] Petrovic M. Fate and distribution of pharmaceuticals in wastewater and sewage sludge of the conventional activated sludge (CAS) and advanced membrane bioreactor (MBR) treatment. Water Research. 2009;**43**(3):831-841

[2] Kasprzyk-hordern B, Dinsdale RM, Guwy AJ. The removal of pharmaceuticals, personal care products, endocrine disruptors and illicit drugs during wastewater treatment and its impact on the quality of receiving waters. Water Research. 2009;**43**(2):363-380

[3] Lin S, Juang R. Adsorption of phenol and its derivatives from water using synthetic resins and low-cost natural adsorbents: A review. Journal of Environmental Management. 2009;**90**(3):1336-1349

[4] Gopal R, Kaur S, Ma Z, Chan C, Ramakrishna S, Matsuura T. Electrospun nanofibrous filtration membrane. Journal of Membrane Science. 2006;**281**(1-2):581-586

[5] Kayaci F, Uyar T. Electrospun zein nanofibers incorporating cyclodextrins. Carbohydrate Polymers. 2012;**90**(1):558-568

[6] Lee KM, Lai CW, Ngai KS, Juan JC. Recent developments of zinc oxide based photocatalyst in water treatment technology: A review. Water Research. 2016;**88**:428-448

[7] Kayaci F, Aytac Z, Uyar T. Surface modification of electrospun polyester nanofibers with cyclodextrin polymer for the removal of phenanthrene from aqueous solution. Journal of Hazardous Materials. 2013;**261**:286-294

[8] Kayaci F, Umu OCO, Tekinay T, Uyar T. Antibacterial electrospun poly(lactic acid) (PLA) nano fi brous webs incorporating triclosan/cyclodextrin inclusion complexes. Journal of Agricultural and Food Chemistry. 2013;**61**:3901–3908

[9] Uyar T, Balan A, Toppare L, Besenbacher F. Electrospinning of cyclodextrin functionalized poly(methyl methacrylate) (PMMA) nanofibers. Polymer (Guildf). 2009;**50**(2):475-480

[10] Chalasani R, Vasudevan S. Cyclodextrin-functionalized Fe_3O_4@TiO_2: Reusable, magnetic nanoparticles for photocatalytic degradation of endocrine-disrupting chemicals in water supplies. ACS Nano. 2013;**7**(5):4093-4104

[11] Zhuo M, Yang T, Fu T, Li Q. High-performance humidity sensors based on electrospinning $ZnFe_2O_4$ nanotubes. RSC Advances. 2015;**5**(84):68299-68304

[12] Del Valle EMM. Cyclodextrins and their uses: A review. Process Biochemistry. 2004; **39**(9):1033-1046

[13] Prieto-Rodriguez L et al. Optimization of mild solar TiO_2 photocatalysis as a tertiary treatment for municipal wastewater treatment plant effluents. Applied Catalysis B: Environmental. 2012;**128**:119-125

[14] Li J, Loh XJ. Cyclodextrin-based supramolecular architectures: Syntheses, structures, and applications for drug and gene delivery. Advanced Drug Delivery Reviews. 2008; **60**(9):1000-1017

[15] Loftsson T, Duehene D. Cyclodextrins and their Pharmaceutical Applicationns. International Journal of Pharmaceuticals. 2007;**329**(1-2):1-11

[16] Nthunya LN, Masheane ML, Malinga SP, Nxumalo EN, Mhlanga SD. Environmentally benign chitosan-based nanofibres for potential use in water treatment. Cogent Chemistry. 2017;**3**(1):1-17

[17] Nthunya LN et al. Greener approach to prepare electrospun antibacterial β-cyclodextrin/ cellulose acetate nanofibers for removal of bacteria from water. ACS Sustainable Chemistry & Engineering. 2017;**5**(1):153-160

[18] Adams FV, Nxumalo EN, Krause RWM, Hoek EMV, Mamba BB. Preparation and characterization of polysulfone/β-cyclodextrin polyurethane composite nanofiltration membranes. Journal of Membrane Science. 2012;**405-406**(July):291-299

[19] Ramaseshan R, Sundarrajan S, Liu Y, Barhate RS, Lala NL, Ramakrishna S. Functionalized polymer nanofibre membranes for protection from chemical warfare stimulants. Nanotechnology. 2006;**17**(12):2947-2953

[20] Li R et al. Preparation and antimicrobial activity of β-cyclodextrin derivative copolymers/cellulose acetate nanofibers. Chemical Engineering Journal. 2014;**248**:264-272

[21] Kamiyama M, Soeda T, Nagajima S, Tanaka K. Development and application of high-strength polyester nanofibers. Polymer Journal. 2012;**44**(10):987-994

[22] Feng CX et al. Carbon nanotubes accelerated poly(vinylidene fluoride) crystallization from miscible poly(vinylidene fluoride)/poly(methyl methacrylate) blend and the resultant crystalline morphologies. European Polymer Journal. 2015;**68**:175-189

[23] Jia YT, Gong J, Gu XH, Kim HY, Dong J, Shen XY. Fabrication and characterization of poly (vinyl alcohol)/chitosan blend nanofibers produced by electrospinning method. Carbohydrate Polymers. 2007;**67**(3):403-409

[24] Hulicova D, Oya A. The polymer blend technique as a method for designing fine carbon materials. Carbon. 2003;**41**(7):1443-1450

[25] Hao X, Kaschta J, Pan Y, Liu X, Schubert DW. Intermolecular cooperativity and entanglement network in a miscible PLA/PMMA blend in the presence of nanosilica. Polymer (United Kingdom). 2016;**82**:57-65

[26] Lodge TP, McLeish TCB. Self-concentrations and effective glass transition temperatures in polymer blends. Macromolecules. 2000;**33**(14):5278-5284

[27] Jiang N, Abe H. Miscibility and morphology study on crystalline/crystalline partially miscible polymer blends of 6-arm poly(l-lactide) and poly(3-hydroxybutyrate-co-3-hydroxyvalerate). Polymer (United Kingdom). 2015;**60**:260-266

[28] Fenouillot F, Cassagnau P, Majesté JC. Uneven distribution of nanoparticles in immiscible fluids: Morphology development in polymer blends. Polymer (Guildf). 2009;**50**(6):1333-1350

[29] Roman C, Es García-Morales M, Gupta J, Mcnally T. On the phase affinity of multi-walled carbon nanotubes in PMMA:LDPE immiscible polymer blends. Polymer (Guildf). 2017;**118**:1-11

[30] Panapitiya NP et al. Stabilization of immiscible polymer blends using structure directing metal organic frameworks (MOFs). Polymer (Guildf). 2014;**55**(8):2028-2034

[31] Wang Z, Liu X, Macosko CW, Bates FS. Nanofibers from water-extractable melt-blown immiscible polymer blends. Polymer (United Kingdom). 2016;**101**:269-273

[32] Zvetkov VL, Djoumaliisky S, Simeonova-Ivanova E. The non-isothermal DSC kinetics of polyethylene tereftalate-epoxy compatible blends. Thermochimica Acta. 2013;**553**:16-22

[33] Taguet A, Cassagnau P, Lopez-Cuesta JM. Structuration, selective dispersion and compatibilizing effect of (nano)fillers in polymer blends. Progress in Polymer Science. 2014;**39**(8):1526-1563

[34] Wang S, Yu J, Yu J. Compatible thermoplastic starch/polyethylene blends by one-step reactive extrusion. Polymer International. 2005;**54**(2):279-285

[35] Celebioglu A, Vempati S, Ozgit-Akgun C, Biyikli N, Uyar T. Water-soluble non-polymeric electrospun cyclodextrin nanofiber template for the synthesis of metal oxide tubes by atomic layer deposition. RSC Advances. 2014;**4**(106):61698-61705

[36] Teo WE, Ramakrishna S. A review on electrospinning design and nanofibre assemblies. Nanotechnology. 2006;**17**(14):R89-R106

[37] Bhardwaj N, Kundu SC. Electrospinning: A fascinating fiber fabrication technique. Biotechnology Advances. 2010;**28**(3):325-347

[38] Huang D et al. Preparation of anatase F doped TiO_2 sol and its performance for photodegradation of formaldehyde. Journal of Materials Science. 2007;**42**(19):8193-8202

[39] Huang ZM, Zhang YZ, Kotaki M, Ramakrishna S. A review on polymer nanofibers by electrospinning and their applications in nanocomposites. Composites Science and Technology. 2003;**63**(15):2223-2253

[40] Li D, Xia Y. Electrospinning of nanofibers: Reinventing the wheel. Advanced Materials. 2004;**16**(14):1151-1170

[41] Pattamaprom C, Hongrojjanawiwat W, Koombhongse P, Supaphol P, Jarusuwannapoo T, Rangkupan R. The influence of solvent properties and functionality on the electrospinnability of polystyrene nanofibers. Macromolecular Materials and Engineering. 2006;**291**(7):840-847

[42] Sapountzi E et al. One-step fabrication of electrospun photo-cross-linkable polymer nanofibers incorporating multiwall carbon nanotubes and enzyme for biosensing. Journal of the Electrochemical Society. 2015;**162**(10):B275-B281

[43] Uyar T, Nur Y, Hacaloglu J, Besenbacher F. Electrospinning of functional poly(methyl methacrylate) nanofibers containing cyclodextrin-menthol inclusion complexes. Nanotechnology. 2009;**20**(12):125703

[44] Celebioglu A, Uyar T. Cyclodextrin nanofibers by electrospinning. Chemical communications (Cambridge, England). 2010;**46**(37):6903-6905

[45] Celebioglu A, Sen HS, Durgun E, Uyar T. Molecular entrapment of volatile organic compounds (VOCs) by electrospun cyclodextrin nanofibers. Chemosphere. 2016;**144**:736-744

[46] Uyar T, Havelund R, Hacaloglu J, Besenbacher F, Kingshott P. Functional electrospun polystyrene nanofibers incorporating α-, β-, and γ-cyclodextrins: Comparison of molecular filter performance. ACS Nano. 2010;**4**(9):5121-5130

[47] Bai J, Yang Q, Li M, Wang S, Zhang C, Li Y. Preparation of composite nanofibers containing gold nanoparticles by using poly(N-vinylpyrrolidone) and β-cyclodextrin. Materials Chemistry and Physics. 2008;**111**(2-3):205-208

[48] Cashion MP, Li X, Geng Y, Hunley MT, Long TE. Gemini surfactant electrospun membranes. Langmuir. 2010;**26**(2):678-683

[49] Mckee MG, Layman JM, Cashion MP, Long TE. Phospholipid nonwoven electrospun membranes. Science. 2006;**311**(5759):353-355

[50] Canbolat MF, Celebioglu A, Uyar T. Drug delivery system based on cyclodextrin-naproxen inclusion complex incorporated in electrospun polycaprolactone nanofibers. Colloids Surfaces B Biointerfaces. 2014;**115**:15-21

[51] Uyar T et al. Cyclodextrin functionalized poly(methyl methacrylate) (PMMA) electrospun nanofibers for organic vapors waste treatment. Journal of Membrane Science. 2010;**365**(1-2):409-417

[52] Celebioglu A, Uyar T. Electrospinning of polymer-free nanofibers from cyclodextrin inclusion complexes. Langmuir. 2011;**27**(10):6218-6226

[53] Celebioglu A, Uyar T. Electrospinning of nanofibers from non-polymeric systems: Electrospun nanofibers from native cyclodextrins. Journal of Colloid and Interface Science. 2013;**404**:1-7

[54] Manasco JL, Saquing CD, Tang C, Khan SA. Cyclodextrin fibers via polymer-free electrospinning. RSC Advances. 2012;**2**(9):3778-3784

[55] Celebioglu A, Uyar T. Electrospinning of nanofibers from non-polymeric systems: Polymer-free nanofibers from cyclodextrin derivatives. Nanoscale. 2012;**4**(2):621-631

[56] Celebioglu A, Aytac Z, Umu OCO, Dana A, Tekinay T, Uyar T. One-step synthesis of size-tunable Ag nanoparticles incorporated in electrospun PVA/cyclodextrin nanofibers. Carbohydrate Polymers. 2014;**99**:808-816

[57] Yoon K, Hsiao BS, Chu B. High flux ultrafiltration nanofibrous membranes based on polyacrylonitrile electrospun scaffolds and crosslinked polyvinyl alcohol coating. Journal of Membrane Science. 2009;**338**(1-2):145-152

[58] Wang ZG, Wang JQ, Xu ZK. Immobilization of lipase from *Candida rugosa* on electrospun polysulfone nanofibrous membranes by adsorption. Journal of Molecular Catalysis B: Enzymatic. 2006;**42**(1-2):45-51

[59] Wang YH, Tian TF, Liu XQ, Meng GY. Titania membrane preparation with chemical stability for very hash environments applications. Journal of Membrane Science. 2006;**280**(1-2):261-269

[60] Ma H, Hsiao BS, Chu B. Functionalized electrospun nanofibrous microfiltration membranes for removal of bacteria and viruses. Journal of Membrane Science. 2014;**452**:446-452

[61] Teng M, Li F, Zhang B, Taha AA. Electrospun cyclodextrin-functionalized mesoporous polyvinyl alcohol/SiO_2 nanofiber membranes as a highly efficient adsorbent for indigo carmine dye. Colloids and Surfaces A: Physicochemical and Engineering Aspects. 2011;**385**(1-3):229-234

[62] Zhang W, Chen M, Diao G. Electrospinning β-cyclodextrin/poly(vinyl alcohol) nanofibrous membrane for molecular capture. Carbohydrate Polymers. 2011;**86**(3):1410-1416

[63] Chen P, Liang HW, Lv XH, Zhu HZ, Bin Yao H, Yu SH. Carbonaceous nanofiber membrane functionalized by β-cyclodextrins for molecular filtration. ACS Nano. 2011;**5**(7):5928-5935

[64] Bazhban M, Nouri M, Mokhtari J. Electrospinning of cyclodextrin functionalized chitosan/PVA nanofibers as a drug delivery system. Chinese Journal of Polymer Science. 2013;**31**(10):1343-1351

[65] Uyar T, Hacaloglu J, Besenbacher F. Electrospun polystyrene fibers containing high temperature stable volatile fragrance/flavor facilitated by cyclodextrin inclusion complexes. Reactive and Functional Polymers. 2009;**69**(3):145-150

[66] Kayaci F, Ertas Y, Uyar T. Enhanced thermal stability of eugenol by cyclodextrin inclusion complex encapsulated in electrospun polymeric nanofibers. Journal of Agricultural and Food Chemistry. 2013;**61**(34):8156-8165

[67] Uyar T, Kingshott P, Besenbacher F. Electrospinning of cyclodextrin-pseudopolyrotaxane nanofibers. Angewandte Chemie International Edition. 2008;**47**(47):9108-9111

[68] Fatin SO, Lim HN, Tan WT, Huang NM. Comparison of photocatalytic activity and cyclic voltammetry of zinc oxide and titanium dioxide nanoparticles toward degradation of methylene blue. International Journal of Electrochemical Science. 2012;**7**:9074-9084

[69] Celebioglu A, Uyar T. Green and one-step synthesis of gold nanoparticles incorporated into electrospun cyclodextrin nanofibers. RSC Advances. 2013;**3**(26):10197

[70] Zhang X, Li X, Deng N. Enhanced and selective degradation of pollutants over cyclodextrin/TiO_2 under visible light irradiation. Industrial and Engineering Chemistry Research. 2012;**51**(2):704-709

[71] Kim ID et al. Dye-sensitized solar cells using network structure of electrospun ZnO nanofiber mats. Applied Physics Letters. 2007;**91**(16):163109

[72] Lu P, Wu F, Deng N. Enhancement of TiO_2 photocatalytic redox ability by β-cyclodextrin in suspended solutions. Applied Catalysis B: Environmental. 2004;**53**(2):87-93

[73] Yoon S, Nichols WT. Cyclodextrin directed self-assembly of TiO_2 nanoparticles. Applied Surface Science. 2013;**285**(part B):517-523

[74] Feng X, Shankar K, Varghese OK, Paulose M, Latempa TJ, a Grimes C. Vertically aligned single crystal TiO_2 nanowire arrays grown directly on transparent conducting oxide coated glass: Synthesis details and applications. Nano Letters. 2008;**8**(11):3781-3786

[75] Feng J, Miedaner A, Ahrenkiel P, Himmel ME, Curtis C, Ginley D. Self-assembly of photoactive TiO_2-cyclodextrin wires. Journal of the American Chemical Society. 2005;**127**:14968-14969

[76] Tachikawa T, Tojo S, Fujitsuka M, Majima T. One-electron oxidation pathways during β-cyclodextrin-modified TiO_2 photocatalytic reactions. Chemistry: A European Journal. 2006;**12**(29):7585-7594

[77] Zhao B, Chen H. Synthesis novel multi-petals ZnO nano-structure by a cyclodextrin assisted solution route. Materials Letters. 2007;**61**(27):4890-4893

[78] Yang X, Shao C, Guan H, Li X, Gong J. Preparation and characterization of ZnO nanofibers by using electrospun PVA/zinc acetate composite fiber as precursor. Inorganic Chemistry Communications. 2004;**7**(2):176-178

[79] Yang Y, Li X, Chen J, Chen H, Bao X. ZnO nanoparticles prepared by thermal decomposition of α and β-cyclodextrin coated zinc acetate. Chemical Physics Letters. 2003;**373**(1-2):22-27

[80] Rakshit S, Vasudevan S. Resonance energy transfer from beta-cyclodextrin-capped ZnO:MgO nanocrystals to included nile red guest molecules in aqueous media. ACS Nano. 2008;**2**(7):1473-1479

[81] Yin J, Deng B. Polymer-matrix nanocomposite membranes for water treatment. Journal of Membrane Science. 2015;**479**:256-275

[82] Zhao S, Wang Z, Wang J, Yang S, Wang S. PSf/PANI nanocomposite membrane prepared by in situ blending of PSf and PANI/NMP. Journal of Membrane Science. 2011;**376**(1-2):83-95

[83] Ng LY, Mohammad AW, Leo CP, Hilal N. Polymeric membranes incorporated with metal/metal oxide nanoparticles: A comprehensive review. Desalination. 2013;**308**:15-33

[84] Yu X et al. High performance thin-film nanofibrous composite hemodialysis membranes with efficient middle-molecule uremic toxin removal. Journal of Membrane Science. 2017;**523**(September):173-184

[85] Song X, Liu Z, Sun DD. Nano gives the answer: Breaking the bottleneck of internal concentration polarization with a nanofiber composite forward osmosis membrane for a high water production rate. Advanced Materials. 2011;**23**(29):3256-3260

[86] Wu H, Tang B, Wu P. Preparation and characterization of anti-fouling β-cyclodextrin/polyester thin film nanofiltration composite membrane. Journal of Membrane Science. 2013;**428**:301-308

[87] Pendergast MM, Hoek EMV. A review of water treatment membrane nanotechnologies. Energy & Environmental Science. 2011;**4**(6):1946

[88] Ncube P, Krause RWM, Mamba BB. Detection of chloroform in water using an azo dye modified β-cyclodextrin-pichlorohydrin copolymer as a fluorescent probe. Physics and Chemistry of the Earth. 2014;**67-69**:79-85

[89] Adams FV, Nxumalo EN, Krause RWM, Hoek EMV, Mamba BB. Application of polysulfone/cyclodextrin mixed matrix membranes in the removal of natural organic matter from water. Physics and Chemistry of the Earth. 2014;**67-69**:71-78

[90] Allabashi R, Arkas M, Hermann G, Tsiourvas D. Removal of some organic pollutants in water employing ceramic membranes impregnated with cross-linked silylated dendritic and cyclodextrin polymers. Water Research. 2007;**41**(2):476-486

[91] Choi SH, Chung JW, Priestley RD, Kwak SY. Functionalization of polysulfone hollow fiber membranes with amphiphilic β-cyclodextrin and their applications for the removal of endocrine disrupting plasticizer. Journal of Membrane Science. 2012;**409-410**:75-81

[92] Mbuli BS, Mhlanga SD, Mamba BB, Nxumalo EN. Fouling resistance and physicochemical properties of polyamide thin-film composite membranes modified with functionalized cyclodextrins. Advances in Polymer Technology. 2017;**36**(2):249-260

[93] Heng M et al. Tunable solvent permeation properties of thin film nanocomposite membrane by constructing dual-pathways using cyclodextrins for organic solvent nanofiltration. ACS Sustainable Chemistry & Engineering. 2015;**3**(9):1925-1933

[94] Mbuli BS et al. Modification of polyamide thin-film composite membranes with amino-cyclodextrins and diethylamino-cyclodextrins for water desalination. Separation and Purification Technology. 2013;**120**:328-340

[95] Mbuli BS et al. Preparation and characterization of thin film composite membranes modified with amine-functionalized β-cyclodextrins. Journal of Applied Polymer Science. 2013;**129**(2):549-558

[96] Krause RWM, Mamba BB, Dlamini LN, Durbach SH. Fe-Ni nanoparticles supported on carbon nanotube-co-cyclodextrin polyurethanes for the removal of trichloroethylene in water. Journal of Nanoparticle Research. 2010;**12**(2):449-456

[97] Liu H, Cai X, Wang Y, Chen J. Adsorption mechanism-based screening of cyclodextrin polymers for adsorption and separation of pesticides from water. Water Research. 2011;**45**(11):3499-3511

[98] Wang ZG, Wan LS, Liu ZM, Huang XJ, Xu ZK. Enzyme immobilization on electrospun polymer nanofibers: An overview. Journal of Molecular Catalysis B: Enzymatic. 2009;**56**(4):189-195

[99] Gopal R et al. Electrospun nanofibrous polysulfone membranes as pre-filters: Particulate removal. Journal of Membrane Science. 2007;**289**(1-2):210-219

[100] Celebioglu A, Umu OCO, Tekinay T, Uyar T. Antibacterial electrospun nanofibers from triclosan/cyclodextrin inclusion complexes. Colloids Surfaces B Biointerfaces. 2014;**116**:612-619

[101] Dong C, Qian LY, Zhao GL, He BH, Xiao HN. Preparation of antimicrobial cellulose fibers by grafting β-cyclodextrin and inclusion with antibiotics. Materials Letters. 2014;**124**:181-183

[102] Subramanian S, Singireddy A, Krishnamoorthy K, Rajappan M. Nanosponges: A novel class of drug delivery system review. Journal of Pharmacy & Pharmaceutical Sciences. 2012;**15**(1):103-111

[103] Lezcano M, Al-Soufi W, Novo M, Rodríguez-Núñez E, Vázquez Tato J. Complexation of several benzimidazole-type fungicides with α- and β-cyclodextrins. Journal of Agricultural and Food Chemistry. 2002;**50**(1):108-112

[104] Taylor MJ, Tanna S, Sahota T. In vivo study of a polymeric glucose-sensitive insulin delivery system using a rat model. Journal of Pharmaceutical Sciences. 2010;**99**(10):4215-4227

[105] de Miranda JC, Martins TEA, Veiga F, Ferraz HG. Cyclodextrins and ternary complexes: Technology to improve solubility of poorly soluble drugs. Brazilian Journal of Pharmaceutical Sciences. 2011;**47**(4):665-681

[106] Gupta SM, Tripathi M. A review on the synthesis of TiO_2 nanoparticles by solution route. Central European Journal of Chemistry. 2012;**10**(2):279-294

[107] Baudin C, Pean C, Perly B, Gosselin P. Inclusion of organic pollutants in cyclodextrins and derivatives. International Journal of Environmental Analytical Chemistry. 2000(April);**77**:233-242

[108] Dardeer HM. Importance of cyclodextrins into inclusion complexes 2-preparation of cyclodextrins 3-structure of cyclodextrins 4-cyclodextrin complexes. International Journal of Advanced Research. 2014;**2**(4):414-428

[109] Alsbaiee A, Smith BJ, Xiao L, Ling Y, Helbling DE, Dichtel WR. Rapid removal of organic micropollutants from water by a porous β-cyclodextrin polymer. Nature. 2016;**529**(7585):190-194

[110] Marques HMC. A review on cyclodextrin encapsulation of essential oils and volatiles. Flavour and Fragrance Journal. 2010;**25**(5):313-326

[111] Sapkal NP, Kilor VA, Bhursari KP, Daud AS. Evaluation of some methods for preparing gliclazide-β-cyclodextrin inclusion complexes. Tropical Journal of Pharmaceutical Research. 2007;**6**(4):833-840

[112] Chen Y, Liu Y. Cyclodextrin-based bioactive supramolecular assemblies. Chemical Society Reviews. 2010;**39**(2):495-505

[113] Wu H, Kong J, Yao X, Zhao C, Dong Y, Lu X. Polydopamine-assisted attachment of β-cyclodextrin on porous electrospun fibers for water purification under highly basic condition. Chemical Engineering Journal. 2015;**270**:101-109

[114] Celebioglu A, Demirci S, Uyar T. Cyclodextrin-grafted electrospun cellulose acetate nanofibers via 'click' reaction for removal of phenanthrene. Applied Surface Science. 2014;**305**:581-588

[115] Mura P. Analytical techniques for characterization of cyclodextrin complexes in the solid state: A review. Journal of Pharmaceutical and Biomedical Analysis. 2015;**113**:226-238

[116] Mura P. Analytical techniques for characterization of cyclodextrin complexes in aqueous solution: A review. Journal of Pharmaceutical and Biomedical Analysis. 2014;**101**:238-250

[117] Rezende BA et al. Complexation with β-cyclodextrin confers oral activity on the flavonoid dioclein. International Journal of Pharmaceutics. 2009;**367**(1-2):133-139

[118] Maestrelli F, Cecchi M, Cirri M, Capasso G, Mennini N, Mura P. Comparative study of oxaprozin complexation with natural and chemically-modified cyclodextrins in solution

and in the solid state. Journal of Inclusion Phenomena and Macrocyclic Chemistry. 2009;**63**(1-2):17-25

[119] Bettinetti GP, Sorrenti M, Rossi S, Ferrari F, Mura P, Faucci MT. Assessment of solid-state interactions of naproxen with amorphous cyclodextrin derivatives by DSC. Journal of Pharmaceutical and Biomedical Analysis. 2002;**30**(4):1173-1179

[120] Al-Marzouqi AH, Jobe B, Dowaidar A, Maestrelli F, Mura P. Evaluation of supercritical fluid technology as preparative technique of benzocaine-cyclodextrin complexes-comparison with conventional methods. Journal of Pharmaceutical and Biomedical Analysis. 2007;**43**(2):566-574

[121] Ikeda Y, Kimura K, Hirayama F, Arima H, Uekama K. Controlled release of a water-soluble drug, captopril, by a combination of hydrophilic and hydrophobic cyclodextrin derivatives. Journal of Controlled Release. 2000;**66**(2-3):271-280

[122] Lantz AW, Wetterer SM, Armstrong DW. Use of the three-phase model and headspace analysis for the facile determination of all partition/association constants for highly volatile solute-cyclodextrin-water systems. Analytical and Bioanalytical Chemistry. 2005;**383**(2):160-166

[123] Loftsson T, Duchêne D. Cyclodextrins and their pharmaceutical applications. International Journal of Pharmaceutics. 2007;**329**(1-2):1-11

[124] Irie T, Uekama K. Pharmaceutical applications of cyclodextrins. III. Toxical issues and safety evaluation. Journal of Pharmaceutical Sciences. 1997;**86**(2):147-162

[125] Kaukonen AM, Kilpelinen I, Mannermaa JP. Water-soluble α-cyclodextrins in paediatric oral solutions of spironolactone: Solubilization and stability of spironolactone in solutions of α-cyclodextrin derivatives. International Journal of Pharmaceutics. 1997;**159**(2):159-170

[126] Tötterman AM, Schipper NG, Thompson DO, Mannermaa JP. Intestinal safety of water-soluble beta-cyclodextrins in paediatric oral solutions of spironolactone: Effects on human intestinal epithelial Caco-2 cells. The Journal of Pharmacy and Pharmacology. 1997;**49**(1):43-48

[127] Jambhekar SS, Breen P. Cyclodextrins in pharmaceutical formulations I: Structure and physicochemical properties, formation of complexes, and types of complex. Drug Discovery Today. 2016;**21**(2):356-362

[128] Szente L, Szejtli J. Highly soluble cyclodextrin derivatives: Chemistry, properties, and trends in development. Advanced Drug Delivery Reviews. 1999;**36**(1):17-28

[129] Gould S, Scott RC. 2-Hydroxypropyl-β-cyclodextrin (HP-βCD): A toxicology review. Food and Chemical Toxicology. 2005;**43**(10):1451-1459

β-Cyclodextrins as Encapsulating Agents of Essential Oils

Ana Paula Capelezzo, Laura Cassol Mohr,
Francieli Dalcanton,
Josiane Maria Muneron de Mello and
Márcio Antônio Fiori

Additional information is available at the end of the chapter

http://dx.doi.org/10.5772/intechopen.73568

Abstract

Many scientific studies have made advances in the ability to encapsulate natural extracts by cyclodextrins. These studies have addressed the physical and chemical conditions of the encapsulation reactions, employed several types of essential oils and characterized the microcapsules as to their ability to release encapsulated active principles. The essential oils studies with cyclodextrin encapsulation processes have been highly varied. However, the most studied are the essential oils with antimicrobial and antioxidant capacities. The essential antimicrobial and antioxidant oils are easily degraded. In the presence of oxygen, they are oxidized, and at low temperatures, they are volatilized and decomposed. Thus, cyclodextrins are coatings capable of protecting these essential oils from environmental conditions and agents capable of promoting oil degradation, in addition to controlling their release. In this chapter of the book, we review scientific papers that examine the encapsulation of antimicrobial essential oils and antioxidant essential oils with β-cyclodextrins.

Keywords: β-cyclodextrin, essential oil, microencapsulation techniques, inclusion complex, protection

1. Introduction

Essential oils are complex and multicomponent mixtures produced from plant secondary metabolites and it can be extracted from different parts of the plants. Its composition depends on factors such as harvest season, part of the plant where the oil was extracted, geographic origin, and extraction method. They are used in the most diverse areas such as pharmaceuticals, cosmetics, agriculture, food, and textiles, among others.

Essential oils' antimicrobial activity has been extensively studied, succeeding both against Gram-positive bacteria and Gram-negative bacteria, as well against fungi. They also exhibit antiparasitic, antiviral, and antioxidant properties. However, its use is conditioned to processes and/or products that do not undergo thermal processing, as these essential oils are largely volatile, oxidizable, and thermosensitive.

Thus, the encapsulation techniques present themselves as an effective alternative in the protection of essential oils, releasing them at the desired time and place. There are several encapsulation techniques, among which we can highlight spray drying, spray cooling, extrusion, solvent evaporation, coacervation, and the use of supercritical fluids. What differs them from each other are the equipment used and the process conditions, the encapsulation efficiency, the particle size obtained, and the cost.

One of the key factors to be considered in the encapsulation process is the coating material. This determines the particles stability, the efficiency, and the degree of the core protection. Examples of coating materials are synthetic and biodegradable polymers, inorganic materials such as clays and silicates, proteins such as gelatin and casein, polysaccharides, and sugars, with emphasis on cyclodextrins. These are widely used in the industry due to properties such as inertia and toxicity.

The encapsulation process can form macroparticles, microparticles, and nanoparticles, and obtaining them is dependent on the choice of techniques and parameters involved in the process. In general, the compound to be encapsulated is suspended in a solution, and then the coating material is dissolved and precipitated by overlaying the core.

Therefore, encapsulating an essential oil ensures that it maintains its properties of interest while being protected from external factors such as mechanical stress, temperature, and oxidation. In the case of thermal protection, this is an extremely important advantage in which the inclusion complex can be used in processes and/or products that make use of thermal sources.

2. Essential oils

The use of plants in daily life has been a constant throughout all stages of evolution. They have been used as an unlimited source of food for humans and animals, fibers for clothing, and as useful medicines. Among the compounds obtained from vegetal material, the essential oils stand out and deserve particular attention due to their peculiar characteristics [1, 2].

2.1. Definitions

Essential oils are oily aromatic liquid compounds containing complex mixtures of volatile compounds, which are the secondary metabolites of plants and play an important role in their defense. They are extracted from the vegetal material (flowers, shoots, seeds, leaves, branches, peels, fruits, and roots) of a large number of plants, usually representing only a small fraction of the plant composition (less than 5% of dry material) [2–5].

These bioactive compounds have promising potential to maintain and promote health and to prevent microbial growth, and have been applied in diverse areas, such as in pharmaceuticals, food, textiles, biomedical applications, cosmetics, and agriculture industries. They usually possess low solubility and absorption and are chemically unstable and susceptible to oxidative deterioration and loss of volatile compounds, especially when exposed to oxygen, light, moisture, and heat, resulting in decreased bioavailability and efficacy [6–8].

The essential oil constituents are a family of organic compounds with a low molecular weight, and they can be divided into four groups according to their chemical structures: terpenes, terpenoids, phenylpropenes, and "others." Terpenes are hydrocarbons produced from the combination of several isoprene units (C_5H_8), and they are synthesized in the cytoplasm of vegetal cells. The main representatives of this group are the monoterpenes ($C_{10}H_{16}$) and sesquiterpenes ($C_{15}H_{24}$), but longer chains, such as diterpenes ($C_{20}H_{32}$) and triterpenes ($C_{30}H_{40}$), are also part of this group. Limonene is a classic example of a terpene. Terpenoids are terpenes that undergo biochemical modifications through enzymes that add oxygen molecules and move or remove methyl groups. Terpenoids can be subdivided into alcohols, esters, aldehydes, ketones, ethers, phenols, and epoxides. Examples of terpenoids are thymol, carvacrol, linalool, menthol, and geraniol [4, 9]. Phenylpropenes constitute a subfamily among the various groups of organic compounds called phenylpropanoids that are synthesized in plants from the amino acid precursor phenylalanine. Phenylpropenes constitute a relatively small part of essential oils, and those that have been more carefully studied are eugenol, vanillin, and cinnamaldehyde [4].

The proportion of these constituents is different in each essential oil and is a function of several factors, including the species, the part of the plant from which the oil was extracted, the harvesting season, geographical origin, and the method of extraction. All these factors directly influence the oil composition and, consequently, the bioactive properties, conferring different biological functionalities to them [10–18].

2.2. Antimicrobial activity and mechanism of action

Antimicrobial activity can be considered the most investigated activity of essential oils, especially when associated with food preservation and the consequent increase in shelf life, because these bioactive compounds have the capacity to slow down growth and even eliminate contaminating pathogens from food products. Therefore, essential oils meet the current requirements of more concerned and demanding consumers who prefer to consume food without synthetic preservatives, expanding their application in this segment of the population [19].

In addition, foodborne illness is a growing public health problem throughout the world; only in the United States, 31 species of pathogens are estimated to cause 9.4 million cases of foodborne illness per year [20]. This demands new strategies and more effective control and has motivated several studies with essential oils. Another characteristic of these compounds is the safety of their use in food. Many essential oils are considered by the Food and Drug Administration (FDA) as Generally Recognized as Safe (GRAS), meaning that they can be used in food products without the need for approval via technical analysis [21].

Some investigations have confirmed the antimicrobial activity of several essential oils. Teixeira et al. [22] studied the antimicrobial activity of 17 different essential oils against 7 different types of bacterial strains. All essential oils inhibited the growth of at least four of the bacteria tested. Pesavento et al. [23] tested the antimicrobial activity of the essential oils of oregano, rosemary, and thymol against *Staphylococcus aureus* and *Listeria monocytogenes* bacteria in meat, verifying that the insertion of essential oil decreased the microbial growth but altered the flavor of the food. Piletti et al. [19] evaluated the antimicrobial activity of eugenol against *Staphylococcus aureus* and *Escherichia coli* bacteria. The authors observed that eugenol has greater inhibitory activity toward *Staphylococcus aureus*, because they are Gram-positive bacteria and, therefore, are more susceptible to essential oils compared with Gram-negative bacteria, such as *Escherichia coli*.

According to studies presented by Affonso et al. [24], clove oil presents pronounced antimicrobial activity when tested against *S. aureus*, *E. coli*, *Campylobacter jejuni*, *Salmonella enteritidis*, and *Listeria monocytogenes*, significantly decreasing the growth rate, because it is effective against Gram-negative and Gram-positive bacteria except for *Pseudomonas aeruginosa*.

Knezevic et al. [25] confirmed the antimicrobial activity of essential oils of *Eucalyptus camaldulensis* against *Acinetobacter baumannii* bacteria, demonstrating the possibility of using this oil together with conventional antibiotics and confirming synergistic interactions between the two compounds in order to develop new strategies for infection treatment and reduce the dose of antibiotics used.

The antimicrobial activity of essential oils is related to their hydrophobicity, a characteristic that favors interaction with the lipids of the cell membranes and with the mitochondria of the microbial cells. These interactions generally alter the permeability of bacterial cells, causing disturbances in the structures and resulting in coarse fractures that cause ion, molecule, and cellular content leakage, leading to microorganism death or inhibition of their growth [3].

In general, essential oils act to inhibit bacterial cell growth and the production of toxic bacterial metabolites. Most essential oils have a more pronounced effect on Gram-positive bacteria than on Gram-negative species, and this effect is likely due to differences in the cell wall composition of these bacteria [9, 26, 27].

According to Muñoz-Bonilla and Fernández-García [28], Gram-positive bacteria have only one outer layer, which facilitates penetration of external molecules, promoting interaction with the cytoplasmic membrane and making them more fragile compared with Gram-negative bacteria. Gram-negative bacteria have an additional membrane with a phospholipid bilayer structure responsible for protection of the inner cytoplasmic membrane, which confers greater resistance to this class of bacteria. The hydrophilic wall hinders the penetration of hydrophobic compounds, for example, essential oils, into the cell [29, 30].

Mechanisms that explain the action of essential oils on bacterial cells have been studied, but it is still not possible to say with certainty how the essential oils act on a microbial cell. These bioactive compounds have many components, and the antimicrobial action cannot be confirmed by the action of only a single component or by the activity on a single cell site [31].

The typical hydrophobic characteristic of essential oils is responsible for the breakdown of bacterial structures, which leads to increased permeability due to the inability of separation between the essential oils and the bacterial cell membrane. This fact alters cellular functions, making it difficult to maintain the energetic state, altering solute transport, promoting cellular component leakage, and deregulating cellular metabolism [9]. Furthermore, because they contain phenolic compounds, the essential oils can disturb the cell membrane and inhibit the cell functional properties and are even capable of spilling cellular materials. The chemical composition of essential oils and/or their volatile compounds has a major impact on their antimicrobial mechanism, because phenolic compounds contain hydroxyl groups, which operate effectively against foodborne pathogenic bacteria [31].

2.3. Miscellaneous properties

Essential oils or their components not only have antibacterial properties [22, 23, 25, 32, 33] but also have antiparasitic [34, 35], antiviral [36, 37], antifungal [38–40], and antioxidant properties [32, 41, 42].

Alves-Silva et al. [43] determined the chemical composition and antimicrobial, antifungal, and antioxidant activities through four different antioxidant tests of three aromatic herb essential oils, coriander (*Coriandrum sativum*), celery (*Apium graveolens*), and bush-basil (*Ocimum minimum*), widely used in Portugal. The results showed that the essential oils of coriander, bush-basil, and celery obtained from plants grown in Portugal have significant antioxidant and antimicrobial activity, and the high antimicrobial activity is due to the high percentage of the main constituents or synergy between the different oil components that provide a biocidal effect against bacteria.

Even with so many well-researched studies, the application of essential oils still has some limitations. When used as a food preservative, the problem of essential oil constituents is that they often cause negative organoleptic changes if added in amounts sufficient to provide an antimicrobial effect, which generally requires high concentrations [44]. Additionally, in many foods, the hydrophobicity of essential oil constituents is detrimental due to interactions with fat-containing foods [4].

There is also another aggravating factor which makes impossible for these compounds to be used in other products that wish to make use of its main characteristic. The compounds that promote antimicrobial and antioxidant activity in the essential oils are highly volatile, thermally unstable, and photodegradable, and in the presence of oxygen, they undergo oxidation. Thus, when they are not protected by a barrier are not very stable and at high temperatures they lose their biological activity and their applications can be compromised [45–48].

Chemical component groups of essential oils are readily converted by oxidation, isomerization, cyclization, or dehydrogenation, which are reactions that can be enzymatically or chemically triggered, and these processes are usually associated with a loss of quality. For example, terpenoids tend to be volatile and thermolabile and can be readily oxidized or hydrolyzed, depending on their structure. Further, the maintenance of essential oil composition depends strongly on the conditions under which it is processed and how it is handled and stored upon production. Certain factors are crucial for maintaining the stability of essential oils, such as temperature, light, and oxygen availability. Therefore, these factors need to be carefully considered [49].

In this way, it is possible to infer that the conditions in which these essential oils are kept are fundamental to their characteristics. Rowshan et al. [50] studied the thermal stability of the *Thymus daenensis* essential oil by storing it at room temperature, under refrigeration (4°C) and frozen (−20°C). The authors verified that the oil composition was a function of temperature and that when frozen, the oil composition underwent only minor changes, and the primary quality was maintained, demonstrating the oil degradation effect under high temperature. The ambient temperature crucially influences the essential oil stability in several respects, and as a rule, chemical reactions are accelerated with increasing temperature because the reaction rates are increased by heat [49].

Turek and Stintzing [51] evaluated the impact of different storage conditions on four essential oils (lavender, pine, rosemary, and thyme) to verify the influence of light and temperature on their composition. The authors obtained interesting results, stating that parameters such as pH, conductivity, and the chemical profile of the essential oils are severely altered when exposed to light and temperature, which modifies their quality. Their work also reinforced that each essential oil responds differently to these external parameters.

One option for minimizing the exposure problems of essential oils is to make use of these compounds in encapsulated form, by means of a protective shell, to limit the degradation/loss of biological activity during processing and storage and to control compound release at the time and site desired [52–54].

Microencapsulation presents a great potential for improvement and development of structures for the conservation of natural products. In the last decade, there has been great progress in the development of microencapsulated compounds in the food and pharmaceutical industries, as they offer greater degradation resistance and compound stability [53, 55–59].

3. Encapsulation

3.1. Approaches

Encapsulation is the process of constructing a functional barrier between the core and the coating material to avoid chemical and physical reactions and to maintain the biological, functional, and physico-chemical properties of the core materials. The proper choice of encapsulation technique and the coating material depends on the end use of the product and the processing conditions involved. The coating material determines the stability of the particles, the process efficiency, and the degree of core protection [8].

Since bioactive compounds has some limitations in their use, for example, induce negative organoleptic change, are highly volatile, thermally unstable, photodegradable, and therefore, they can be easily deteriorated, the use of a barrier that limits these exchanges is interesting. When encapsulated, these compounds are protected against a number of factors, such as temperature, moisture, light, oxidation, undesirable reactions with other compounds and mechanical stress during handling, processing, and storage of the final product. This leads to a prolonged shelf life and maintenance of metabolic activity for long periods of time during storage, which maintains the biological and functional characteristics of essential oils [60–62].

3.2. Procedure

The encapsulation technique can be used for solid, liquid, or gaseous material packaging using fine polymer coatings to form macrocapsules (>5000 μm), microcapsules (0.2–5000 μm), or nanocapsules (<0.2 μm) [63, 64]. Nanoencapsulation is the coating of one or more substances within another material at the nanoscale. Microencapsulation is similar to nanoencapsulation, but it involves larger particles and is an already consolidated technique, with a longer study time compared to the nanoencapsulation process. On the other hand, macroencapsulation involves a larger scale than microencapsulation [65].

In general, the compound to be encapsulated is suspended in a solution containing the encapsulating agent, and then, this agent is dissolved and precipitated by coating the suspended material, or the compound to be encapsulated and the encapsulating agent are dissolved in a single solvent and simultaneously precipitated (coprecipitation). In this situation, various particles of the compound are within the layer of the encapsulating agent, with the capsule formation, which may be microcapsules and/or microspheres, for example [66]. The encapsulating agent protects the core by isolating it and allows release through a specific stimulus at the time and place desired [64]. **Figure 1** shows a schematic picture of microcapsules and microspheres.

Microcapsules are particles consisting of a substantial central inner core containing the active substance covered by a layer of the encapsulating agent, constituting the capsule membrane, while microspheres are matrix systems in which the nucleus is uniformly dispersed and/or dissolved in a polymer network. The microspheres may be homogeneous or heterogeneous

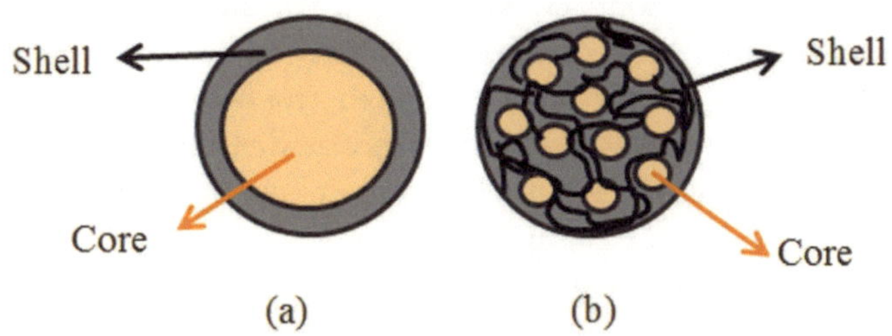

Figure 1. Types of microparticles: (a) microcapsules and (b) microspheres. Source: Adapted [67].

depending on whether the core is in the molecular state (dissolved) or in particle form (suspended), respectively [64, 68].

The encapsulating agent should not react with the core, and it should have the ability to seal and hold the core within the capsule, protecting it from adverse conditions. Interactions between the wall material and the core can affect the release rate as well as the core volatility and particle size [63, 69].

The mechanisms involved in the core release are diffusion, where the active compound is released slowly by permeating the wall of the coating without compromising its physical integrity, or through a release trigger, which involves a change in pH, mechanical stress, temperature, enzymatic activity, time, or osmotic force, among other triggers, that promotes capsule breakdown and instantly releases the active compound [70].

3.3. Techniques

The selection of the encapsulation technique and the coating material depends on the final application of the product, considering the physical and chemical stability, the concentration of the compounds in the encapsulation process, the required particle size, the release mechanism and the manufacturing costs [62, 64].

The encapsulation technique efficiency depends on several parameters. Retention of the active agent within the membrane shell is regulated by factors related to the chemical nature of the core, including its molecular weight, chemical functionality, polarity, and volatility, and by the properties of the coating material and the chosen encapsulation technique [62].

The key steps in an encapsulation method are incorporation of the bioactive compounds; droplet formation; removal of the solvent; collection of the capsules, and drying [71]. Different encapsulation methods have been developed to meet different types of core and shell materials, as well as to generate particles with various sizes, thickness, and shell permeability, thus adjusting the active principle release rate [62]. Some of the main encapsulation methods are spray drying, coacervation, solvent evaporation, extrusion, lyophilization, and encapsulation using supercritical fluids [72]. These methods are described in the following sections.

3.3.1. Spray drying

Very simply, this process consists of (i) preparing, (ii) homogenizing, and (iii) atomizing the suspension and (iv) drying the atomized particles [8]. The encapsulation efficiency using the spray drying technique is dependent on the preparation of a stable emulsion/suspension of oil in water and spraying it into small droplets on the drying bed [63]. Therefore, the bulk ratio of hydrophilic and lipophilic phases, the oil droplet size distribution, the dry matter content, and the emulsion viscosity need to be optimized prior to use in this technique. The emulsions must have sufficient viscosity to be pumped and sprayed, and they should not be sticky and hygroscopic after drying, which will ensure particle stability during storage [7].

This technique requires well-adjusted operating conditions as well as a suitable composition of the solution containing the bioactive compounds. The first includes factors such as inlet air temperature, atomizing air flow, liquid flow rate, vacuum aspirator velocity, and solid concentration, among others [73]. The success in obtaining particles using the spray drying method depends on factors such as the choice of polymer and the size distribution of the oil droplets in the emulsion [74, 75].

The main advantages of the spray drying technique are the combination of particle formation and drying in a single step, the possibility of using a wide variety of encapsulating agents, potentially large-scale production, simple equipment, low operating costs, high quality capsules with good yield, quick solubility of the capsules, small size, high capsule stability, and continuous operation [61, 64, 73, 76, 77]. The disadvantages are variations in the particle size and shape distribution, high temperatures, and rapid drying rates that normally do not allow encapsulation of thermosensitive compounds [61, 64, 73, 77].

Gallo et al. [76] studied the influence of the operation conditions of the spray drying method on the physical properties of *Rhamnus purshiana* extract powder. Roccia et al. [73] evaluated the influence of spray-drying operation conditions on the qualities of powdered sunflower oil. Fernandes et al. [78], Goñi et al. [79], and Oliveira et al. [80] microencapsulated essential oil of rosemary, eugenol, and passion fruit seed oil, respectively, using the dying spray technique. Gallardo et al. [81] microencapsulated linseed oil by spray-drying for application in functional foods.

3.3.2. Coacervation

Coacervation is one of the oldest and most widely used encapsulation techniques, and it involves electrostatic attraction between two oppositely charged polymers and the formation of coacervates over a narrow pH range. This technique involves the addition of a coacervating agent to a homogeneous polymer solution. The coacervation agent desolvates the polymer solution in a coacervate (polymer rich phase) and coacervation medium (poor polymer phase). During the encapsulation process, the bioactive compound is encapsulated within the polymer rich phase [7, 8].

The coacervation technique may be simple or complex. What differentiates them is the method of phase separation. In simple coacervation, the polymer is salted by the action of electrolytes, such as sodium sulfate, or desolvated by the addition of a water-miscible nonsolvent, such as ethanol, or by increasing/decreasing the temperature. These conditions promote macromolecule-macromolecule interactions, allowing the production of microcapsules containing hydrophobic substances, such as essential oils. Simple coacervation offers important advantages over complex coacervation in terms of cost savings and flexible operations. To induce phase separation, simple coacervation uses cheap inorganic salts, while complex coacervation is more sensitive even at small pH changes. In addition, complex coacervation uses expensive hydrocolloids [8, 82].

On the other hand, complex coacervation involves complexation between two oppositely charged polymers (commonly a polysaccharide and a protein). Complex coacervation involves three basic steps. The first step consists of the formation of an oil/water (o/w) emulsion, and the second step consists of separation of the liquid phase rich in the insoluble polymer; this phase results from the electrostatic attraction between opposing charges of the polymers caused by a reduction in the pH of the solution. The last step is coating stabilization (coating hardening, using thermal, cross-linking or desolvation techniques, to form self-sustained microcapsules). The formation of the coacervation coating is conducted by the difference in surface tension between the coacervation phase, the water and the hydrophobic material [7, 8, 62].

The main advantage of complex coacervation is that it has a high level payload (up to 99%). In addition, this method is simple, low cost, and solvent free. Therefore, complex coacervation can be used to manufacture microcapsules at an industrial scale [83].

Several investigations have utilized the coacervation method to microencapsulate bioactive compounds, such as propolis extract, sweet orange oil, essential oil of mustard seeds (*Sinapis alba*) and pepper essential oil [52, 84–86].

3.3.3. Solvent evaporation

This technique is based on the evaporation of the internal phase of an emulsion by shaking. Generally, the coating material is dissolved in a volatile organic solvent. The core material is then dissolved or dispersed in the encapsulating agent solution to form a suspension, an emulsion, or a solution. Thereafter, the organic phase is emulsified under agitation in a dispersion phase consisting of a nonsolvent of the encapsulant, which is immiscible with the organic solvent and contains an appropriate emulsifying agent. Once the emulsion is stabilized, the shaking is maintained, and the solvent evaporates after diffusing through the continuous phase, resulting in solid microspheres. The microspheres are recovered by filtration or centrifugation, and washed and dried [87, 7]. Patil et al. [88] encapsulated clove oil in methylcellulose microcapsules using the solvent evaporation method.

3.3.4. Extrusion technique

Oil encapsulation by extrusion consists basically of (i) injection and (ii) melt extrusion, followed by (iii) centrifugal extrusion (coextrusion). The main advantages of the encapsulation of essential oils by extrusion are stability against oxidation, prolonged shelf life, and lower rates of essential oil evaporation. However, this is an expensive process, and the particles do not have a uniform distribution [8]. Soliman et al. [89] microencapsulated essential oils of clove (*Eugenia caryophyllata*), thyme (*Thymus vulgaris*), and cinnamon (*Cinnamomum zeylanicum*) using this technique.

3.3.5. Freeze drying

Freeze drying is a method that involves dehydration of the frozen material under a vacuum sublimation process; the removal of water occurs without subjecting the sample to high temperatures. This method provides products of excellent quality because it minimizes the changes associated with high temperature. However, its high cost and long process time reduce its applicability [64]. Examples of studies that used this technique include Calvo et al. [90] who microencapsulated extra virgin olive oil in the presence of maltodextrin, carboxymethylcellulose, and lecithin; Ezhilarasi et al. [91], who studied microencapsulation of garcinia fruit extract by spray drying and its effect on bread quality; Piletti et al. [19], who encapsulated eugenol essential oil into β-cyclodextrin molecules through lyophilization; and Hill et al. [21], who encapsulated cinnamon bark extract, trans-cinnamaldehyde, clove extract, eugenol, and a 2:1 mixture (trans-cinnamaldehyde: eugenol) with β-cyclodextrin using the lyophilization method.

3.3.6. Technology employing supercritical fluids

The encapsulation technology employing supercritical fluids has been developed to minimize the disadvantages associated with traditional encapsulation methods, and it has a great relevance for the pharmaceutical, cosmetic, and food industries. It has several inherent advantages: non-toxicity and easy removal of the solvent without degradation of the product, and the process uses a wide variety of materials that produce controlled particle sizes and morphologies. Generally, it is preferred for essential oils that are sensitive to high temperatures, oxygen, and chemicals. Technology using supercritical fluids is considered a green technology because of the use of supercritical carbon dioxide in most cases. Supercritical carbon dioxide (CO_2) has properties that are ideal for bioactive compound encapsulation. The characteristic properties of supercritical CO_2 are lower viscosity, higher diffusivity, lower surface tension, faster process, and high solubility of the active compound [7, 8].

The supercritical apparatus consists of a high pressure stainless steel impregnation cell, a magnetic stirrer plate, a temperature controlled water bath, a high pressure CO_2 pump, and a pressure transducer. The impregnation cell contains two chambers separated by a mesh. The lower chamber is filled with essential oil, and the upper chamber is filled with microparticles or matrices in which the oil needs to be impregnated [7]. The essential oils of lavandin, oregano, canola, and passion fruit seed oil have been encapsulated using this method [80, 92–94].

For the encapsulation process, selection of the encapsulating material is a very important step. This material should be chosen according to its bioactivity, non-toxicity, intended application, and method of particle formation [71]. Biodegradable polymers, such as PLA, PLGA, and PCL, have primarily been used as a coating in the medical field, especially in tissue engineering and drug release. To a lesser extent, inorganic materials such as silicates, clays, and polyphosphates can also be used. Further, proteins (gelatin, casein, and soy proteins), lipids (waxes, paraffin, and oils), and synthetic polymers (acrylic polymers, polyvinylpyrrolidone) may be used. However, the most widespread materials used as encapsulating agents are polysaccharides and sugars (gums, starches, and celluloses), especially cyclodextrins, which are widely used mainly in the food industry due to their interesting properties; specifically, they are inert and non-toxic [7, 62].

4. Cyclodextrins

Cyclodextrin had its origin around 1981, when Villier discovered a new starch derivative obtained from bacterial degradation, which presented properties similar to those of cellulose, and distinguished two types of crystals of cellulosin: the cyclodextrins α and β [95]. Twelve years later, when studying the bacterial digestion of starch, Szejtli [95] identified two crystalline products with the same characteristics as Villier's cellulosins. Deepening his studies, he perfected the process of obtaining these crystals and isolated the bacterium that produced them, deeming it *Bacillus macerans*. The crystalline products were called crystallized α-dextrin and β-dextrin. Later, λ-dextrin was also isolated, and several fractionation schemes for the production of cyclodextrins were developed [96].

Cyclodextrins (CD's) are cyclic oligosaccharides consisting of glucose units linked by α-(1,4) glycosidic bonds derived from the enzymatic degradation of starch by certain bacteria, and they are chemically and physically stable molecules [95, 97]. The most common natural CDs have six, seven, and eight D-glucopyranose units and are named α, β, and γ cyclodextrin, respectively, and they differ from each other by virtue of ring size and solubility [98]. While the central cavity of CDs has a hydrophobic character, the surrounding walls are hydrophilic, and this feature allows CDs to form capsules, acting as a host for lipophilic compounds in their cavities and forming inclusion complexes [97, 99–101].

The binding of bioactive compounds within the host cyclodextrin is not fixed or permanent, but rather a dynamic equilibrium. This way, the formation of inclusion complexes is result of an equilibrium between the free and CD molecules and the bioactive compounds—CD complex [99]. Therefore, some factors may affect inclusion complex formation, such as type of cyclodextrin, cavity size, pH and ionization state, temperature, and method of preparation [102].

CD molecules are cone-like in shape with a cavity 7.9 Å deep. The upper and lower diameters of the CD wells are 4.7 and 5.3 Å, 6.0 and 6.5 Å, and 7.5 and 8.3 Å for α-CD, β-CD, and γ-CD, respectively [45].

Among the CDs, β-CD is the most used, because its apolar cavity can host molecules of molecular masses between 100 and 400 g mol^{-1}, which is the molecular mass range of most molecules of interest. β-CD is also easy to recover industrially through the crystallization process [103], and it has the lowest solubility and an intermediate size (**Table 1**). In addition, β-CD production is the most economically viable, with an industrial cost per kilogram approximately 20 times lower than that of the other CD types [104].

These inclusion complexes are important because they improve the chemical and physical stability and solubility of the compounds encapsulated in water. Due to the solubility of CDs in water and because they have the ability to form reversible inclusion complexes with non-polar molecules in aqueous solution, the water molecules inside the ring are easily replaced by non-polar molecules or molecules with less polarity than water, forming structures that are energetically more stable [105].

The encapsulation can reduce volatilization rates, and promote the gradual release of the encapsulated molecules, which improves their efficacy and bioavailability. Furthermore, they act as protectors against oxidative damage, light degradation, and heat, and other adverse effects linked to the medium in which they are inserted and maintain the initial characteristics of the compound for a long period. These inclusion complexes are relatively more hydrophilic and larger in size than the non-associated active compound, which helps to increase the retention of the encapsulated substance. They are also very interesting because they can mask undesirable flavors and odors that the encapsulated compounds may present [21, 56, 72, 101, 102, 106–110].

	α-CD	β-CD	γ-CD
Glucose number	6	7	8
Molecular mass	972	1135	1297
Aqueous solution (g 100 mL^{-1} at 25°C)	14.2	1.85	23.2
Cavity diameter (Â)	4.7–5.3	6.0–6.5	1.5–8.3
Cavity volume ($Â^3$)	174	262	427
Crystal form	Hexagonal blades	Monoclinic parallelograms	Quadratic prisms
Pk	12.332	12.202	12.081
Melting point (°C)	275	280	275
Surface tension (nM/m)	73	73	73
Rate of acid hydrolysis (h^{-1})	0.11	0.13	0.23

Source: [95].

Table 1. Physical and chemical properties of α-CD, β-CD, and γ-CD.

Marques [102] notes that the goal of encapsulation using cyclodextrin is to reduce the volatility and toxicity of the encapsulated compounds, provide protection of compounds that are sensitive to factors that promote their degradation, and alter the kinetics of migration and release of the encapsulated active components into the external environment.

The use of cyclodextrins is verified in diverse industrial products, such as pharmaceuticals [107, 111, 112], agrochemicals [113–115], and foods [116–119]. In the food area, cyclodextrins are nontoxic and considered GRAS, and thus are used for several purposes [120, 121]. These structures offer increased resistance to degradation of the active compounds and make the host-microcapsule complex more stable [53, 56, 57, 122, 123].

Szente and Szejtli [104] studied the toxicity of CDs and demonstrated that oral administration of high doses of CDs does not cause any harm. Several studies have shown that CDs are nontoxic and do not present intoxication risks, because they are not absorbed in the gastrointestinal tract or through lipophilic biological membranes, and the same results have been obtained with regard to teratogenicity and mutagenicity [124–128]. Antisperger [129] also evaluated the toxicity of CDs when introduced in an amount equivalent to 20% in the diet of rats and dogs and found no toxicity.

5. Cyclodextrins in thermal protection

The thermal degradation is one of the main natural compounds' degradation forms. In most cases, the increase in temperature is undesirable, as it favors the volatilization of less stable compounds, which are responsible for the biological activity. Therefore, the thermal degradation makes it impossible to apply many of the natural compounds studied, due to the alteration of their characteristics when exposed to high temperatures. Due to this situation, several authors have studied the encapsulation of these bioactive compounds with cyclodextrin, in order to provide a barrier, aiming the thermal protection of these natural compounds and preventing bioactive compounds from being lost and thus ensuring the application of these products in different situations.

Abarca et al. [130] prepared an inclusion complex of 2-nonanone (2-NN) with β-cyclodextrin by a co-precipitation method. 2-Nonanone are aliphatic hydrocarbons, aromatic volatiles commonly found in plant tissues, presenting antifungal behavior with low mammalian toxicity, a pleasant fruity/floral odor, resistance to rapid decomposition, adequate volatility, environmental acceptability, and a high potential for commercial development. The TGA and DSC analyses showed that thermal stability increased when 2-NN was encapsulated with β-CD. The antifungal activity of the inclusion complex was tested against *Botrytis cinerea*. All inclusion complexes tested showed potential antifungal activity, but the complex 1:0.5 (β-CD: 2-NN) showed the highest antifungal activity with a radius of 0.6 cm and 80% of growth inhibition.

Babaoglu et al. [101] encapsulated clove essential oil in hydroxypropyl-beta-cyclodextrin using the kneading method (a low-cost and easy-to-operate encapsulation technique) with hydroxypropyl beta-cyclodextrin and oil at a molar ratio of 1:1. The study demonstrated that the stability of the inclusion complex formed was greater and that the encapsulation process also increased the total phenolic content and antioxidant properties compared with the essential oil in free form. The authors indicate that this increase is due to an increase in the solubility of the essential oil molecules in water as a result of inclusion complex formation. Furthermore, the release rate of the essential oil was controlled with encapsulation. However, the authors concluded that this rate could be improved with the use of different proportions of essential oils. With this study, the potential for the use of microencapsulated clove oil in the pharmaceutical and food industries is evident, because this formulation keeps the oil constituents active and avoids losses and degradation.

Inclusion complexes formed with cyclodextrin are already being used as additives in final products, as reported by Wang et al. [131], that performed a work demonstrating this possibility when preparing cyclodextrin microencapsulated ammonium polyphosphate (MCAPP), with the goal of improving the water durability of APP and making a novel functional flame retardants. One of the interesting results found by the authors was that cyclodextrin resulted in the transformation of hydrophilic to hydrophobic of the flame retardant surface. Then, MCAPP was incorporated into the ethylene vinyl acetate copolymer (EVA), extensively used for the several applications like electrical insulation, cable jacketing and repair, water proofing, and corrosion protection, in order to improve flame retardancy of the EVA. The results showed that after the incorporation, the EVA composites presented improvements in mechanical, thermal stability, combustion properties, and flame-retardant properties, mainly because cyclodextrin shell improves the compatibility of the composites and the dispersion of APP in the EVA matrix evidencing that the microencapsulation technology with cyclodextrin contributes to obtain products with better characteristics and greater applicability. This study showed that cyclodextrin encapsulation is not only limited to natural products, but can also act as an encapsulating agent for other products as well, increasing its stability.

Another study that inserted the inclusion complex obtained in a final product was done by Kayaci et al. [132]. Geraniol is a natural component of plant essential oils, generally used as a fragrance/flavor in food industry to treat infectious diseases and/or preserve the food. The authors studied solid inclusion complexes of geraniol/cyclodextrins (α-CD, β-CD, and γ-CD). The results showed that the complexation efficiency between geraniol and γ-CD was higher. After this verification, the authors incorporated this inclusion complex into polyvinyl alcohol (PVA) nanofibers (NF) via electrospinning. The SEM analysis showed a homogeneous distribution of the inclusion complex (geraniol/γ-CD) to the PVA nanofibers. PVA/inclusion complex (geraniol/γ-CD) nanofibers presented higher thermal stability when compared to PVA/geraniol nanofibers only. Geraniol is easily volatilized, a fact that can be observed during electrospinning or during storage. When the PVA/geraniol nanofibers are evaluated, it was verified that after one day of its production, the geraniol had already evaporated completely. In contrast, PVA/inclusion complex (geraniol/γ-CD) nanofibers lost only about 10% of geraniol after two years of manufacturing. This result led the authors to conclude that PVA/inclusion complex (geraniol/γ-CD) nanofibers have potential application in the food packaging sector due to the high surface area and nanoporous structure of nanofibers and also due to the high thermal stability and longer durability of the agent active because it is encapsulated.

Hădărugă et al. [133] studied *Ocimum basilicum* L. essential oil and its β-cyclodextrin (β-CD) complex with respect to stability against the degradative action of air/oxygen and temperature using GC-MS analysis. Compounds associated with the degradation of the essential oil, which appear at high concentrations in degraded feedstocks, were limited and nearly constant in the complex formed by oil and β-CD, even at very high degradation temperatures, indicating improvement of the quality and stability of the complex.

Kalogeropoulos et al. [134] performed a thermal study of *Hypericum perforatum* methanolic extract, which is very rich in flavonoids, encapsulated in β-cyclodextrin (β-CD). Through DSC analysis after thermal oxidation, the authors found that the inclusion complex remained intact at temperatures at which the free extract was oxidized. Therefore, they showed that β-CD protected *Hypericum perforatum* extracts against thermal degradation, suggesting that this inclusion complex can be used as a food supplement or a novel additive to enhance the antioxidant capacity of fresh or thermally processed food.

Hill et al. [21] investigated the complexes formed by oils encapsulated in β-cyclodextrin (BCD) and their antimicrobial activity. The natural products studied were cinnamon bark extract, trans-cinnamaldehyde, clove bud extract, eugenol, and a 2:1 (trans-cinnamaldehyde:eugenol) mixture microencapsulated with the freeze-drying method. The oils and their BCD complexes were analyzed for their antimicrobial activity against *Salmonella enterica* serovar Typhimurium LT2 and *Listeria innocua*. In addition to the antimicrobial analysis, the authors investigated, among other things, the protection of the biological compounds against thermal oxidation, which should be the role played by β-cyclodextrin. The authors investigated the thermal stability of the oils through DSC analysis and compared EOs in their free form and their encapsulated form. As noted, there are two exothermic peaks at approximately 265°C and 260°C that, according to the authors, may be related to hydrolysis or oxidation of trans-cinnamaldehyde and eugenol. These peaks were not detected in the thermogram of the inclusion complex formed by the oils and β-cyclodextrin, suggesting that the EOs were protected at the β-cyclodextrin cavity. The temperature peaks of 100°C were attributed to water evaporation in all the samples, and the exothermic peaks at approximately 300°C for the β-cyclodextrin samples are a result of thermal degradation of the compound itself. Similar results were observed for the extracts and their inclusion complexes formed with β-cyclodextrin, indicating that the encapsulating agent provided thermal protection.

The antimicrobial analysis showed that all the antimicrobials effectively inhibited bacterial growth within the tested concentration range except for free eugenol. The EO-BCD complexes inhibited both bacterial strains at lower active compound concentrations than free oils, likely due to increased solubility in water that led to greater contact between the pathogens and essential oils. Moreover, the results showed that in addition to masking the sensory effect of the attributes of antimicrobial agents, complexation may potentiate their activity.

Wang et al. [103] studied the encapsulation of garlic oil (GO) and obtained an inclusion complex with GO encpasulated by the β-cyclodextrin using the co-precipitation method. The authors also used DSC to evaluate the thermal stability of the complex. The garlic oil is rich in organosulphur compounds that have a variety of antimicrobial and antioxidant activities but are very volatile and have low physicochemical stability.

The BCD thermogram showed a large endothermic peak at approximately 127°C that, according to the authors, is related to elimination of water molecules that are bound to the cyclodextrin molecules. For GO in its free form, the authors verified the existence of two peaks at approximately 186° and 223°C and associated the peaks with GO oxidation. These two exothermic peaks were not found in the GO-BCD complex thermogram, indicating that the biological compound is protected from oxidation within the BCD cavity.

Hădărugă et al. [135] studied the thermal and oxidative stability of Atlantic salmon oil (*Salmo salar* L.) and complexation with β-cyclodextrin. Due to being very unstable, even with low temperature degradation, it is interesting to encapsulate Atlantic salmon oil to ensure the permanence of its characteristics even after some time. The results showed good yields in the preparation of β-CD/Atlantic salmon oil complexes by co-crystallization, thereby increasing the thermal and oxidative stability of this oil.

Li et al. [136] also prepared an inclusion complex of benzyl isothiocyanate (BITC) extracted from papaya seeds with β-cyclodextrin. The thermal properties of BCD, BITC and its inclusion complex (BITC-BCD) were investigated using DSC and TG techniques. The DSC curve of BITC-BCD shows that volatilization of uncoated BITC occurred. The TG curve of BCD showed a slope close to 300°C, which was generally attributed to the onset of BCD decomposition. The BITC is a volatile material and quickly loses mass at 80–165°C. The inclusion complex showed volatility between 140°C and 300°C, indicating that the BCD cavity provides protection against BITC volatilization.

Zhou et al. [137] studied the Baicalein (Ba) encapsulation, an active ingredient extracted from a medicinal herb *Scutellaria baicalensis*, which has anti-inflammatory, antioxidant and anti-tumor activity among other biological activities; however, it presents limited solubility and high instability. In order to overcome the unfavorable physical-chemical properties presented by Ba, the authors performed a study with the various natural forms of cyclodextrin and its derivatives by using the freeze-drying method to obtain a complex that allows thermal protection of the natural compound. The solubility of Ba in the presence of natural cyclodextrins and its derivatives was higher than that of free Ba, with emphasis on the inclusion complex formed by 2,6-di-O-methyl-β-cyclodextrin (DM-β-CD), which showed the solubility constant of 13672.67 L mol^{-1}. The dissolution rate and thermal stability of the inclusion complex were significantly enhanced compared with the Ba pure; thus, DM-β-CD considerably improves the solubility and thermal stability of Ba, which make the chemical application of this drug promising.

Vilanova and Solans [138] studied the inclusion complexes of Vitamin A Palmitate with β-cyclodextrins, without the use of organic solvents. The low stability and low water solubility of some vitamins limit its use as a food additive, so the authors' interest was to use cyclodextrin as an encapsulating agent to overcome these deficiencies, making possible the production of foods enriched with vitamins, in order to prevent diseases related to their deficiency. All results showed a notably increase of Vitamin A Palmitate water solubility and stability in front of temperature, oxygen, and UV light when encapsulated. This works showed that the formation of inclusion complexes is a potential strategy to not only enrich but also to provide stability in surfactant-free food emulsion formulations, which seem to be a promising vehicle to increase the bioavailability of Vitamin A Palmitate in food.

Fernandes et al. [139] evaluated the thermal stability of cyanidin-3-O-glucoside (cy3glc) (major blackberry anthocyanin) and blackberry purees through molecular inclusion with β-cyclodextrin (β-CD). This work evidenced the thermal protection provided by the encapsulating agent, which showed a thermal stabilization of cy3glc, resulting in a decrease of the degradation rate constant (k) and in several alterations in the cy3glc-β-CD DSC thermogram. According to the authors, anthocyanin-loaded β-CD could potentially carry and stabilize anthocyanins, improving their bioavailability, which could be an advantage for efficient utilization in food systems.

All the showed works evidenced the importance of encapsulation to maintain the properties of the studied compounds, allowing their application in different situations. It is evident that cyclodextrin is the most widely used encapsulating agent, as it provides the formation of inclusion complexes with interesting properties.

6. A case study

Eugenol is an essential oil with excellent antimicrobial properties. However, because it is thermosensitive, it has restricted the applicability in processes that require high temperatures. Piletti et al. [19] proposed a method for protecting this oil by encapsulating it in β-cyclodextrin. The authors evaluated the encapsulation of eugenol molecules by means of lyophilization and later evaluated the antimicrobial activity of the complex (eugenol-β-cyclodextrin) against the bacteria *Escherichia coli* and *Staphylococcus aureus* through the diffusion technique in agar. The authors also investigated the thermal and morphological properties of the complex. When evaluating the antimicrobial activity complexes obtained with different concentrations of eugenol (9.68, 10.90, 17.08 mmol L^{-1}) against *Escherichia coli* and *Staphylococcus aureus* bacteria, the authors verified that the antimicrobial activity was maintained even after encapsulation.

However, when using cyclodextrin as an encapsulating agent, the idea was that there would be thermal protection of the essential oil, ensuring that the compound property of interest (antimicrobial activity) was not altered. This was confirmed by the heat treatment of the eugenol-β-cyclodextrin complex in a furnace maintained at 80°C (temperature approximately twice the temperature of free eugenol volatilization) for 2 h and subsequent re-evaluation of antimicrobial activity. **Figures 2** and **3** illustrate the antimicrobial capacity of the complex after the heat treatment against *E. coli* and *S. aureus* bacteria, respectively.

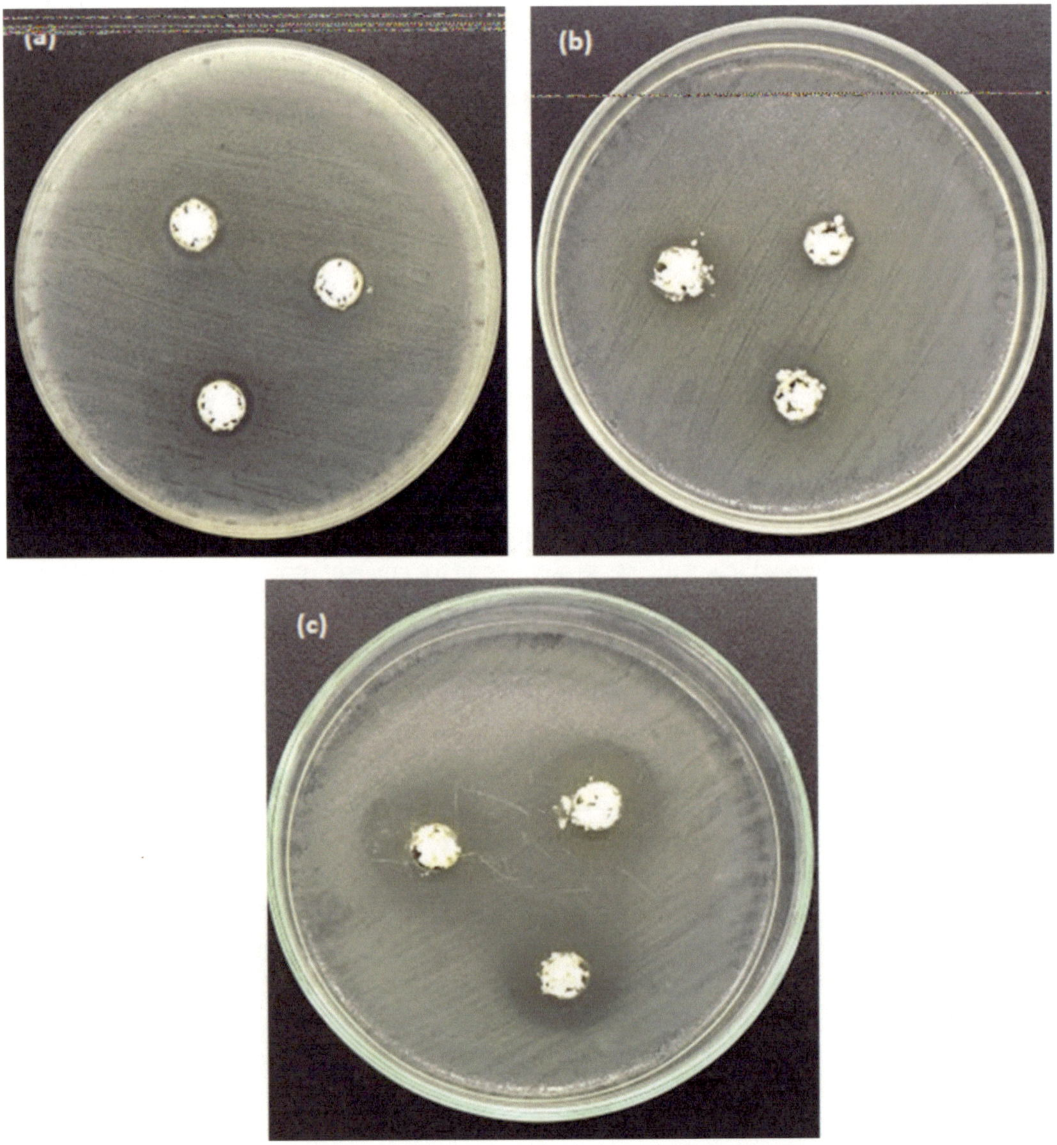

Figure 2. Agar diffusion tests of a eugenol-β-cyclodextrin complex synthesized using different eugenol concentrations in the reaction solution and thermally treated at 80°C for 2 h for *E. coli* bacteria. Eugenol concentration: (a) 9.68; (b) 10.90; and (c) 17.08 mmol L^{-1}.

http://dx.doi.org/10.5772/intechopen.73568

The encapsulated eugenol molecules were thermally protected, remained in the complexes after heat treatment and manifested the antimicrobial activity of this essential oil. Therefore, encapsulation using β-cyclodextrin is a promising method to protect eugenol, preserving its antibacterial action when it is used under conditions higher than its volatilization temperature.

All these studies show the efficiency of β-cyclodextrin as an encapsulating agent and demonstrate its high thermal protection capacity for bioactive natural compounds, which are highly unstable, without damaging the biological property of interest in these compounds.

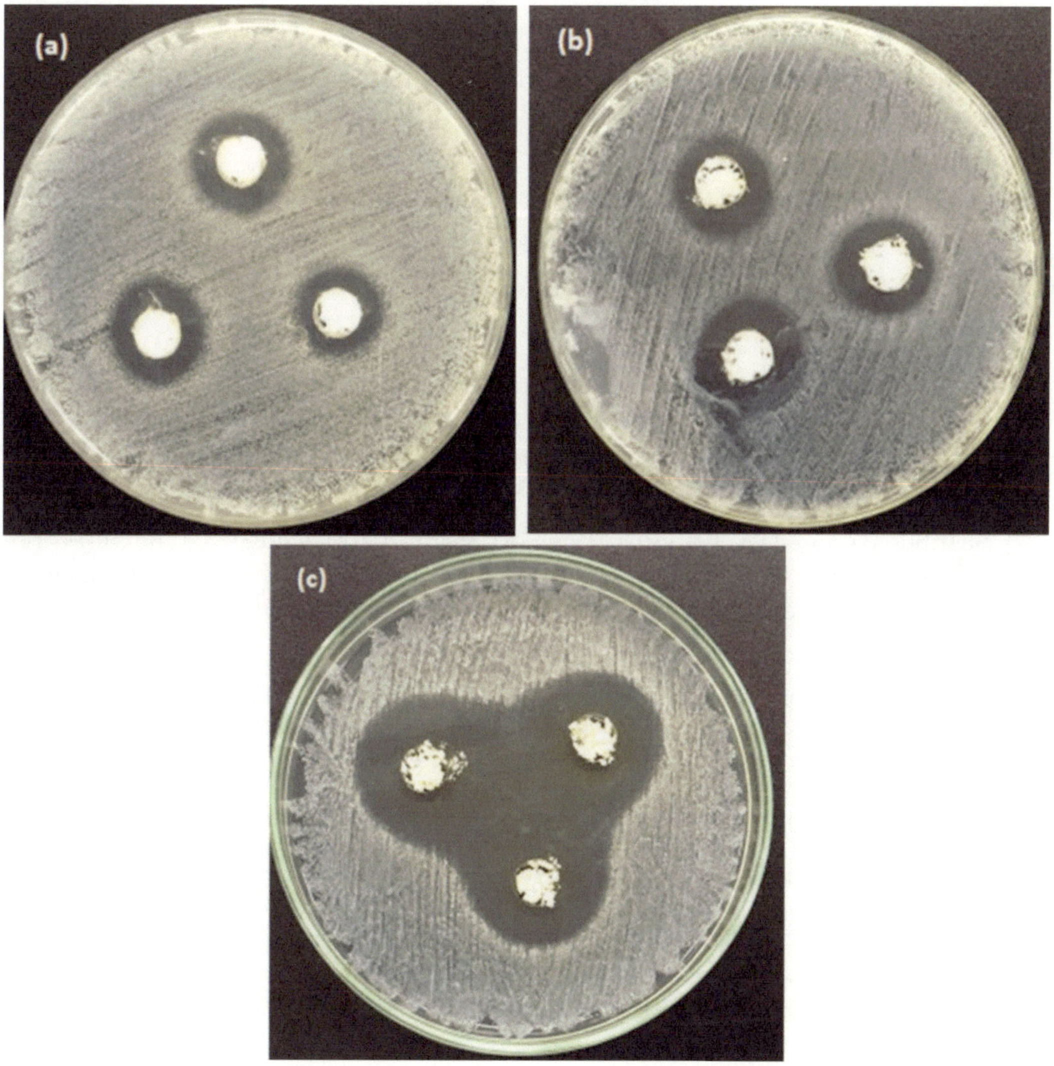

Figure 3. Agar diffusion tests of a eugenol-β-cyclodextrin complex synthesized using different eugenol concentrations in the reaction solution and thermally treated at 80°C for 2 h against *S. aureus* bacteria. Eugenol concentration: (a) 9.68; (b) 10.90; and (c) 17.08 mmol L^{-1}.

Thus, the encapsulation of essential oils using β-cyclodextrin is an alternative to promote the use of these biocomposites as additives, boosting the development of functional materials, providing new applications for them in the diverse areas, such as medical, pharmaceutical, cosmetic, and food, combining the use of technology with the appreciation of natural raw materials.

Author details

Ana Paula Capelezzo[1], Laura Cassol Mohr[1], Francieli Dalcanton[2], Josiane Maria Muneron de Mello[1,2] and Márcio Antônio Fiori[1,2*]

*Address all correspondence to: fiori@unochapeco.edu.br

1 Post-Graduate Program in Environmental Science, Universidade Comunitária da Região de Chapecó (UNOCHAPECÓ), Chapecó, SC, Brazil

2 Post-Graduate Program in Technology and Management of the Innovation, Universidade Comunitária da Região de Chapecó (UNOCHAPECÓ), Chapecó, SC, Brazil

References

[1] Ayala-Zavala JF, Soto-Valdez H, González-León A, Álvarez-Parrilla E, Martıín-Belloso O, González-Aguilar GA. Microencapsulation of cinnamon leaf (*Cinnamomum zeylanicum*) and garlic (*Allium sativum*) oils in b-cyclodextrin. Journal of Inclusion Phenomena and Macrocyclic Chemistry. 2008;**60**:359-368. DOI: 10.1007/s10847-007-9385-1

[2] Sharifi-Rad J, Sureda A, Tenore GC, Daglia M, Sharifi-Rad M, Valussi M, Tundis R, Sharifi-Rad M, Loizzo MR, Ademiluyi AO, Sharifi-Rad R, Ayatollahi SA, Iriti M. Biological activities of essential oils: from plant chemoecology to traditional healing systems. Molecules. 2017;**22**:1-55. DOI: 10.3390/molecules22010070

[3] Burt S. Essential oils: Their antibacterial properties and potential applications in foods—A review. International Journal of Food Microbiology. 2004;**94**:223-253. DOI: 10.1016/j.ijfoodmicro.2004.03.022

[4] Hyldgaard M, Mygind T, Meyer RL. Essential oils in food preservation: mode of action, synergies, and interactions with food matrix components. Frontiers in Microbiology. 2012;**3**:1-24. DOI: 10.3389/fmicb.2012.00012

[5] El Asbahani A, Miladi K, Badri W, Sala M, Aït Addi EH, Casabianca H, El Mousadik A, Hartmann D, Jilale A, Renaud FNR, Elaissari A. Essential oils: From extraction to encapsulation. International Journal of Pharmaceutics. 2015;**483**(1-2):220-243. DOI: 10.1016/j.ijpharm.2014.12.069

[6] Codevilla CF, Bazana MT, Silva CB, Barin JS, Menezes CR. Nanoestruturas contendo compostos bioativos extraídos de plantas. Ciência e Natura. 2015;**37**:142-151. DOI: 10.5902/2179-460X19743

[7] Sagiri SS, Anis A, Pal K. A review on encapsulation of vegetable oils: Strategies, preparation methods and applications. Polymer-Plastics Technology and Engineering. 2016;**55**(3):291-311. DOI: 10.1080/03602559.2015.1050521

[8] Bakry AM, Abbas S, Ali B, Majeed H, Abouelwafa MY, Mousa A, Liang L. Microencapsulation of oils: A comprehensive review of benefits, techniques, and applications. Comprehensive Reviews in Food Science and Food Safety. 2016;**15**:143-182. DOI: 10.1111/1541-4337.12179

[9] Nazzaro F, Fratianni F, De Martino L, Coppola R, De Feo V. Effect of essential oils on pathogenic bacteria. Pharmaceuticals. 2013;**6**:1451-1474. DOI: 10.3390/ph6121451

[10] Mejri J, Abderrabba M, Mejri M. Chemical composition of the essential oil of *Ruta chalepensis* L: Influence of drying, hydro-distillation duration and plant parts. Industrial Crops and Products. 2010;**32**(3):671-673. DOI: 10.1016/j.indcrop.2010.05.002

[11] Haddouchi F, Chaouche TM, Zaouali Y, Ksouri R, Attou A, Benmansour A. Chemical composition and antimicrobial activity of the essential oils from four Ruta species growing in Algeria. Food Chemistry. 2013;**141**(1):253-258. DOI: 10.1016/j.foodchem.2013.03.007

[12] Ehlert PAD, Ming LC, Marques MOM, Fenandes DM, Rocha WA, Luz JMQ, Silva RF. Influência do horário de colheita sobre o rendimento e composição do óleo essencial de erva-cidreira brasileira [Lippia alba (Mill.) N. E. Br.]. Revista Brasileira de Plantas Medicinais. 2013;**15**:72-77. DOI: 10.1590/S1516-05722013000100010

[13] Russo A, Formisano C, Rigano D, Senatore F, Sebastiano D, Cardile V, Rossellie S, Bruno M. Chemical composition and anticancer activity of essential oils of Mediterranean sage (*Salvia officinalis* L.) grown in different environmental conditions. Food and Chemical Toxicology. 2013;**55**:42-47. DOI: 10.1016/j.fct.2012.12.036

[14] Jordán MJ, Lax V, Rota MC, Loráns S, Sotomayor JA. Effect of bioclimatic area on the essential oil composition and antibacterial activity of *Rosmarinus officinalis* L. Food Control. 2013;**30**(2):463-468. DOI: 10.1016/j.foodcont.2012.07.029

[15] Veloso RA, Castro HG, Barbosa LCA, Cardoso DP, Chagas Júnior AF, Scheidt GN. Teor e composição do óleo essencial de quatro acessos e duas cultivares de manjericão (*Ocimum basilicum* L.). Revista Brasileira de Plantas Medicinais. 2014;**16**:364-371. DOI: 10.1590/1983-084X/12_180

[16] Luz JMQ, Silva SM, Habber LL, Marquez MOM. Essential oil of *Melissa officinalis* L. at differents seasons, systems of planting and fertilizations. Revista Brasileira de Plantas Medicinais. 2014;**16**:552-560. DOI: 10.1590/1983-084X/11_130

[17] López-Mejía OA, López-Malo A, Palou E. Antioxidant capacity of extracts from amaranth (*Amaranthus hypochondriacus* L.) seeds or leaves. Industrial Crops and Products. 2014;**53**:55-59. DOI: 10.1016/j.indcrop.2013.12.017

[18] Jiang H, Wang J, Song L, Cao X, Yao X, Tang F, Yue Y. GC×GC-TOFMS analysis of essential Ooils composition from leaves, twigs and deeds of *Cinnamomum camphora* L. Presl and their insecticidal and repellent activities. Molecules. 2016;**21**:423. DOI: 10.3390/molecules21040423

[19] Piletti R, Bugiereck AM, Pereira AT, Gussati E, Dal Magro J, Mello JMM, Dalcanton F, Ternus RZ, Soares C, Riella HG, Fiori MA. Microencapsulation of eugenol molecules by β-cyclodextrine as a thermal protection method of antibacterial action. Materials Science and Engineering C. 2017;**75**:259-271. DOI: 10.1016/j.msec.2017.02.075

[20] Scallan E, Hoekstra RM, Angulo FJ, Tauxe RV, Widdowson MA, Roy SL, Jones JL, Griffin PM. Foodborne illness acquired in the United States-major pathogens. Emerging Infectious Diseases. 2011;**17**:7-15. DOI: 10.3201/eid1701.P11101

[21] Hill LE, Gomes C, Taylor TM. Characterization of beta-cyclodextrin inclusion complexes containing essential oils (trans-cinnamaldehyde, eugenol, cinnamon bark, and clove bud extracts) for antimicrobial delivery applications. LWT—Food Science and Technology. 2013;**51**(1):86-93. DOI: 10.1016/j.lwt.2012.11.011

[22] Teixeira B, Marques A, Ramos C, Neng NR, Nogueira JMF, Saraiva JA, Nunes ML. Chemical composition and antibacterial and antioxidant properties of commercial essential oils. Industrial Crops and Products. 2013;**43**:587-595. DOI: 10.1016/j.indcrop.2012.07.069

[23] Pesavento G, Calonico C, Bilia AR, Barnabei M, Calesini F, Addona R, Mencarelli L, Carmagnini L, Di Martino MC, Lo Nostro A. Antibacterial activity of Oregano, Rosmarinus and Thymus essential oils against Staphylococcus aureus and Listeria monocytogenes in beef meatballs. Food Control. 2015;**54**:188-199. DOI: 10.1016/j.foodcont.2015.01.045

[24] Affonso RS, Rennó MN, GBCA S, TCC F. Aspectos químicos e biológicos do óleo essencial de Cravo da Índia. Revista Virtual de Química. 2012;**4**:46-161. ISSN: 1984-6835

[25] Knezevic P, Aleksic V, Simin N, Svircev E, Petrovic A, Mimica-Dukic N. Antimicrobial activity of *Eucalyptus camaldulensis* essential oils and their interactions with conventional antimicrobial agents against multi-drug resistant *Acinetobacter baumannii*. Journal of Ethnopharmacology. 2016;**178**:125-136. DOI: 10.1016/j.jep.2015.12.008

[26] Chorianopoulos N, Kalpoutzakis E, Aligiannis N, Mitaku S, Nychas G-J, Haroutounian SA. Essential oils of Satureja, Origanum, and Thymus species: Chemical composition and antibacterial activities against foodborne pathogens. Journal of Agricultural of Food Chemistry. 2004;**52**:8261-8267. DOI: 10.1021/jf049113i

[27] Kim SI, Yoon JS, Jung JW, Hong K, Ahn YJ, Kwon HW. Toxicity and repellency of origanum essential oil and its components against *Tribolium castaneum* (Coleoptera: Tenebrionidae) adults. Journal of Asia Pacific Entomology. 2010;**13**(4):369-373. DOI: 10.1016/j.aspen.2010.06.011

[28] Muñoz-Bonilla A, Fernández-García M. Polymeric materials with antimicrobial activity. Progress in Polymer Science. 2012;**37**(2):281-339. DOI: 10.1016/j.progpolymsci.2011.08.005

[29] Calsamiglia S, Busquet M, Cardozo PW, Castillejos L, Ferret A. Invited review: Essential oils as modifiers of rumen microbial fermentation. Journal of Dairy Science. 2007;**90**(6):2580-2595. DOI: 10.3168/jds.2006-644

[30] Ravichandran M, Hettiarachchy NS, Ganesh V, Ricke SC, Singh S. Enhancement of antimicrobial activities of naturally occurring phenolic compounds by nanoscale delivery against Listeria monocytogenes, Escherichia coli O157:H7 and Salmonella Typhimurium in broth and chicken meat system. Journal of Food Safety. 2011;**31**:462-471. DOI: 10.1111/j.1745-4565.2011.00322.x

[31] Bajpai VK, Baek KH, Kang SC. Control of Salmonella in foods by using essential oils: A review. Food Research International. 2012;**45**(2):722-734. DOI: 10.1016/j.foodres.2011.04.052

[32] Nikolic M, Glamoclija J, Ferreira ICFR, Calhelha RC, Fernandes A, Markovic T, Markovic D, Giweli A, Sokovic M. Chemical composition, antimicrobial, antioxidant and antitumor activity of *Thymus serpyllum* L., Thymus algeriensis Boiss. and Reut and *Thymus vulgaris* L. essential oils. Industrial Crops and Products. 2014;**52**:183-190. DOI: 10.1016/j.indcrop.2013.10.006

[33] Abdollahzadeh E, Rezaei M, Hosseini H. Antibacterial activity of plant essential oils and extracts: the role of thyme essential oil, nisin, and their combination to control Listeria monocytogenes inoculated in minced fish meat. Food Control. 2014;**35**(1):177-183. DOI: 10.1016/j.foodcont.2013.07.004

[34] Soares BV, Neves LR, Oliveira MSB, Chaves FCM, Dias MKR, Chagas EC, Tavares-Dias M. Antiparasitic activity of the essential oil of Lippia alba on ectoparasites of Colossoma macropomum (tambaqui) and its physiological and histopathological effects. Aquaculture. 2016;**452**:107-114. DOI: 10.1016/j.aquaculture.2015.10.029

[35] Meira CS, Menezes LRA, Santos TB, Macedo TS, Fontes JEN, Costa EV, Pinheiro MLB, Silva TB, Guimarães ET, Soares MBP. Chemical composition and antiparasitic activity of essential oils from leaves of *Guatteria friesiana* and *Guatteria pogonopus* (Annonaceae). Journal of Essential Oil Research. 2017;**29**(2):156-162. DOI: 10.1080/10412905.2016.1210041

[36] Pilau MR, Alves SH, Weiblen R, Arenhart S, Cueto AP, Lovato LT. Antiviral activity of the Lippia graveolens (Mexican oregano) essential oil and its main compound carvacrol against human and animal viruses. Brazilian Journal of Microbiology. 2011;**42**(4):1616-1624. DOI: 10.1590/S1517-83822011000400049

[37] Dunkic V, Vuko E, Bezic N, Kremer D, Ruscic M. Composition and antiviral activity of the essential oils of *Eryngium alpinum* and *E. amethystinum*. Chemistry & Biodiversity. 2013;**10**(10):1894-1902. DOI: 10.1002/cbdv.201300061

[38] Mota KSL, Pereira FO, Oliveira WA, Lima IO, Lima EO. Antifungal activity of *Thymus vulgaris* L. essential Ooil and its constituent phytochemicals against Rhizopus oryzae: Interaction with Ergosterol. Molecules. 2012;**17**(12):14418-14433. DOI: 10.3390/molecules171214418

[39] Zeng H, Chen X, Liang J. In vitro antifungal activity and mechanism of essential oil from fennel (*Foeniculum vulgare* L.) on dermatophyte species. Journal of Medical Microbiology. 2015;**64**:93-103. DOI: 10.1099/jmm.0.077768-0

[40] Gakuubi MM, Maina AW, Wagacha JM. Antifungal activity of essential oil of *Eucalyptus camaldulensis* Dehnh. against selected *Fusarium* spp. International Journal of Microbiology. 2017;**2017**:1-7. DOI: 10.1155/2017/8761610

[41] Cherrat L, Espina L, Bakkali M, García-Gonzalo D, Pagán R, LaglaouiA. Chemical composition and antioxidant properties of *Laurus nobilis* L. and *Myrtus communis* L. essential oils from Morocco and evaluation of their antimicrobial activity acting alone or in combined processes for food preservation. Journal of the Science of Food and Agriculture. 2014;**94**(6):1197-1204. DOI: 10.1002/jsfa.6397

[42] Aleksic V, Knezevic P. Antimicrobial and antioxidative activity of extracts and essential oils of *Myrtus communis* L. Microbiological Research. 2014;**169**(4):240-254. DOI: 10.1016/j.micres.2013.10.003

[43] Alves-Silva JM, Santos SMD, Pintado ME, Pérez-Álvarez JA, Fernández-López JL, Viuda-Martos M. Chemical composition and in vitro antimicrobial, antifungal and antioxidant properties of essential oils obtained from some herbs widely used in Portugal. Food Control. 2013;**32**(2):371-378. DOI: 10.1016/j.foodcont.2012.12.022

[44] Emiroglu ZK, Yemis GP, Skun BK, Candogan K. Antimicrobial activity of soy edible films incorporated with thyme and oregano essential oils on fresh ground beef patties. Meat Science. 2010;**86**:283-288. DOI: 10.1016/j.meatsci.2010.04.016

[45] Venturini CG, JNC M, Machado G. Propriedades e aplicações recentes das ciclodextrinas. Química Nova. 2008;**31**:360-368. DOI: 10.3109/02652041003681398

[46] Seo EJ, Min SG, Choi MJ. Release characteristics of freeze-dried eugenol encapsulated with beta-cyclodextrin by molecular inclusion method. Journal of Microencapsulation. 2010;**27**:496-505. DOI: 10.3109/02652041003681398

[47] Turasan H, Sarin S, Sumnu G. Encapsulation of rosemary essential oil. LWT—Food Science and Technology. 2015;**64**(1):112-119. DOI: 10.1016/j.lwt.2015.05.036

[48] Calo JR, Crandall PG, O'Bryan CA, Ricke SC. Essential oils as antimicrobials in food systems—A review. Food Control. 2015;**54**:111-119. DOI: 10.1016/j.foodcont.2014.12.040

[49] Turek C, Stintzing FC. Stability of essential oils: A review. Comprehensive Reviews in Food Science and Food Safety. 2013;**12**:40-53. DOI: 10.1111/1541-4337.12006

[50] Rowshan V, Bahmanzadegan A, Saharkhiz MJ. Influence of storage conditions on the essential oil composition of *Thymus daenensis* Celak. Industrial Crops and Products. 2013;**49**:97-101. DOI: 10.1016/j.indcrop.2013.04.029

[51] Turek C, Stintzing FC. Impact of different storage conditions on the quality of selected essential oils. Food Research International. 2012;**46**:341-353. DOI: 10.1016/j.foodres.2011.12.028

[52] Nori MP, Favaro-Trindade CS, Alencar SM, Thomazini M, Balieiro JCC, Castillo CJC. Microencapsulation of propolis extract by complex coacervation. LWT—Food Science and Technology. 2011;**44**(2):429-435. DOI: 10.1016/j.lwt.2010.09.010

[53] Hosseini SF, Zandi M, Rezaei M, Farahmangdghavi F. Two-step method for encapsulation of oregano essential oil in chitosan nanoparticles: Preparation, characterization and in vitro release study. Carbohydrate Polymers. 2013;**95**:50-56. DOI: 10.1016/j.carbpol.2013.02.031

[54] Beirão-Da-Costa S, Duarte C, Bourbon AI, Pinheiro NC, Januário MIN, Vicente AA, Beirão-Da-Costa ML, Delgadillo I. Inulin potential for encapsulation and controlled delivery of Oregano essential oil. Food Hydrocolloids. 2013;**33**(2):199-206. DOI: 10.1016/j.foodhyd.2013.03.009

[55] Choi MJ, Soottitantawat A, Nuchuchua O, Min SG, Ruktanonchai U. Physical and light oxidative properties of eugenol encapsulated by molecular inclusion and emulsion–diffusion method. Food Research International. 2009;**42**(1):148-156. DOI: 10.1016/j.foodres.2008.09.011

[56] Singh M, Dua JS, Menra M, Soni M, Prasad DN. Microencapsulation and its various aspects: a review. International Journal of Advanced Research. 2016;**4**:2094-2108

[57] Wu Y, Luo Y, Wang Q. Antioxidant and antimicrobial properties of essential oils encapsulated in zein nanoparticles prepared by liquid-liquid dispersion method. LWT—Food Science and Technology. 2012;**48**(2):283-290. DOI: 10.1016/j.lwt.2012.03.027

[58] Kamimura JA, Santos EH, Hill LE, Gomes CL. Antimicrobial and antioxidant activities of carvacrol microencapsulated in hidroxypropyl-beta-ciclodextryn. LWT—Food Science and Technology. 2014;**57**:701-709. DOI: 10.1016/j.lwt.2014.02.014

[59] Pinho E, Grootveld M, Soares G, Henriques M. Cyclodextrins as encapsulation agents for plant bioactive compounds. Carbohydrate Polymers. 2014;**101**:121-135. DOI: 10.1016/j.carbpol.2013.08.078

[60] Wang S, Marcone MF, Barbut S, Lim LT. Fortification of dietary biopolymers-based packaging material with bioactive plant extracts. Food Research International. 2012;**49**(1):80-91. DOI: 10.1016/j.foodres.2012.07.023

[61] Vemmer M, Patel AV. Review of encapsulation methods suitable for microbial biological control agents. Biological Control. 2013;**67**(3):380-389. DOI: 10.1016/j.biocontrol.2013.09.003

[62] Martins IM, Barreiro MF, Coelho M, Rodrigues AE. Microencapsulation of essential oils with biodegradable polymeric carriers for cosmetic applications. Chemical Engineering Journal. 2014;**245**:191-200. DOI: 10.1016/j.cej.2014.02.024

[63] Gharsallaoui A, Roudaut G, Chambin O, Voilley A, Saurel R. Applications of spray-drying in microencapsulation of food ingredients: An overview. Food Research International. 2007;**40**(9):1107-1121. DOI: 10.1016/j.foodres.2007.07.004

[64] Silva PT, Fries LCM, Menezes CR, Holkem AT, Schwan CL, Wigmann EF, Bastos JO, Silva CB. Microencapsulation: concepts, mechanisms, methods and some applications in food technology. Ciência Rural. 2014;**44**:1304-1311. DOI: 10.1590/0103-8478cr20130971

[65] Suganya V, Anuradha V. Microencapsulation and Nanoencapsulation: A review. International Journal of Pharmaceutical and Clinical Research. 2017;**9**:233-239. DOI: 10.25258/ijpcr.v9i3.8324

[66] Jyothi NVN, Prasanna PM, Sakarkar SN, Prabha KS, Ramaiah PS, Srawan GY. Microencapsulation techniques, factors influencing encapsulation efficiency. Journal of Microencapsulation. 2010;**27**(3):187-197. DOI: 10.3109/02652040903131301

[67] Nesterenko A, Alric I, Silvestre F, Durrieu V. Vegetable proteins in microencapsulation: A review of recent interventions and their effectiveness. Industrial Crops and Products. 2013;**42**:469-479. DOI: 10.1016/j.indcrop.2012.06.035

[68] Gaonkar A, Niraj V, Khare A, Sobel R, editors. Microencapsulation in the Food Industry: A Practical Implementation Guide. 1st ed. USA: Elsevier; 2014. 590 p

[69] Nazzaro F, Orlando P, Fratianni F, Coppola R. Microencapsulation in food science and biotechnology. Current Opinion in Biotechnology. 2012;**23**(2):182-186. DOI: 10.1016/j.copbio.2011.10.001

[70] Fathi M, Mozafari MR, Mohebbi M. Nanoencapsulation of food ingredients using lipid based delivery systems. Trends in Food Science & Technology. 2012;**23**(1):13-27. DOI: 10.1016/j.tifs.2011.08.003

[71] Dalmoro A, Barba AA, Lamberti G, D'Amore M. Intensifying the microencapsulation process: Ultrasonic atomization as an innovative approach. European Journal of Pharmaceutics and Biopharmaceutics. 2012;**80**(3):471-477. DOI: 10.1016/j.ejpb.2012.01.006

[72] Risch SJ, Reineccius GA. Encapsulation and Controlled Release of Food Ingredients. Washington, DC: American Chemical Society (ACS); 1995. 214 p. DOI: 10.1021/bk-1995-0590

[73] Roccia P, Martínez ML, Llabot JM, Ribotta PD. Influence of spray-drying operating conditions on sunflower oil powder qualities. Powder Technology. 2014;**254**:307-313. DOI: 10.1016/j.powtec.2014.01.044

[74] Turchiuli C, Munguia MTJ, Sanchez MH, Ferre HC, Dumoulin E. Use of different supports for oil encapsulation in powder by spray drying. Powder Technology. 2014;**255**:103-108. DOI: 10.1016/j.powtec.2013.08.026

[75] Carneiro HCF, Tonon RV, Grosso CRF, Hubinger MD. Encapsulation efficiency and oxidative stability of flaxseed oil microencapsulated by spray drying using different combinations of wall materials. Journal of Food Engineering. 2013;**115**(4):443-451. DOI: 10.1016/j.jfoodeng.2012.03.033

[76] Gallo L, Llabot JM, Allemandi D, Bucalá V, Piña J. Influence of spray-drying operating conditions on *Rhamnus purshiana* (*Cáscara sagrada*) extract powder physical properties. Powder Technology. 2011;**208**(1):205-214. DOI: 10.1016/j.powtec.2010.12.021

[77] Mahdavi SA, Jafari SM, Ghorbani M, Assadpoor E. Spray-drying microencapsulation of anthocyanins by natural biopolymers: A review. Drying Technology. 2014;**32**(5):509-518. DOI: 10.1080/07373937.2013.839562

[78] Fernandes RVB, Carmo EL, Borges SV, Botrel DA, Silva YF, Souza HJB. Comportamento de óleo essencial de alecrim microencapsulado por spray drying em diferentes umidades relativas. Ciência Agrícola. 2016;**14**(1):73-82. DOI: 0103-8699

[79] Goñi ML, Gañán NA, Strumia MC, Martini RE. Eugenol-loaded LLDPE films with antioxidant activity by supercritical carbon dioxide impregnation. Journal of Supercritical Fluids. 2016;**111**:28-35. DOI: 10.1016/j.supflu.2016.01.012

[80] Oliveira DA, Mezzomo N, Gomes C, Ferreira SRS. Encapsulation of passion fruit seed oil by means of supercritical antisolvent process. Journal of Supercritical Fluids. 2017;**129**:96-105. DOI: 10.1016/j.supflu.2017.02.011

[81] Gallardo G, Guida L, Martinez V, López MC, Bernhardt D, Blasco R, Pedroza-Islas R, Hermida LG. Microencapsulation of linseed oil by spray drying for functional food application. Food Research International. 2013;**52**(2):473-482. DOI: 10.1016/j.foodres.2013.01.020

[82] Sutaphanit P, Chitprasert P. Optimisation of microencapsulation of holy basil essential oil in gelatin by response surface methodology. Food Chemistry. 2014;**150**:313-320. DOI: 10.1016/j.foodchem.2013.10.159

[83] Xiao Z, Liu W, Zhu G, Zhou R, Niu Y. Production and characterization of multinuclear microcapsules encapsulating lavender oil by complex coacervation. Flavour and Fragrance Journal. 2013;**29**(3):166-172. DOI: 10.1002/ffj.3192

[84] Jun-Xia X, Hai-Yan Y, Jian Y. Microencapsulation of sweet orange oil by complex coacervation with soybean protein isolate/gum Arabic. Food Chemistry. 2011;**125**(4):1267-1272. DOI: 10.1016/j.foodchem.2010.10.063

[85] Peng C, Zhao S-Q, Zhang J, Huang G-Y, Chen L-Y, Zhao F-Y. Chemical composition, antimicrobial property and microencapsulation of Mustard (*Sinapis alba*) seed essential oil by complex coacervation. Food Chemistry. 2014;**165**:560-568. DOI: 10.1016/j.foodchem.2014.05.126

[86] Dima C, Cotârlet M, Alexe P, Dima S. Microencapsulation of essential oil of pimento [*Pimenta dioica* (L) Merr.] by chitosan/k-carrageenan complex coacervation method. Innovative Food Science and Emerging Technologies. 2014;**22**:203-211. DOI: 10.1016/j.ifset.2013.12.020

[87] Hwisa NT, Katakam P, Chandu BR, Adiki SK. Solvent evaporation techniques as promising advancement in microencapsulation. Vedic Research International Biological Medicinal Chemistry. 2013;**1**(1):8-22. DOI: 10.14259/bmc.v1i1.29

[88] Patil DK, Agrawal DS, Mahire RR, More DH. Synthesis, characterization and controlled release studies of ethyl cellulose microcapsules incorporating essential oil using an emulsion solvent evaporation method. American Journal of Essential Oils and Natural Products. 2016;**4**(1):23-31. DOI: 2321 9114

[89] Soliman EA, El-Moghazy AY, El-Din MSM, Massoud MA. Microencapsulation of essential oils within alginate: formulation and in vitro evaluation of antifungal activity. Journal of Encapsulation and Adsorption Sciences. 2013;**3**(1):48-55. DOI: 10.4236/jeas.2013.31006

[90] Calvo P, Castaño AL, Lozano M, González-Gómez D. Influence of the microencapsulation on the quality parameters and shelf-life of extra-virgin olive oil encapsulated in the presence of BHT and different capsule wall components. Food Research International. 2012;**45**(1):256-261. DOI: 10.1016/j.foodres.2011.10.036

[91] Ezhilarasi PN, Indrani D, Jena BS, Anandharamakrishnan C. Freeze drying technique for microencapsulation of Garcinia fruit extract and its effect on bread quality. Journal of Food Engineering. 2013;**117**(4):513-520. DOI: 10.1016/j.jfoodeng.2013.01.009

[92] Almeida AP, Rodríguez-Rojo S, Serra AT, Vila-Real H, Simplicio AL, Delgadilho I, Da Costa SB, Da Costa LB, Nogueira ID, Duarte CMM. Microencapsulation of oregano essential oil in starch-based materials using supercritical fluid technology. Innovative Food Science & Emerging Technologies. 2013;**20**:140-145. DOI: 10.1016/j.ifset.2013.07.009

[93] Varona S, Rodríguez-Rojo A, Martín A, Cocero MJ, Duarte CMM. Supercritical impregnation of lavandin (*Lavandula hybrida*) essential oil in modified starch. The Journal of Supercritical Fluids. 2011;**58**(2):313-319. DOI: 10.1016/j.supflu.2011.06.003

[94] Ciftci ON, Temelli F. Formation of solid lipid microparticles from fully hydrogenated canola oil using supercritical carbon dioxide. Journal of Food Engineering. 2016;**178**:137-144. DOI: 10.1016/j.jfoodeng.2016.01.014

[95] Szejtli J. Introduction and general overview of cyclodextrin chemistry. Chemical Reviews. 1998;**98**(5):1743-1753. DOI: 10.1021/cr970022c

[96] Teixeira B, Marques A, Pires C, Ramos C, Batista I, Saraiva JA, Nunes ML. Characterization of fish protein films incorporated with essential oils of clove, garlic and origanum: Physical, antioxidant and antibacterial properties. LWT—Food Science and Technology. 2014;**59**(1):533-539. DOI: 10.1016/j.lwt.2014.04.024

[97] Aguiar UN, De Lima SG, Rocha MS, De Freitas RM, Oliveira TM, Silva RM, Moura LCB, De Almeida LTG. Preparação e caracterização do complexo de inclusão do óleo essencial de croton zehntneri com β-ciclodextrina. Química Nova. 2014;**37**(1):50-55. DOI: 10.1590/S0100-40422014000100010

[98] Lyra MAM, Alves LDS, Fontes DAF, Soares-Sobrinho JL, Rolim-Neto PJ. Ferramentas analíticas aplicadas à caracterização de complexos de inclusão fármacociclodextrina. Revista de Ciências Farmacêuticas Básica e Aplicada. 2010;**31**(2):117-124. DOI: 1808-4532

[99] Cevallos PAP, Buera MP, Elizalde BE. Encapsulation of cinnamon and thyme essential oils components (cinnamaldehyde and thymol) in β-cyclodextrin: Effect of interactions with water on complex stability. Journal of Food Engineering. 2010;**99**(1):70-75. DOI: 10.1016/j.jfoodeng.2010.01.039

[100] Rakmai J, Cheirsilp B, Mejuto JC, Torrado-Agrasar A, Simal-Gándara J. Physico-chemical characterization and evaluation of bio-efficacies of black pepper essential oil encapsulated in hydroxypropyl-betacyclodextrin. Food Hydrocolloids. 2017;**65**:157-164. DOI: 10.1016/j.foodhyd.2016.11.014

[101] Babaoglu HC, Bayrak A, Ozdemir N, Ozgun N. Encapsulation of clove essential oil in hydroxypropyl beta-cyclodextrin for characterization, controlled release, and antioxidant activity. Journal of Food Processing and Preservation. Forthcoming. DOI: 10.1111/jfpp.13202

[102] Marques HMC. A review on cyclodextrin encapsulation of essential oils and volatiles. Flavour and Fragrance Journal. 2010;**25**(5):313-326. DOI: 10.1002/ffj.2019

[103] Wang J, Cao Y, Sun B, Wang C. Physicochemical and release characterisation of garlic oil-ß-cyclodextrin inclusion complexes. Food Chemistry. 2011;**127**(4):1680-1685. DOI: 10.1016/j.foodchem.2011.02.036

[104] Szente L, Szejtli J. Cyclodextrins as food ingredientes. Trends in Food Science and Technology. 2004;**15**(3-4):137-142. DOI: 10.1016/j.tifs.2003.09.019

[105] Bobbio PA, Bobbio FO. Química do processamento de alimentos. 2nd ed. São Paulo: Varela; 1995. 151 p. DOI: 8585519126

[106] Antunes MD, Dannenberg GS, Fiorentini AM, Pinto VZ, Lim L-T, Zavareze ER, Dias ARG. Antimicrobial electrospun ultrafine fibers from zein containing eucalyptus essential oil/cyclodextrin inclusion complex. International Journal of Biological Macromolecules. 2017;**104**:874-882. DOI: 10.1016/j.ijbiomac.2017.06.095

[107] Hattrem MN, Kristiansen KA, Aachmann FL, Dille MJ, Draget KI. Ibuprofen-in-cyclodextrin-in-W/O/W emulsion—Improving the initial and long-term encapsulation efficiency of a model active ingredient. International Journal of Pharmaceutics. 2015;**487**(1-2):1-7. DOI: 10.1016/j.ijpharm.2015.03.059

[108] Khatibi SA, Misaghi A, Moosavy M, Basti AA, Koohi MK, Khosravi P, Haghirosada F. Encapsulation of Zataria Multiflora Bioss. essential oil into nanoliposomes and in vitro antibacterial activity against Escherichia coli O157:H7. Journal of Food Processing and Preservation. 2017;**41**(3):1-10. DOI: 10.1111/jfpp.12955

[109] Devi KP, Nisha SA, Sakthivel R, Pandian SK. Eugenol (an essencial oil of clove) acts as an antibacterial agent against Samonella typhi by disrupting the cellular membrane. Journal of Ethnopharmacology. 2010;**130**(1):107-115. DOI: 10.1016/j.jep.2010.04.025

[110] Chatterjee D, Bhattacharjee P. Comparative evaluation of the antioxidant efficacy of encapsulated and un-encapsulated eugenol-rich clove extracts in soybean oil: Shelf-life and frying stability of soybean oil. Jounal of Food Engineering. 2013;**117**(4):545-550. DOI: 10.1016/j.jfoodeng.2012.11.016

[111] Machín R, Isasi JR, Vélaz I. Cyclodextrin hydrogels as potential drug delivery systems. Carbohydrate Polymers. 2012;**87**(3):2024-2030. DOI: 10.1016/j.carbpol.2011.10.024

[112] Zhao Y, Sun C, Shi F, Firempong CK, Yu J, Xu X, Zhang W. Preparation, characterization, and pharmacokinetics study of capsaicin via hydroxypropyl-beta-cyclodextrin encapsulation. Pharmaceutical Biology. 2016;**54**(1):130-138. DOI: 10.3109/13880209.2015.1021816

[113] Liu H, Cai X, Wang Y, Chen J. Adsorption mechanism-based screening of cyclodextrin polymers for adsorption and separation of pesticides from water. Water Research. 2011;**45**(11):3499-3511. DOI: 10.1016/j.watres.2011.04.004

[114] Ge X, He J, Qi F, Yang Y, Huang Z, Lu R, Huang L. Inclusion complexation of chloropropham with -cyclodextrin: Preparation, characterization and molecular modeling. Spectrochimica Acta Part A. 2011;**81**:397-403. DOI: 10.1016/j.saa.2011.06.028

[115] Garrido J, Cagide F, Melle-Franco M, Borges F, Garrido EM. Microencapsulation of herbicide MCPA with native b-cyclodextrin and its methyl and hydroxypropyl derivatives: An experimental and theoretical investigation. Journal of Molecular Structure. 2014;**1061**:76-81. DOI: 10.1016/j.molstruc.2013.12.067

[116] Teixeira BN, Ozdemir N, Hill LE, Gomes CL. Synthesis and characterization of nano-encapsulated black pepper oleoresin using hydroxypropyl beta-cyclodextrin for antioxidant and antimicrobial applications. Journal of Food Science. 2013;**78**(12):1913-1920. DOI: 10.1111/1750-3841.12312

[117] Tao F, Hill LE, Peng Y, Gomes CL. Synthesis and characterization of ß-cyclodextrin inclusion complexes of thymol and thyme oil for antimicrobial delivery applications. LWT—Food Science and Technology. 2014;**59**(1):247-255. DOI: 10.1016/j.lwt.2014.05.037

[118] Shao P, Zhang J, Fang Z, Sun P. Complexing of chlorogenic acid with b-cyclodextrins: Inclusion effects, antioxidative properties and potential application in grape juice. Food Hydrocolloids. 2014;**41**:132-139. DOI: 10.1016/j.foodhyd.2014.04.003

[119] Gong L, Li T, Chen F, Duan X, Yuan Y, Zhang D, Jiang Y. An inclusion complex of eugenol into β-cyclodextrin: Preparation, and physicochemical and antifungal characterization. Food Chemistry. 2016;**196**:324-330. DOI: 10.1016/j.foodchem.2015.09.052

[120] Porte A, Porte LHM, De Oliveira LM. Cromatografia gasosa quiral na resolução de enantiômeros envolvidos em flavours de frutas. Química Nova. 2014;**37**(10):1670-1679. DOI: 10.5935/0100-4042.20140262

[121] Santos EH, Kamimura JA, Hill LE, Gomes CL. Characterization of carvacrol beta-cyclodextrin inclusion complexes as delivery systems for antibacterial and antioxidant applications. LWT—Food Science and Technology. 2015;**60**(1):583-592. DOI: 10.1016/j.lwt.2014.08.046

[122] De Souza RC, Valarini Júnior O, Pinheiro KH, Klososki SJ, Pimentel TC, Cardozo Filho L, Barão CE. Prebiotic green tea beverage added inclusion complexes of catechin and β-cyclodextrin: Physicochemical characteristics during storage. LWT—Food Science and Technology. 2017;**85**:212-217. DOI: 10.1016/j.lwt.2017.07.022

[123] Wei Y, Zhang J, Memon AH, Liang H. Molecular model and in vitro antioxidant activity of a water-soluble and stable phloretin/hydroxypropyl-β-cyclodextrin inclusion complex. Journal of Molecular Liquids. 2017;**236**:68-75. DOI: 10.1016/j.molliq.2017.03.098

[124] Thompson DO. Cyclodextrins—enabling excipients: Their present and future use in pharmaceuticals. Critical Reviews in Therapeutic Drug Carrier Systems. 1997;**14**(1):101-104. DOI: 10.1615/CritRevTherDrugCarrierSyst.v14.i1.10

[125] Hirayama F, Uekama K. Cyclodextrin-based controlled drug release system. Advanced Drug Delivery Reviews. 1999;**36**(1):125-141. DOI: 10.1016/S0169-409X(98)00058-1

[126] Rowe RC, Sheskey PJ, Owen SC. Handbook of Pharmaceutical Excipients. 50th ed. American Pharmaceutial Association: London, Chicago; 2006. 918 p

[127] Cunha-Filho MSS, Sá-Barreto LCL. Utilização de ciclodextrinas na formação de complexos de inclusão de interesse farmacêutico. Revista de Ciências Farmacêuticas Básica e Aplicada. 2007;**28**(1):1-9. DOI: 1808-4532

[128] Liang H, Yuan Q, Vriesekoop F, Lv F. Effects of cyclodextrins on the antimicrobial activity of plant-derived essential oil compounds. Food Chemistry. 2012;**135**(3):1020-1027. DOI: 10.1016/j.foodchem.2012.05.054

[129] Antisperger G. New aspects in cyclodextrin toxicology. In: 6th International Symposium on Cyclodextrins; Abril; Chicago. Chicago: Editions de Santé; 1992. p. 227-283

[130] Abarca RL, Rodríguez FJ, Guarda A, Galotto MJ, Bruna JE. Characterization of beta-cyclodextrin inclusion complexes containing an essential oil componente. Food Chemistry. 2016;**196**:968-975. DOI: 10.1016/j.foodchem.2015.10.023

[131] Wang B, Qian X, Shi Y, Yu B, Hong N, Song L, Hu Y. Cyclodextrin microencapsulated ammonium polyphosphate: Preparation and its performance on the thermal, flame retardancy and mechanical properties of ethylene vinyl acetate copolymer. Composites: Part B. 2015;**69**:22-30. DOI: 10.1016/j.compositesb.2014.09.015

[132] Kayaci F, Sen HS, Durgun E, Uyar T. Functional electrospun polymeric nanofibers incorporating geraniol–cyclodextrin inclusion complexes: High thermal stability and enhanced durability of geraniol. Food Research International. 2014;**62**:424-431. DOI: 10.1016/j.foodres.2014.03.033

[133] Hădărugă DI, Hădărugă NG, Costescu CI, David I, Gruia AT. Thermal and oxidative stability of the *Ocimum basilicum* L. essential oil/β-cyclodextrin supramolecular system. Beilstein Journal of Organic Chemistry. 2014;**10**:2809-2820. DOI: 10.3762/bjoc.10.298

[134] Kalogeropoulos N, Yannakopoulou K, Gioxari A, Chiou A, Makris DP. Polyphenol characterization and encapsulation in β-cyclodextrin of a flavonoid-rich *Hypericum perforatum* (St John's wort) extract. LWT—Food Science and Technology. 2010;**43**(6):882-889. DOI: 10.1016/j.lwt.2010.01.016

[135] Hădărugă DI, Ünlüsayin M, Gruia AT, Birău C, Rusu G, Hădărugă NG. Thermal and oxidative stability of Atlantic salmon oil (*Salmo salar* L.) and complexation with β-cyclodextrin. Beilstein Journal of Organic Chemistry. 2016;**12**:179-191. DOI: 10.3762/bjoc.12.20

[136] Li W, Liu X, Yang Q, Zhang N, Du Y, Zhu H. Preparation and characterization of inclusion complex of benzyl isothiocyanate extracted from papaya seed with β-cyclodextrin. Food Chemistry. 2015;**184**:99-104. DOI: 10.1016/j.foodchem.2015.03.091

[137] Zhou Q, Wei X, Dou W, Chou G, Wang Z. Preparation and characterization of inclusion complexes formed between baicalein and cyclodextrins. Carbohydrate Polymers. 2013;**95**:733-739. DOI: 10.1016/j.carbpol.2013.02.038

[138] Vilanova N, Solans C. Vitamin A Palmitate–b-cyclodextrin inclusion complexes: Characterization, protection and emulsification properties. Food Chemistry. 2015;**175**: 529-535. DOI: 10.1016/j.foodchem.2014.12.015

[139] Fernandes A, Rocha MAA, LMNBF S, Brás J, Oliveira J, Mateus N, Freitas V. Blackberry anthocyanins: β-Cyclodextrin fortification for thermal and gastrointestinal stabilization. Food Chemistry. 2018;**245**:426-431. DOI: 10.1016/j.foodchem.2017.10.109

Chapter 8

Advanced Polymer-Surfactant Systems via Self-Assembling

Laura Romero-Zerón and Xingzhi Jiang

Additional information is available at the end of the chapter

http://dx.doi.org/10.5772/intechopen.74618

Abstract

This chapter summarizes the formulation of supramolecular polymers built via noncovalent and β-cyclodextrin (β-CD) host-guest interactions. The self-assembling polymeric (SAP-AP) systems were formulated by mixing associative polymers with an anionic surfactant and β-CD. These SAP-AP systems were characterized by rheological analysis and several techniques to establish their stability under mechanical shear, high ionic strength, and high temperature. The experimental results demonstrate that the SAP-AP systems display enhanced viscoelastic properties, shear stability, superior structural strength, and tolerance to high-salinity brines relative to the corresponding baseline polymers.

Keywords: self-assembling, β-CD host-guest complexations, noncovalent bonding, structural regeneration, self-healing, self-repairing, associating polymers, supramolecular systems, functional polymers

1. Introduction

The chemistry of polymers for applications in enhanced oil recovery (EOR) has been advanced to improve their stability and functionality at elevated temperatures and in brines containing high salinity and hardness concentration (i.e. harsh reservoir conditions). Therefore, a variety of functional moieties have been attached to the polymer structure including: salt-tolerant and hydrolysis-resistant moieties such as allyl sulfonic acid, 2-acrylamido-2-methylpropane sulfonate (AMPS), and/or n-vinyl pyrrolidone (n-VP) monomers; hydrophobic groups like n-alkyl (i.e. $\geq C_6$ carbon numbers) acrylamide, and styrene; ring structures and large-rigid side groups to improve the shear stability such as styrene sulfonic acid, n-alkyl maleimide, acrylamide-base long-chain alkyl acid, and 3-acrylamide-3-methyl butyric acid, among others [1–6].

This work evaluated the generation of advanced polymer-surfactant systems built via noncovalent and host-guest interactions based on β-cyclodextrin that would be stable under harsh reservoir conditions. The aim was to formulate efficient, chemically stable, and cost-effective systems as an alternative to expensive chemical synthesis.

Self-assembly allows the coassembly of two or more types of building blocks, resulting in increasingly structurally complex nanoassemblies that may have physical and chemical properties that are distinct from those of the original monostructures [7]. Host-guest interactions refer to the formation of supramolecular inclusion complexes based on macrocyclic molecules (i.e. host molecules) consisting of two or more entities connected via noncovalent interactions in a highly controlled and cooperative manner. These host-guest inclusions are relatively stable and provide reliable and robust connections for the fabrication of stimuli-responsive supramolecular systems [8]. Cyclodextrins (CDs) are the most used and affordable hosts in the field of inclusion chemistry [7, 9, 10].

Supramolecular β-CD-based polymer systems retain their structural stability and functionality (i.e. self-healing) after exposure to externally applied stimuli or shear forces increasing the life span of these materials [8, 7]. The self-healing capability is highly dependent on the noncovalent connections in the polymer backbone and on the decomplexation and complexation of the supramolecular system [7, 8, 11, 12]. Therefore, supramolecular chemistry built on weak and reversible noncovalent interactions has emerged as a powerful and versatile strategy to design and fabricate materials with extraordinary reversibility and adaptivity with potential applications in diverse fields [8].

In this chapter, we first discuss the steps followed during the formulation of the SAP-AP systems and their rheological analysis. Secondly, we explore the effect of ionic strength on the rheological properties of the SAP-AP systems. Next, we discuss the combined effect of shear and ionic strength on the structural stability of the self-assembling systems. Finally, we summarize the short- and long-term thermal performance of the SAP-AP systems.

2. Formulation of self-assembling systems

The associating polymers (APs) were provided by SNF Holding Co. (Riceboro, Georgia, USA). Two of these APs, designated as AP1 and AP2, display high grade of anionicity (hydrolysis degree: 25 – 30 mol% at room temperature) and high molecular weights (≈16 – 20 million Dalton). AP1 has a low hydrophobic content, while AP2 has a medium hydrophobic content. The third associating polymer, designated as AP3, has low anionicity (hydrolysis degree: 15 mol%), low molecular weight (8 – 12 million Dalton), and high hydrophobic content [6, 13, 14]. The relative hydrophobic contents of AP1, AP2, and AP3 are 1, 2, and 5 – 6, respectively [13]. APs are tolerant to high salinities at moderate temperatures [14]. The anionic surfactant, a primary alcohol alkoxy sulfate 30% active, was supplied by Sasol North America (Houston, Texas) [15]. β-Cyclodextrin (β-CD) Technical Grade Trappsol® was acquired from Cyclodextrins Technology Development Inc. (CDT, Inc. Alachua, Florida, USA). The assay of the β-CD powder was 98% (molecular weight: 1135 g/mol).

The SAP-AP systems were formulated in synthetic reservoir brines [16] and the effect of different concentrations of brine on the rheological properties of the SAP-AP systems was determined. **Table 1** shows the compositions of the synthetic brines.

The formulation of the SAP-AP systems was based on a molar ratio of surfactant to β-CD of 2:1 established from our previous research [17–20]. In the initial formulation of the SAP-AP system, the concentration of polymer was kept fixed at 0.5 wt% in 2.1 wt% brine. **Table 2** presents the experimental design applied for the SAP-AP formulation. All the experiments were duplicated or even triplicated, and the results presented are the average of several measurements.

Self-assembly was monitored through rheology by observing the changes in the elastic (G') and viscous behavior (G''), loss factor ($\tan\delta = G''/G'$), and complex viscosity, $|\eta^*|$, relative to the baseline associating polymers [8–10, 12, 21]. The rheological analysis was conducted using a Bohlin Gemini HR Nano Rheometer manufactured by Malvern (Worcestershire,

Components	Total concentration (wt% or TDS)				
	1.40	**2.10**	**4.21**	**6.31**	**8.41**
NaCl	1.15	1.72	3.45	5.17	6.9
$MgCl_2$	0.03	0.04	0.09	0.13	0.18
$CaCl_2$	0.22	0.33	0.65	0.98	1.30
Na_2SO_4	0.01	0.01	0.02	0.03	0.04

Table 1. Synthetic brine compositions.

Surfactant (ppm)	β-CD concentration (ppm)					
	0	**30**	**50**	**70**	**90**	**110**
0	Baseline-*AP	β-CD-AP	β-CD-AP	β-CD-AP	β-CD-AP	β-CD-AP
30	**S 30-AP	***SAP-AP AP 30				
50	S 50-AP		SAP-AP S/β-CD50			
70	S 70-AP			SAP-AP S/β-CD70		
90	S 90-AP				SAP-AP S/β-CD90	
110	S 110-AP					SAP-AP S/β-CD110

*AP: associating polymer: AP1, AP2, or AP3 mixed at a fixed concentration of 0.5 wt%.
**S: surfactant.
***SAP-AP: self-assembling polymeric system.

Table 2. Experimental design: SAP-AP formulations.

UK) equipped with parallel-plate measuring geometry (gap between the plates of 1000 μm) and solvent trap to avoid evaporation and/or drying effects. First, amplitude sweeps were run to determine the limit of the linear viscoelastic (LVE) range of the samples at 25°C; followed by frequency sweeps to establish the time-dependent deformation behavior [22]. G', G'', $\tan\delta$ (G''/G'), and $|\eta^*|$ were plotted as a function of the angular frequency (ω) in logarithmic scales on both axis.

2.1. Effect of β-CD addition

Figure 1 displays the results of the oscillatory tests for β-CD/polymer blends at different β-CD concentrations and fixed concentration of polymer (0.5 wt%). **Figure 1(a–c)** corresponds to formulations using polymers AP1, AP2, and AP3, respectively. **Figure 1(a–c)** indicates that the addition of different concentrations of β-CD did not improve the frequency-dependent function of the baseline polymers; on the contrary, the addition of β-CD makes these polymers more inflexible and rigid. In all cases, tanδ increases, while G' and G'' decrease relative to the baseline polymers.

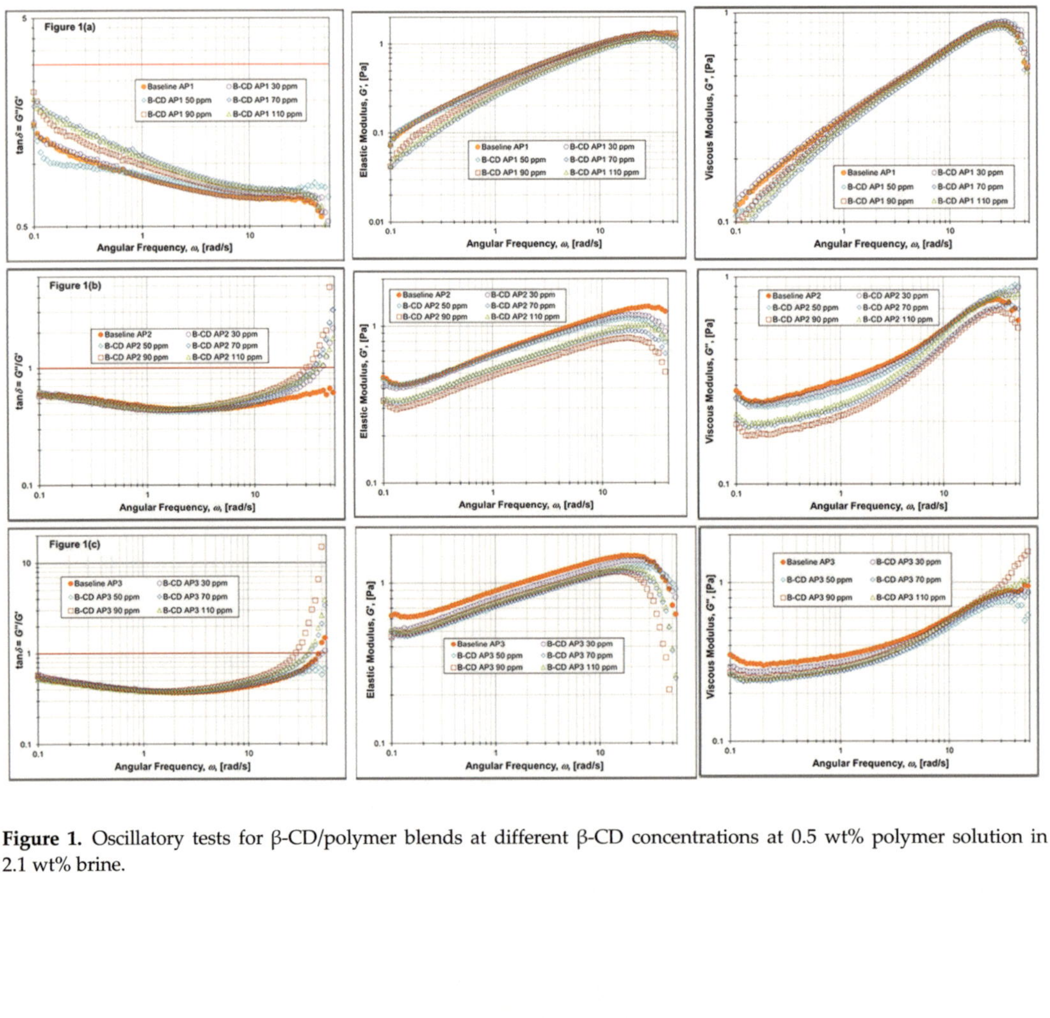

Figure 1. Oscillatory tests for β-CD/polymer blends at different β-CD concentrations at 0.5 wt% polymer solution in 2.1 wt% brine.

2.2. Effect of surfactant addition

Figure 2(a–c) demonstrates the effect of the addition of surfactant on the viscoelastic properties of polymers AP1, AP2, and AP3. These plots reveal interactions (noncovalent associations) among the surfactant and polymers AP1, AP2, and AP3. The addition of surfactant increases the elasticity (tanδ decreases and G' increases) and viscosity of the samples (G'' and $|\eta^*|$ increase) relative to the respective baseline polymers. However, these plots also indicate that there is no a clear relationship between surfactant concentration and the improvement of the viscoelastic properties. For instance, the surfactant concentrations that render the best viscoelastic properties for surfactant-AP1 ranged from 30 to 70 ppm; for surfactant-AP2 was 70 ppm, and for surfactant-AP3 ranged from 30 to 90 ppm.

2.3. Effect of the simultaneous addition of surfactant and β-CD

Figure 3(a–c) demonstrates that the simultaneous addition of surfactant and β-CD produces strong noncovalent interactions and robust self-assembling. Self-aggregation significantly increases the viscoelastic properties of the different systems, specifically for the SAP-AP systems formulated using polymers AP2 (**Figure 3(b)**) and AP3 (**Figure 3(c)**). Furthermore, the

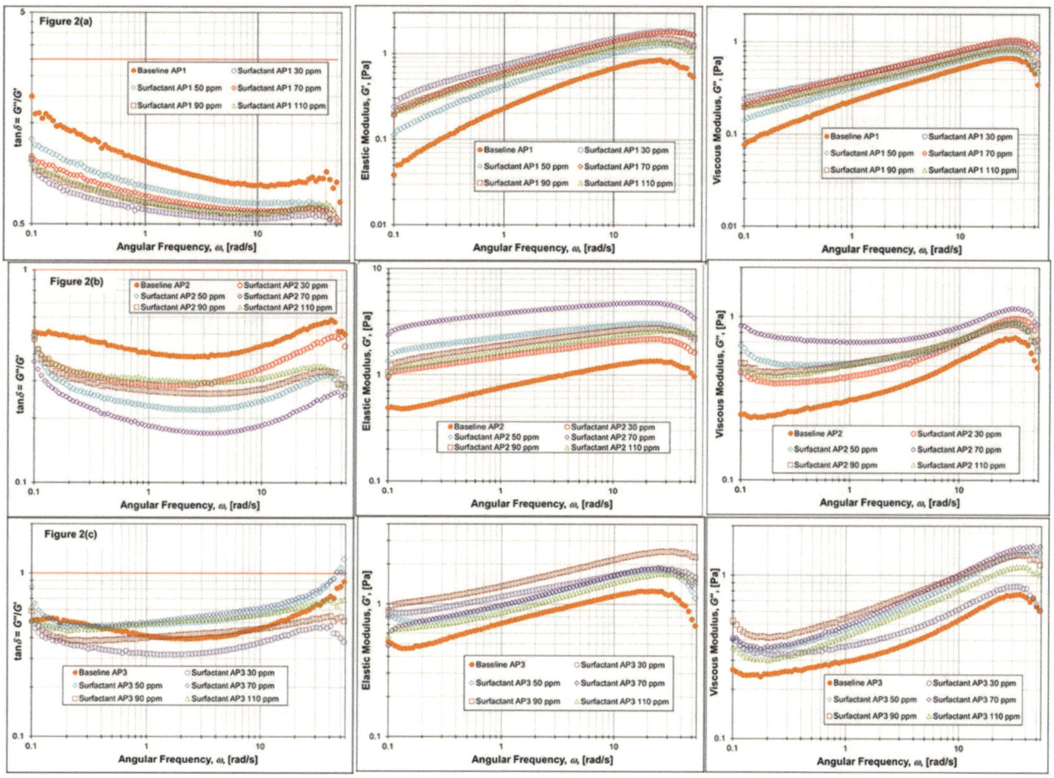

Figure 2. Oscillatory tests for surfactant/AP blends at different surfactant concentrations at 0.5 wt% polymer solution in 2.1 wt% brine.

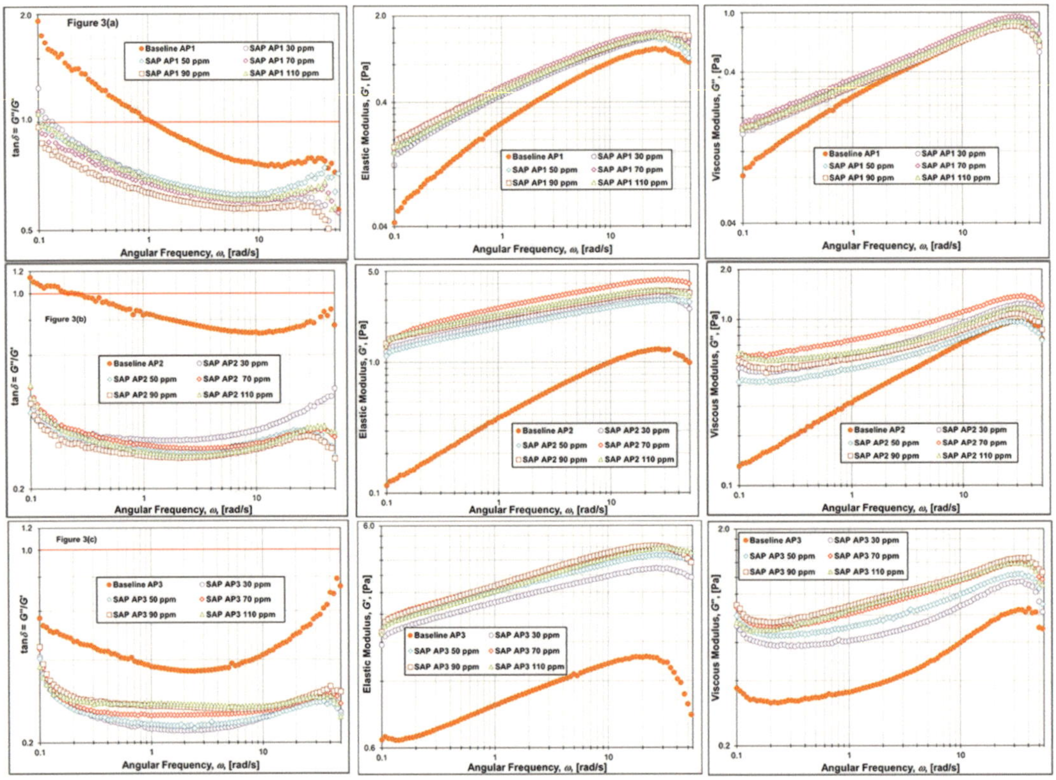

Figure 3. Oscillatory tests for surfactant/β-CD/AP blends at different surfactant/β-CD concentrations at 0.5 wt% polymer solution in 2.1 wt% brine.

experimental results show a clear and consistent relationship between surfactant/β-CD concentration and the improved viscoelastic properties of the SAP-AP systems. The self-assembling systems displaying the most enhanced viscoelasticity were obtained at a surfactant concentration of 70 ppm and β-CD concentration of 70 ppm (**Figure 3(a–c)**). The viscoelastic properties achieved by the SAP-AP systems at a surfactant/β-CD concentration of 70 ppm overlapped the viscoelastic performance corresponding to surfactant/β-CD concentrations of 90 and 110 ppm; therefore, a surfactant/β-CD concentration of 70 ppm was selected as the optimum concentration from the technical and cost-effective standpoint.

All the self-assembling systems display a decrease of the loss factor (tanδ), which indicates that these SAP-AP systems are more elastic (i.e. improved reversible deformation behavior) relative to the baseline AP polymers. This observation agrees with the significant increase of G' shown by the SAP-AP systems, especially for the SAP-AP2 with a percentage increase of $G' = 310\%$ and the SAP-AP3 with a percentage increase of $G' = 220\%$. Likewise, the loss modulus (G'') and the complex viscosity ($|\eta^*|$) increased significantly. For instance, the SAP-AP2 shows a percentage increase of $G'' = 61\%$ and of $|\eta^*| = 253\%$; while the SAP-AP3 displays a percentage increase of $G = 694\%$ and of $|\eta^*| = 414\%$.

The remarkable gain of viscoelastic properties achieved by these SAP-AP systems is attributed to self-association through β-CD host-guest interactions and other noncovalent interactions (i.e. hydrogen bonding and hydrophobic interactions, among others). These observations agree with previous research in which self-aggregation through intermolecular noncovalent associations increased solution viscosity [7].

In this work, only 70 ppm (0.007 wt%) of surfactant and 70 ppm (0.007 wt%) of β-CD were added to the AP polymers dissolved in brine. β-CD rapidly forms inclusion complexes with various hydrophobic guest moieties and polymeric chains [7]. The hydrophobic pendant groups from the associating polymers and the hydrophobic tails of the surfactant are typical guest moieties that can be spontaneously included into the β-CD cavity through both faces: the primary and secondary faces. While the hydrophilic end (i.e. polar part) of the anionic surfactant can interact with the hydrophilic pendant groups (i.e. amide groups contained in the associating polymers) through hydrogen bonding (i.e. H-bridges) or with the carboxylate pendant groups of the polymer backbone via electrostatic interactions through divalent cation-bridges (i.e. Ca^{2+} or Mg^{2+}) present in the brine.

Figure 4 displays a hypothetical network structure formed through self-assembly. This 3D supramolecular structure is responsive and reversible because these physical bonds are not rigid [22]. Furthermore, the resulting high-order macromolecule displays increased hydrophilicity

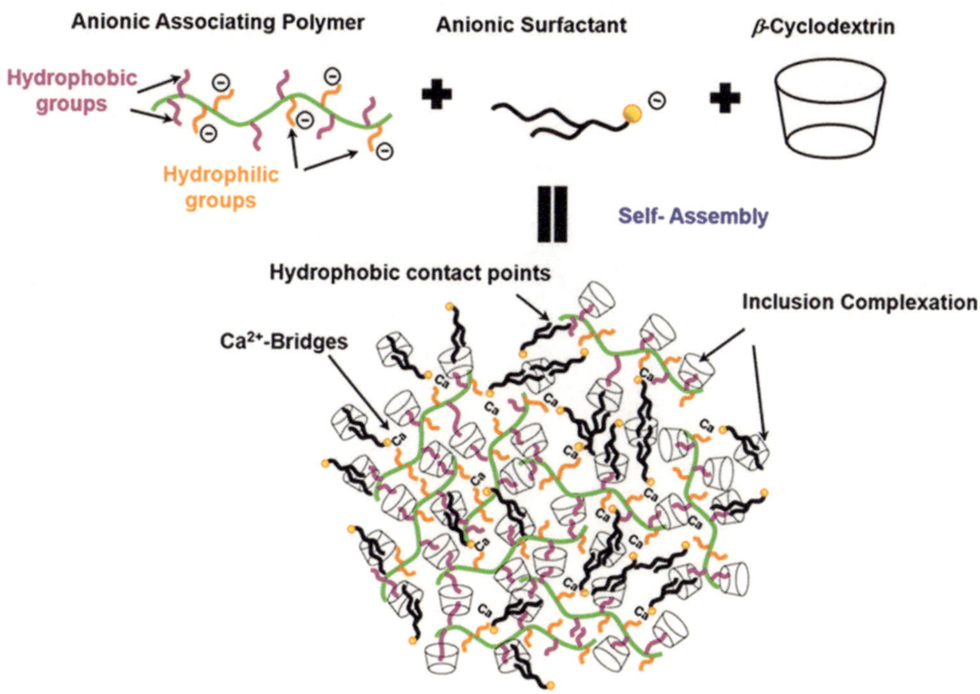

Figure 4. Hypothetical network structure via self-assembly among β-CD, associating polymer, and anionic surfactant in brine solution.

because the molecular locations containing the β-CD complexations become more hydrophilic in nature [7].

2.4. Effect of polymer concentration

Associating polymers contain hydrophobic pendant groups, which are important contact points between the hydrophobic tails of the surfactant and the cavity of the β-CD leading to the self-association and formation of supramolecular three-dimensional (3D) networks. Therefore, the effect of polymer concentration on the properties of the SAP-AP systems at a fixed concentration of surfactant and β-CD (i.e. optimum concentration: 70 ppm surfactant; 70 ppm β-CD) was established. The concentrations of polymer evaluated were 0.25, 0.5, and 0.75 wt%.

Figure 5(a–c) displays the results of the oscillatory tests for the optimum SAP-AP systems at different concentrations of the respective polymers AP1, AP2, and AP3. As polymer concentration increases, G' and G'' increase. A higher concentration of associating polymers increases the number of hydrophobic contact points, which promotes more intermolecular hydrophobic interactions, host-guest complexations, and other noncovalent associations (i.e. H- and Ca^{2+}-bridging). Additionally, higher polymer concentration enhances chain overlapping, which also contributes to the formation of a network of higher structural strength [13].

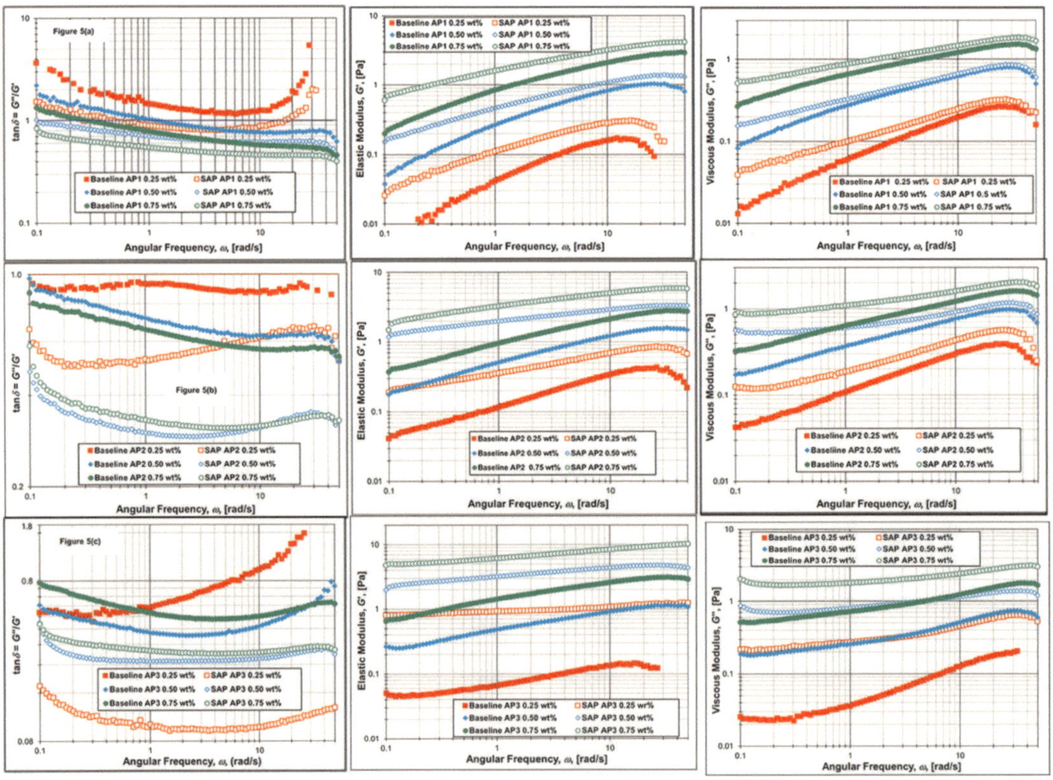

Figure 5. Oscillatory tests for the optimum SAP-AP systems at different concentrations of polymers: AP1, AP2, and AP3.

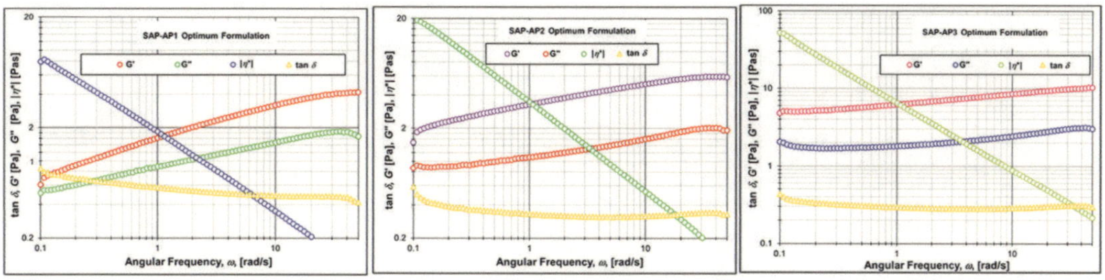

Figure 6. Frequency-dependent rheological performance of the optimum SAP-AP systems for polymers AP1, AP2, and AP3 in saline solution.

These results guided the selection of the optimum formulation of the SAP-AP system as follows, 0.75 wt% associating polymer (AP), 0.007 wt% (70 ppm) of surfactant, and 0.007 wt % (70 ppm) of β-CD prepared in saline solution. **Figure 6(a–c)** presents the frequency-dependent rheological performance of the optimum SAP-AP systems for polymers AP1, AP2, and AP3, respectively.

The frequency-dependent rheological data (**Figure 6**) indicates that these systems follow the typical behavior of three-dimensional network structures showing $G' > G''$ in the range of frequency studied [22]. As explained by Mezger, "the curves of G' and G'' often occur in the form of almost parallel straight lines throughout the entire frequency range showing a slight slope only" … "The shape of the curves [also indicates that these] … network structures are exhibiting a relatively constant structural strength in the whole frequency range" [22].

3. Effect of ionic strength on the SAP-AP systems

Salts significantly affect the viscosity of polymer solutions. The screening of the negatively charged moieties (i.e. carboxyl groups) in the polymer structure in the presence of mono- and divalent cations causes viscosity loss due to polymer coiling, polymer precipitation, and phase separation [2, 3, 5, 6, 23–25].

Figure 7(a–c) displays the results of the frequency sweeps of polymers AP1 and AP2 and their corresponding SAP-AP systems at the following brine concentrations: 1.4, 2.1, 4.2, 6.3, and 8.4 wt%. While for polymer AP3 and its corresponding SAP-AP3 system, the effect of ionic strength was evaluated at the following brine concentrations: 2.1, 4.2, 6.3, and 8.4 wt%.

In the case of polymer AP1, **Figure 7(a)** shows that it is strongly affected by salinity and hardness. As brine concentration increases, the tanδ-curve shifts from lower values toward medium range values, which means that the polymer flow behavior changes to a more viscoelastic liquid. G'', G', and $|\eta^*|$ decrease as salinity increases. This rheological behavior indicates that an increase in ionic strength weakens the hydrophobic interactions in the associating polymer resulting from the electrostatic screening of the charged segments [13] causing the coiling/folding of the polymer backbone.

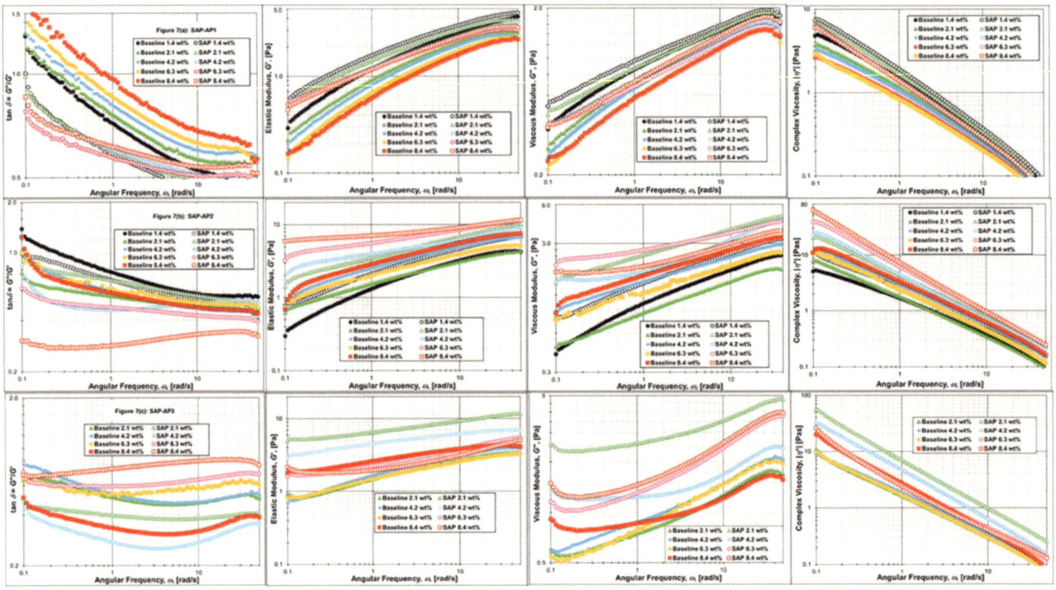

Figure 7. Frequency sweeps of polymers AP1, AP2, and AP3 and the corresponding SAP-AP systems at several brine concentrations at T = 25°C.

On the contrary, for the SAP-AP1 system, **Figure 7(a)** demonstrates that as salinity and hardness concentration increase, $G' > G''$ (tan$\delta < 1$) throughout the entire frequency range; thus, in these systems, the elastic behavior dominates, which is characteristic of supramolecular aggregates [22]. Although as salinity concentration increases, G'', G', and $|\eta^*|$ decreases, the SAP-AP1 system displays higher structural strength relative to the AP1 polymer. Self-association seems to prevent electrostatic screening of the charged sections of the polymer molecule. Therefore, the SAP-AP1 system is less affected by electrostatic effects, which enhances its structural strength and functionality. Furthermore, the bulky size and shape of the β-CD molecules increase the steric hindrance that might diminish the electrostatic effects in these network structures [3, 13].

Figure 7(a) also shows that the SAP-AP1 prepared in the highest brine concentration of 8.4 wt% seems to collapse showing the same rheological behavior of the SAP-AP1 prepared in 6.3 wt% brine. These observations indicate that for the experimental conditions of this work, a brine concentration of 6.3 wt% seems to be the threshold before electrostatic effects become significant for this system.

The effect of salinity on the rheological behavior of polymer AP2 presented in **Figure 7(b)** is noticeably different than that for polymer AP1. In this case, it seems that increasing brine salinity and hardness strengthens the intermolecular hydrophobic interactions. The tanδ-values are shifted from tan$\delta > 1$ (1.4 wt% brine) to tan$\delta < 1$ (2.1 – 8.4 wt% brine) and the polymer attained a flow behavior of a physical network. G'-, G''-, and $|\eta^*|$-values increase throughout the entire range of angular frequency as salinity and hardness concentrations increase. These results suggest that a higher content of hydrophobic groups in the AP2 polymer makes it less susceptible to electrostatic effects. **Figure 7(b)** makes evident that the SAP-AP2 system displays a better rheological performance than the baseline AP2. Again, an increase in the ionic strength causes a decrease in the tanδ-values. The SAP-AP2 prepared in 8.4 wt% brine shows the lowest tanδ-value. G', G'', and $|\eta^*|$ also increase with salinity and hardness concentration. The SAP-AP2 formulated in 6.3 and 8.4 wt% brine displays the highest values of complex viscosity, $|\eta^*|$. These observations demonstrate that an increase in the ionic strength reinforces the inter- and intramolecular forces building up the supramolecular SAP-AP2 system, which improves its viscosifying power [13].

Figure 7(c) presents the effect of salinity and hardness concentration on the rheological behavior of polymer AP3 and the SAP-AP3 system. **Figure 7(c)** demonstrates a negligible effect of brine salinity in the range from 2.1 to 6.3 wt% on the rheological properties of polymer AP3. In contrast, above this salinity range (i.e. brine 8.4 wt%), the rheological behavior of AP3 is significantly affected. For instance, the tanδ-curve shifts toward very low values and $G''/G' < 1$ in the entire range of angular frequency, which suggests the reinforcement of the intermolecular hydrophobic interactions generating a stronger physical network structure. These observations seem to indicate that the lower anionicity and larger hydrophobic content of the AP3 polymer improve its salt tolerance.

The viscoelasticity of the SAP-AP3 system is greater than the viscoelasticity of polymer AP3 at all salinity concentrations (see **Figure 7(c)**). At the lowest salinity concentration of 2.1 wt%, the SAP-AP3 system shows far superior rheological properties (i.e. enhanced viscosifying power and elasticity) in the entire angular frequency range. G', G'', and $|\eta^*|$ are significantly higher compared to the baseline. It seems that this salinity concentration (i.e. 2.1 wt%) greatly reinforces the strength of the inter- and intramolecular interactions taking place in this system. The G'-curve is almost parallel to the x-axis, which is typical of a highly stable structural network [22]. However, increasing the ionic strength beyond 2.1 wt% negatively affects the viscoelasticity of this system.

Overall, the SAP-AP systems are less sensitive to electrostatic effects compared to the AP polymers. The enhanced salt tolerance may result from the bulky size of the β-CD host-guest complexations that "sterically [hinders] the polymer chain so that the hydrodynamic radius does not fully collapse to a random coil configuration at high salinity" [6].

These experimental observations also indicate that the hydrophobic content in the associating polymers plays a vital role in the strength, rheological behavior, and ionic strength sensitivity of the SAP-AP systems. For instance, the SAP-AP system derived from the baseline polymer AP2, which has a medium content of hydrophobic groups, shows the formation of a stable supramolecular network highly functional in brines with high salinity and hardness concentration (i.e. brine 8.4 wt%). For this SAP-AP2 system, an increase in the ionic strength increases its elasticity and viscosifying power. This functionality is important for applications in enhanced oil recovery.

4. Effect of shear and ionic strength on the SAP-AP systems

Previous research has demonstrated that the mechanical degradation of polymers is path independent; thus, the effect of shear on the mechanical stability of polymers can be evaluated using any kind of degrading geometry [26]. In this work, the mechanical stability of the SAP-AP systems was determined through thixotropic behavior analysis using oscillatory rheology as recommended in Ref. [22]. **Table 3** summarizes the dynamic-mechanical conditions employed during the thixotropic analysis. The percentage of regeneration method was used to analyze the thixotropic behavior of the samples. In this method, the percentage of regeneration that takes place at the end of the third interval is read off and the percentage is calculated in relation to the reference value G'-at-rest at the end of the first interval which was taken as the 100% value [22]. These analyses were conducted for the baseline polymers and for the SAP-AP systems prepared at different salinity concentrations: 1.4, 2.1, 4.2, 6.3, and 8.4 wt% (see **Table 1**). The objective was to determine the simultaneous effect of shear and ionic strength on the structural and shear stability of the different systems.

Figures 8–10 summarize the time-dependent function of G' and G'' for Step 1 and Step 3 for polymers AP1 and SAP-AP1, AP2 and SAP-AP2, and AP3 and SAP-AP3, respectively. Likewise, **Tables 4** and **5** summarize the structural strength of the baseline polymers and the corresponding SAP-AP systems in terms of the G'- or G''-values taken at the end of Step 1, and the G'- or G''-values taken at the end of Step 3, and the percentage (%) of structural regeneration of the respective samples.

Steps	Oscillation test	ω (rad/s)	Strain (%)	Number of samples	Test time (s)
1	Single frequency/strain controlled	6.283	20 within the LVE	200	≈960
2	Single frequency/strain controlled	6.283	100 outside the LVE	100	≈480
3	Single frequency/strain controlled	6.283	20 within the LVE	200	≈960

LVE refers to the linear viscoelastic range.

Table 3. Preset of the dynamic-mechanical conditions for each of the intervals used during the thixotropic behavior analysis.

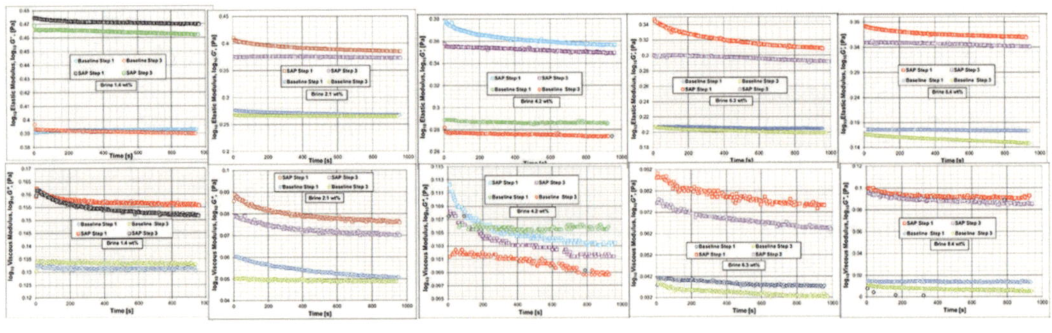

Figure 8. $\log_{10} G'$ and $\log_{10} G''$ for steps 1 and 3 vs. time for polymer AP1 and SAP-AP1.

http://dx.doi.org/10.5772/intechopen.74618

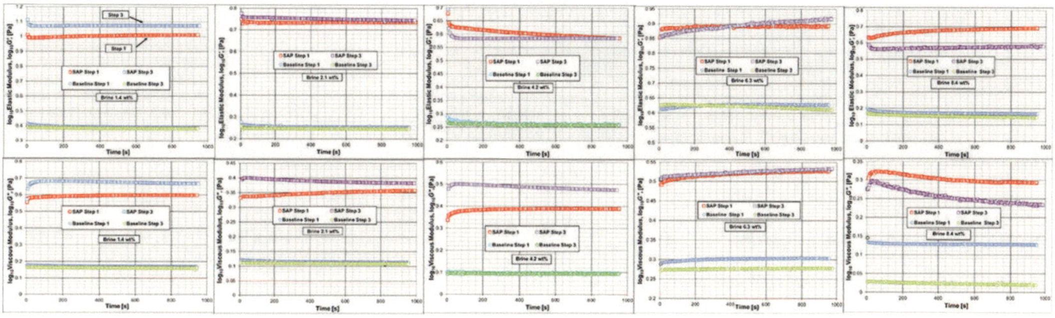

Figure 9. $\log_{10} G'$ and $\log_{10} G''$ for steps 1 and 3 vs. time for polymer AP2 and SAP-AP2.

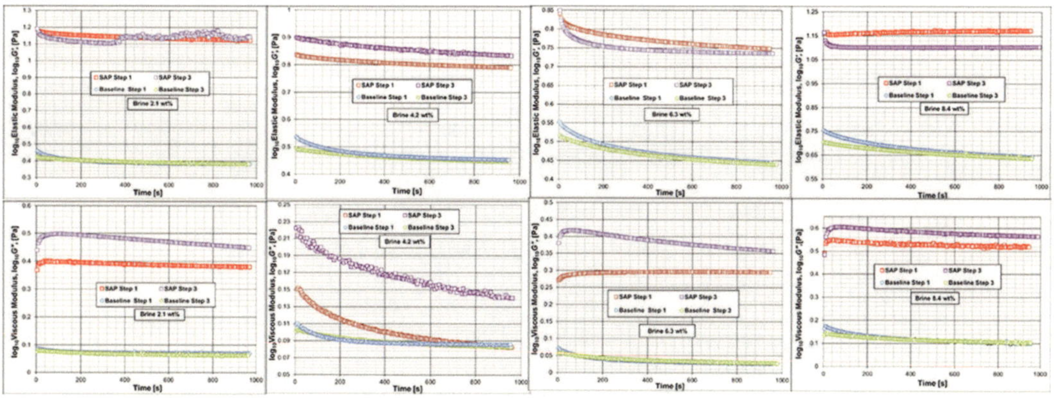

Figure 10. $\log_{10} G'$ and $\log_{10} G''$ for steps 1 and 3 vs. time for polymer AP3 and SAP-AP3.

Figure 8 demonstrates that the optimum SAP-AP1 shows higher structural strength (> G'-values and > G''-values) relative to polymer AP1. The second important result is that there is no complete regeneration of the initial structural strength. The first two rows of **Table 5** show that the percentage of structural regeneration for polymer AP1 in terms of G' > 95% for all brine salinities (1.4, 4.2, 6.3 wt%) except for the 8.4 wt% brine with a structural regeneration of 91%. In terms of G'' (see **Table 5**), both systems: AP1 and SAP-AP1 show a structural regeneration > 96%, excluding the baseline polymer AP1 in 1.4% brine that shows a gain in structural strength with a regeneration of G'' > 100%.

The minor loss of structural strength observed for both systems might be related to macromolecular shear degradation under the high-shear forces applied during Step 2 causing the breaking of carbon/carbon bonds of some of the polymer chains [5].

Figure 9 presents the time-dependent behavior of polymer AP2 and the SAP-AP2 system in terms of G' and G'' for the low-shear interval (Step 1) and for the regeneration interval (Step 3). Experimental trends observed from **Figure 9** and **Tables 4** and **5** are as follows:

	Brine concentration (wt%)									
	1.4		2.1		4.2		6.3		8.4	
	G' (Pa)	Reg (%)	G' (Pa)	Reg (%)	G' (Pa)	Reg (%)	G' (Pa)	Reg (%)	G' (Pa)	Reg (%)
Polymer AP1										
Step 1	2.47		1.92		1.82		1.59		1.49	
Step 3	2.34	**98.84**	1.85	**96.51**	1.80	**98.82**	1.57	**98.59**	1.36	**91.03**
SAP-AP1										
Step 1	3.05		2.50		2.28		2.04		2.18	
Step 3	3.01	**98.49**	2.42	**96.75**	2.25	**98.34**	1.99	**97.72**	2.17	**99.7**
Polymer AP2										
Step 1	2.45		1.86		1.82		—		1.07	
Step 3	2.41	**98.16**	1.72	**92.47**	1.82	**100**	—	—	1.04	**97.70**
SAP-AP2										
Step 1	10.26		6.05		3.61		—		4.92	
Step 3	11.82	**>100**	6.20	**>100**	3.61	**100**	—	—	3.79	**77.01**
Polymer AP3										
Step 1	—		2.41		2.84		2.79		4.38	
Step 3	—	—	2.41	**100**	2.79	**98.27**	2.75	**98.82**	4.33	**98.68**
SAP-AP3										
Step 1	—		13.22		6.17		5.60		14.85	
Step 3	—	—	13.86	**>100**	6.81	**>100**	5.45	**97.23**	12.68	**85.39**

The objective of bold numbers in Table 4 is to emphasize the percentage of regeneration in each system.

Table 4. G'-values and percentage of regeneration.

- The SAP-AP2 system formulated at different ionic strengths displays significantly higher structural strength, G' and G'', than the baseline AP2 polymer.
- As the ionic strength increases, the structural strength of the AP2 polymer and the SAP-AP2 system decreases.
- The baseline polymer AP2 shows structural regeneration G' and $G'' > 92\%$ but $< 100\%$ in the range of salinity concentration studied. While the SAP-AP2 systems prepared in 1.4, 2.1, and 4.2 wt% brines display gain in structural strength with regenerations ≥ 100%. In 8.4 wt% brine, the structural regeneration SAP-AP2 system falls to 77% and to 87.26% in terms of G' and G'', respectively.

The gain in structural strength is significant for the SAP-AP2 prepared in low brine salinity with a 15% gain of the G'-value and 17% gain of the G''-value (see **Figure 9**). Although at brine salinity concentrations up to 4.2 wt%, the gain in structural strength is ≥ 100%, it decreases as salinity increases, which demonstrates the negative effect of increased ionic strength on the shear stability of the SAP-AP2 system.

	Brine concentration (wt%)									
	1.4		2.1		4.2		6.3		8.4	
	G'' (Pa)	Reg (%)	G'' (Pa)	Reg (%)	G'' (Pa)	Reg (%)	G'' (Pa)	Reg (%)	G'' (Pa)	Reg (%)
Polymer AP1										
Step 1	1.311		1.14		1.28		1.05		1.03	
Step 3	1.371	**>100**	1.10	**96.92**	1.25	**97.66**	1.04	**98.89**	1.02	**99.03**
SAP-AP1										
Step 1	1.47		1.23		1.27		1.14		1.18	
Step 3	1.46	**99.25**	1.21	**98.17**	1.26	**99.2**	1.12	**98.29**	1.17	**99.07**
Polymer AP2										
Step 1	1.476		1.39		1.27		—		1.34	
Step 3	1.447	**98.2**	1.28	**98.54**	1.27	**100**	—	**—**	1.29	**96.26**
SAP-AP2										
Step 1	3.96		2.72		2.45		—		1.97	
Step 3	4.64	**>100**	2.91	**>100**	2.99	**>100**	—	**—**	1.72	**87.26**
Polymer AP3										
Step 1	—		1.172		1.22		1.07		1.26	
Step 3	—	**—**	1.155	**98.55**	1.21	**98.27**	1.07	**100**	1.27	**>100**
SAP-AP3										
Step 1	—		2.39		1.209		1.97		3.30	
Step 3	—	**—**	3.16	**>100**	1.381	**>100**	2.28	**>100**	3.67	**>100**

Table 5. G''-values and percentage of regeneration.

The mechanical stability shown by the SAP-AP2 system might result from the increased rigidity of the polymeric network due to self-aggregation with the bulky structure of the β-CD and the long-chain alkyl branched anionic surfactant [3, 5]. Supramolecular aggregates based on physical bonding are considerably more rigid compared to the superstructures of the thread-like macromolecules of synthetic polymers [22].

Figure 10 indicates that the SAP-AP3 system displays higher structural strength in terms of G' and G'' relative to the baseline polymer. In low-salinity brines (≤ 4.2 wt%), the SAP-AP3 systems show a gain in the structural strength (see G'- and G''-curves in **Figure 10**). However, for brine salinities ≥ 4.2 wt%, there is no complete structural regeneration as shown by the G'-curve.

The gain in structural strength displayed by the SAP-AP2 and the SAP-AP3 systems in salinity concentrations ≤ 4.2 wt% demonstrates the self-healing character of these self-assemblies. As stated by Yang, "supramolecular self-healing materials [rely] on the use of noncovalent bonds to generate reversibility and dynamic networks, which are able to heal the damaged sites" [10]. According to Mezger, (this self-healing or self-repairing effect occurs because) "the structural development aims at achieving a balance or equilibrium of the active forces or energies ...

based on the very fast connection, disconnection, and re-connection" of the physical bonds such as the decomplexation and complexation of host-guest interactions [7, 22], that results in improved structural strength. "Therefore, when tension is imparted on these networks, the force is distributed homogeneously across the whole network [due to the dynamic disassembling and re-assembling of the physical interactions]" protecting the macromolecules in the network from permanent shear degradation [7].

5. Thermal stability of the SAP-AP systems

Thermal degradation of polymer leads to chemical changes of the polymer structure. For instance, in the case of polymers derived from polyacrylamides, high temperatures induce the hydrolyzation of the acrylamide group to the acrylate moiety. These hydrolysis reactions are strongly correlated to temperature; thus, the higher the temperature, the higher the hydrolysis [2]. In this study, both short- and long-term thermal stabilities of the AP polymers and the corresponding SAP-AP systems were evaluated.

5.1. Short-term thermal stability

The AP polymers and the corresponding optimum SAP-AP systems were subjected to dynamic-mechanical thermo-analysis or DMTA. In this analysis, the dynamic temperature sweep was conducted by using a linear heating rate (8.65°C/min) from 9 to 81°C (±0.2°C) as an upward ramp immediately followed by a linear cooling rate (7.3°C/min) from 81 to 9°C (±0.2°C) as a downward ramp. In these tests, the angular frequency (ω) was fixed at 7 rad/s, while the strain was fixed at 20% (within the LVE range). **Figure 11** displays the temperature-dependent functions of G', G'', and tanδ for polymer AP1 and the SAP-AP1 system for low (2.1 wt%)- and high (8.4 wt%)-concentration salinity; while **Figure 12** presents the temperature-dependent functions of G', G'', and tanδ for polymers AP2 and AP3 and their corresponding SAP-AP systems in 8.4 wt % brine concentration.

At low-ionic strength (i.e. 2.1 wt% brine), the SAP-AP1 system shows higher structural strength in terms of G' and G'' relative to polymer AP1. The SAP-AP1 system displays a tan$\delta < 1$ (i.e. $G' > G''$) in the entire range of temperature that is consistent with the behavior of network structures. At low temperatures, the tanδ-curve of the AP1 polymer shows tan$\delta < 1$, as temperature increases the crossover point ($G' = G''$) is reached and the tanδ-curve is shifted to a value of tan$\delta > 1$ at higher temperatures, which corresponds to the behavior of viscoelastic liquids. The cooling curves of tanδ, G', and G'' as a function of temperature demonstrate a decreasing structural strength of the polymer systems relative to the corresponding heating curves. According to Mezger, as temperature increases, "the [polymer] molecules are able to move along one another which results in an increasing number of disentanglements. With increasing temperature, more motion occurs between the molecules which again cause an increase in the amount of the frictional forces [that produces] frictional heat that afterward is lost for the sample in the form of thermal energy … This process can be observed as a decreasing G'' value" [22]. The cooling temperature curve displays indeed lower G''-values. Likewise, for the SAP-AP1 system, with increasing temperature, dissociation and/or disassembling of the

http://dx.doi.org/10.5772/intechopen.74618

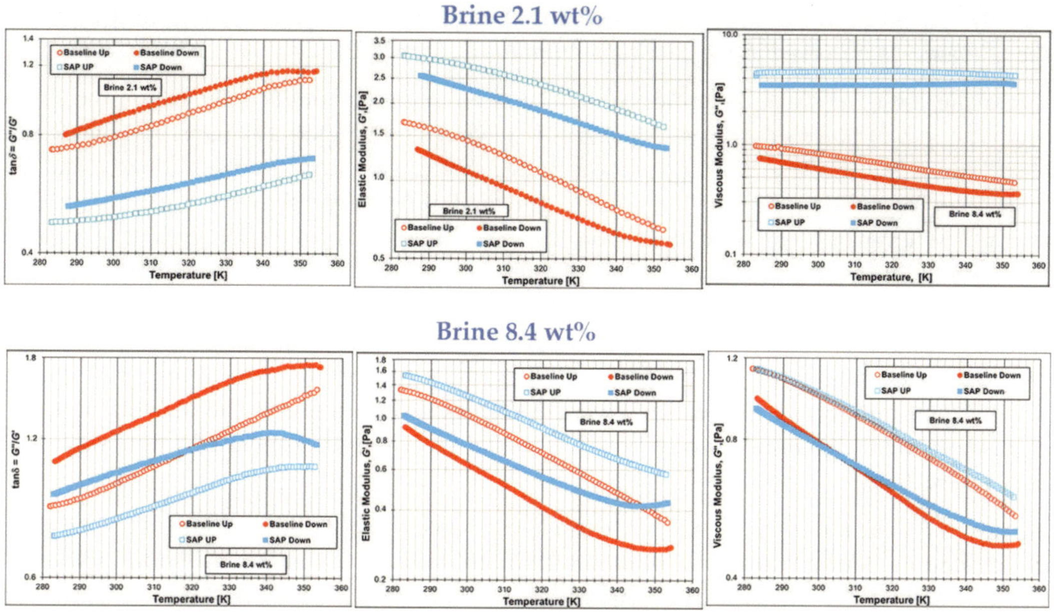

Figure 11. Temperature-dependent functions of G', G'', and tanδ for polymer AP1 and SAP-AP1 system.

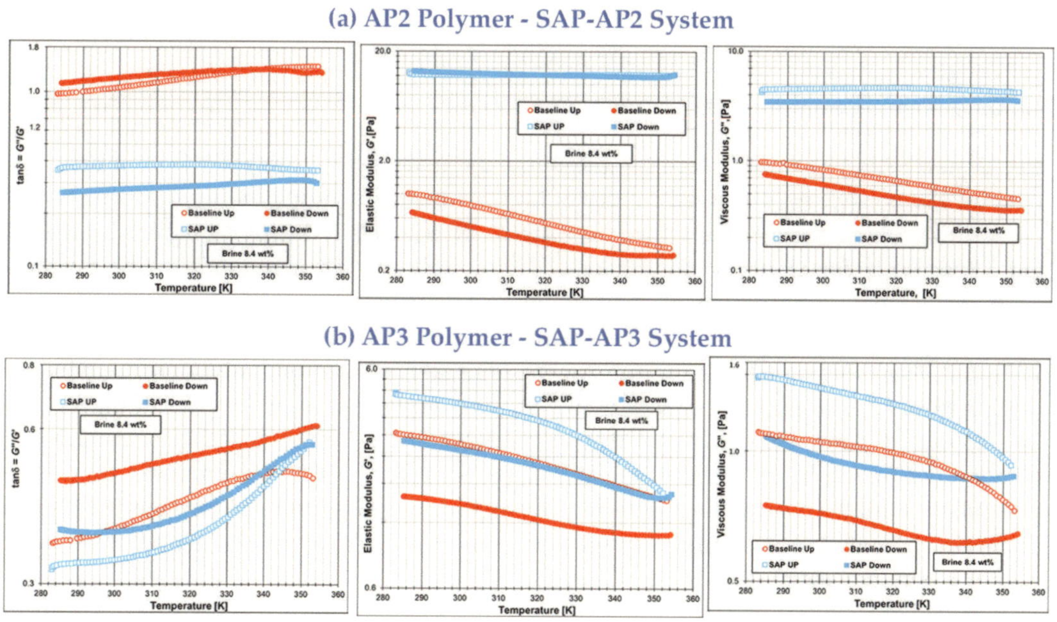

Figure 12. Temperature-dependent functions of G', G'', and tanδ for polymers AP2 and AP3 and systems SAP-AP2 and SAP-AP3.

physical interactions takes place. Therefore, "the superstructure [network] yields more and more, breaking increasingly into smaller parts, turning into a state of flexibility. Later, during the cooling process, the physical bonds are reformed, and the network structure regains rigidity. However, due to losses of frictional heat in the form of thermal energy, a decrease of the G''-values is observed in the cooling temperature curve" [22].

Increasing the ionic strength (brine 8.4 wt%) and temperature significantly affects the structural strength of polymer AP1 and the SAP-AP1 system (see **Figure 11**). Similarly, the downward curve (i.e. cooling process) shows a decreasing structural strength relative to the corresponding heating curves. The SAP-AP1 temperature curve shows a higher structural strength in terms of G'-values compared to the AP1 polymer; however, in terms of G''-values, there is no difference between the AP1 polymer and the SAP-AP1 system. Furthermore, the $\tan\delta$-curve of the cooling process for the SAP-AP1 system shows a shift from $\tan\delta$-values > 1 at elevated temperatures to $\tan\delta$-values < 1 at lower temperatures. This makes evident that at elevated temperatures, the SAP-AP1 network system changes its flow behavior, showing $G'' > G'$, performing as a viscoelastic liquid. As temperature decreases, the network becomes more rigid and the $\tan\delta$-values become < 1, showing again the consistency of a rigid supramolecular structure.

At elevated ionic strength (i.e. 8.4 wt% brine), the baseline polymers AP2 and AP3 and their corresponding SAP-AP systems show similar behavior to the baseline AP1 and SAP-AP1, discussed above. Nevertheless, it is important to mention that the SAP-AP2 system shows the highest structural strength in terms of G' and G'' and the least hysteresis between the heating and the cooling curves among the systems. These observations make evident that the optimum SAP-AP2 displays enhanced structural strength at elevated temperatures and ionic strengths compared to the baseline.

5.2. Long-term thermal stability

The AP and SAP-AP systems were subjected to a thermal stability test at 90°C for a period of 8 weeks. The presence of dissolved oxygen at high temperatures might induce the formation of free radicals which degrade the polymer molecule by cleavage reducing its molecular weight and viscosifying functionality [2]. Besides, if dissolved oxygen is present in the polymer solutions together with very low concentrations of dissolved iron, it might also cause substantial polymer degradation [4, 26–30]. Therefore, to prevent chemical degradation, the AP and the SAP-AP samples were placed in a glove chamber and bubbled with nitrogen at a pressure ranging from 10 to 20 psi for a period of 30 min. Two duplicated set of samples of the AP and the SAP-AP systems were prepared in two different brine salinity concentrations: 2.1 and 8.4 wt% and placed in the oven at 90 ± 0.5°C for 8 weeks. Every 2 weeks, a set of samples were taken out of the oven and subjected to rheological analysis. **Figure 13** displays the G'- and G''-curves as a function of angular frequency and time for the AP and SAP-AP systems at brine salinity concentrations of 2.1 and 8.4 wt%.

The G'- and G''-plots in **Figure 13** demonstrate that the AP polymers and the SAP-AP systems are significantly degraded at a temperature of 90°C after 2 weeks of testing. A dramatic drop of G' and G'' is observed for all the samples in both low and high salinities.

In low-salinity brine, samples AP1, AP2, SAP-AP1, and SAP-AP2 show precipitation of solids and color change at week # 4; while in samples AP3 and SAP-AP3, the precipitation of solids was observed at week # 8. In high-salinity brine, precipitation of solids took place faster at week # 2.

http://dx.doi.org/10.5772/intechopen.74618

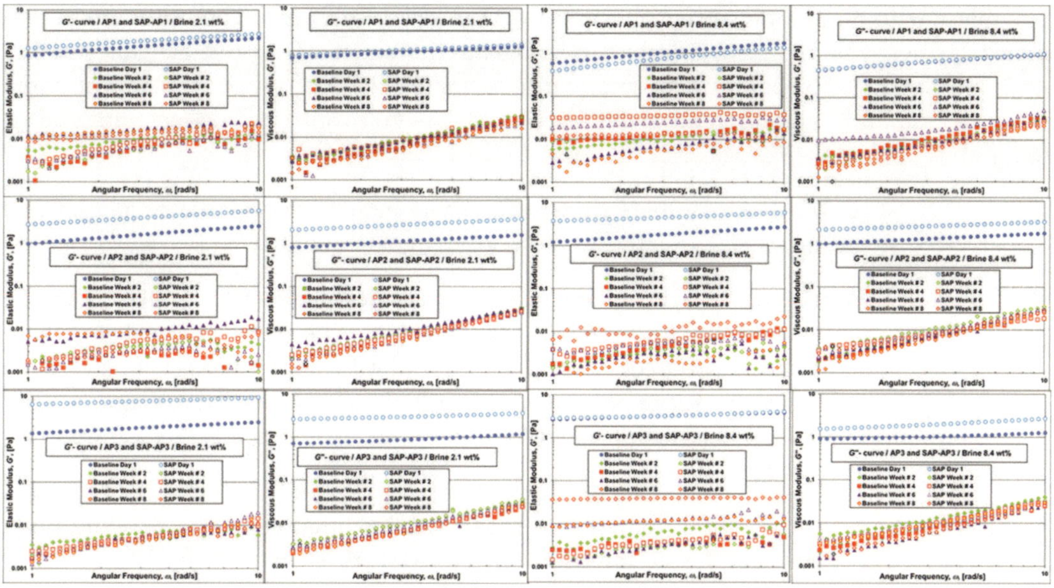

Figure 13. G'- and G''-curves vs. angular frequency, time, and brine salinity.

These observations clearly indicate the combined damaging effect of salinity and elevated temperature on the stability of the AP and SAP-AP systems.

Although the introduction of hydrophobic groups into the molecular chain is an effective way to improve polymer salt tolerance and thermal stability [3], these hydrophobic moieties such as AMPS or n-VP are still susceptible to hydrolysis at high temperatures (i.e. > 85°C). As explained by Levitt and Pope, "…hydrolysis of the [amide] group from the acrylamide moiety and/or β, β, dimethyl taurine from the AMPS moiety [takes place] forming additional acrylate moiety in the hydrolysed polymer molecule" [6]. At high salinity and hardness concentration, these acrylate moieties (i.e. hydrolyzed poly(AM-co-AMPS)) associate strongly with cations (i.e. Ca^{2+}, M^{2+}) and polymer precipitation takes place, which results in a rapid drop of viscosity of the polymer solution that becomes turbid [6, 31]. Precipitation of polymer due to interactions with multivalent cations increases with temperature. Furthermore, thermal degradation of polymers results in a "reduction of molecular weight because of free-radical induced scission of the acrylic backbone" [6].

This long-term thermal stability testing demonstrates that the AP and the SAP-AP systems are not stable at elevate temperatures; therefore, these SAP-AP systems are recommended for low-temperature (< 90°C) applications.

6. Conclusions

In summary, we formulated advanced polymer-surfactant systems via self-assembling driven by host-guest chemistry and other physical noncovalent bonding by mixing associating polymers, with an anionic surfactant, and β-CD in brine solutions. The optimum supramolecular systems were subjected to rheological characterization and several stability testing techniques to establish their tolerance to increasing ionic strength concentration, shear degradation, and thermal stability. The most significant findings are outlined as follows:

- The formulated SAP-AP systems show remarkable gain of viscoelastic properties and elevated structural strength relative to the associating baseline polymers.
- The SAP-AP network structures are less sensitive to electrostatic effects than the corresponding baseline polymers AP1, AP2, and AP3. The increase in salt tolerance may be aided by the steric effect of the β-CD host-guest complexations acting as large rigid groups within the supramolecular network that prevents the collapse and/or coiling of the polymer chain at elevated ionic strength. This functionality is important for EOR applications.
- The SAP-AP systems show remarkable shear stability relative to the baseline polymers. If the high-shear forces imposed on the network structures are lifted and as soon as the equilibrium of the shear forces and the flow resistance forces is reached, the SAP-AP systems show high structural regeneration, which is > 95% for the case of the SAP-AP1 system; while the SAP-AP2 and SAP-AP3 systems display a gain of structural strength with regeneration > 100%. The gain in structural strength evidences the self-healing performance of these SAP-AP self-assemblies. The overall trend is that increasing the hydrophobic content of the associating polymers increases the shear stability and the structural strength of the supramolecular formulations.
- The short-term thermal stability testing demonstrates that all SAP-AP systems display higher structural strength in terms of G' and G'' when compared to the corresponding baseline polymers in the entire range of temperature and ionic strength evaluated.
- The long-term thermal stability testing carried out at 90°C for a period of 8 weeks demonstrates that the AP polymers and the optimum SAP-AP systems are not stable at elevated temperatures; therefore, these systems are recommended for low-temperature ($\ll$ 90°C) applications.

Acknowledgements

The authors would like to extend their appreciation to Grace Hicks and Tayler Hunt from the Chemical Engineering Department, University of New Brunswick, for their technical involvement during the rheological evaluation of the SAP-AP systems. The authors are grateful to Sasol Chemicals and SNF Floerger for providing surfactant and polymer samples. Financial support from the University of New Brunswick through the Sabbatical Research Grant and the Canada Foundation for Innovation (CFI) is also acknowledged.

Author details

Laura Romero-Zerón* and Xingzhi Jiang

*Address all correspondence to: laurarz@unb.ca

Chemical Engineering Department, University of New Brunswick, Fredericton, New Brunswick, Canada

References

[1] Kang PS, Lim JS, Huh C. A novel approach in estimating shear-thinning rheology of HPAM and AMPS polymers for enhanced oil recovery using artificial neural network. In: The Twenty-third International Offshore and Polar Engineering Conference (ISOPE-I-13-171); Jun 30–Jul 5, 2013. Anchorage, Alaska: 2013. pp. 81-85

[2] Vermolen E, Van Haasterecht MJ, Masalmeh SK, Faber MJ, Boersma DM, Gruenenfelder MA. Pushing the envelope for polymer flooding towards high-temperature and high-salinity reservoirs with polyacrylamide based ter-polymers. In: SPE Middle East Oil and Gas Show; September 25–28, 2011; Manama, Bahrain: 2011. pp. 1-9

[3] Wu Y, Mahmoudkhani A, Watson P, Fenderson TR, Nair M. Development of new polymers with better performance under conditions of high temperature and high salinity. In: SPE EOR Conference at Oil and Gas West Asia; April 16–18, 2012; Muscat, Oman: 2012. pp. 1-11

[4] Sheng JJ, Leonhardt B, Azri N. Status of polymer-flooding technology. Journal of Canadian Petroleum Technology. 2015;**54**(02):16-26

[5] Zaitoun A, Makakou P, Blin N, Al-Maamari RS, Al-Hashmi AA, Abdel-Goad M. Shear stability of EOR polymers. SPE Journal. 2012;**17**(02):335-339

[6] Levitt D, Pope GA. Selection and screening of polymers for enhanced-oil recovery. In: SPE Symposium on Improved Oil Recovery; April 19–23, 2008; Tulsa, Oklahoma: pp. 1-18

[7] Tan S, Ladewig K, Fu Q, Blencowe A, Qiao GG. Cyclodextrin-based supramolecular assemblies and hydrogels: Recent advances and future perspectives. Macromolecular Rapid Communications. 2014;**35**(13):1166-1184

[8] Hu J, Liu S. Engineering responsive polymer building blocks with host–guest molecular recognition for functional applications. Accounts of Chemical Research. 2014;**47**(07):2084-2095

[9] Adler-Abramovich L, Gazit E. The physical properties of supramolecular peptide assemblies: From building block association to technological applications. Chemical Society Reviews. 2014;**43**(20):6881-6893

[10] Yang L, Tan X, Wang Z, Zhang X. Supramolecular polymers: Historical development, preparation, characterization, and functions. Chemical Reviews. 2015;**115**(15):7196-7239

[11] Dong R, Zhou Y, Huang X, Zhu X, Lu Y, Shen J. Functional supramolecular polymers for biomedical applications. Advanced Materials. 2015;**27**(03):498-526

[12] Hart LR, Harries JL, Greenland BW, Colquhoun HM, Hayes W. Supramolecular approach to new inkjet printing inks. Applied Materials & Interfaces. 2015;**7**(16):8906-8914

[13] Perttamo EK. Characterization of associating polymer (AP) solutions. Influences on flow behavior by the degree of hydrophobicity and salinity [thesis]. The University of Bergen; 2013

[14] SNF Floerger. Enhancing Polymer Flooding Performance. 30 Years of Experience in EOR [Internet]. 2017. Availble from: http://www.snfoil.com/images/phocagallery/brochures/SNF_in_chemical_enhanced_oil_recovery.pdf [Accessed: August 2, 2017]

[15] Sasol. Product Application Areas. Surfactants for Enhanced Oil Recovery Bulletin [Internet]. 2017. Available from: http://sasolnorthamerica.com/Images/Interior/productsearch-documents/eor%20bulletin%20sept%2025%202013.pdf [Accessed: August 2, 2017]

[16] Wassmuth FR, Hodgins LA, Schramm LL, Kutay SM. Screening and coreflood testing of gel foams to control excessive gas production in oil wells. SPE Reservoir Evaluation & Engineering. 2001;**4**(03):187-194

[17] Wei B, Romero-Zerón L, Rodrigue D. Novel self-assembling polymeric system based on a hydrophobic modified copolymer: Formulation, rheological characterization, and performance in enhanced heavy oil recovery. Polymers for Advanced Technologies. 2014;**25**(07): 732-741

[18] Wei B, Romero-Zerón L, Rodrigue D. Evaluation of two new self-assembly polymeric systems for enhanced heavy oil recovery. Industrial and Engineering Chemistry Research. 2014;**53**(43):16600-16611

[19] Wei B, Romero-Zerón L, Rodrigue D. Improved viscoelasticity of xanthan gum through self-association with surfactant: β-cyclodextrin inclusion complexes for applications in enhanced oil recovery. Polymer Engineering and Science. 2015;**55**(03):523-532

[20] Wei B, Romero-Zerón L, Rodrigue D. Formulation of a self-assembling polymeric network system for enhanced oil recovery applications. Advances in Polymer Technology. 2014;**33**(03):21413 (1–12)

[21] Herbst F, Döhler D, Michael P, Binder WH. Self-healing polymers via supramolecular forces. Macromolecular Rapid Communications. 2013;**34**(03):203-220

[22] Mezger TG. The Rheology Handbook: For Users of Rotational and Oscillatory Rheometers. 3rd ed. Hanover, Germany: Vincentz Network GmbH & Co KG; 2011

[23] Ayirala SC, Uehara-Nagamine E, Matzakos AN, Chin RW, Doe PH, van den Hoek PJ. A designer water process for offshore low salinity and polymer flooding applications. In: SPE Improved Oil Recovery Symposium; April 24–28, 2010; Tulsa, Oklahoma; 2010. pp. 1-12

[24] Han B, Lee J. Sensitivity analysis on the design parameters of enhanced oil recovery by polymer flooding with low salinity waterflooding. In: The Twenty-fourth International Ocean and Polar Engineering Conference; June 15–20, 2014; Busan, Korea: 2014. pp. 147-151

[25] Khorsandi S, Qiao C, Johns RT. Displacement efficiency for low-salinity polymer flooding including wettability alteration. SPE Journal. 2017;**22**(02):417-30

[26] Rivas C, Gathier F. C-EOR projects—offshore challenges. In: The Twenty-Third International Offshore and Polar Engineering Conference (ISOPE); June 30–July 5, 2013; Anchorage, Alaska: 2013

[27] Jouenne S, Anfray J, Levitt D, Souilem I, Marchal P, Lemaitre C, Choplin L, Nesvik J, Waldman T. Degradation (or lack thereof) and drag reduction of HPAM solutions during transport in turbulent flow in pipelines. Oil and Gas Facilities. 2015;**4**(01):80-92

[28] Seright R, Skjevrak I. Effect of dissolved iron and oxygen on stability of hydrolyzed polyacrylamide polymers. SPE Journal. 2015;**20**(03):433-441

[29] Manichand RN, Let MS, Kathleen P, Gil L, Quillien B, Seright RS. Effective propagation of HPAM solutions through the Tambaredjo reservoir during a polymer flood. SPE Production & Operations. 2013;**28**(04):358-368

[30] Jouenne S, Klimenko A, Levitt D. Polymer flooding: Establishing specifications for dissolved oxygen and iron in injection water. SPE Journal. 2017;**22**(02):438-446

[31] Levitt DB. The optimal use of enhanced oil recovery polymers under hostile conditions [Doctoral dissertation]. Austin: The University of Texas; 2009

Chapter 9

Supramolecular Polymer-Surfactant System for Heavy Oil Recovery

Laura Romero-Zerón and Xingzhi Jiang

Additional information is available at the end of the chapter

http://dx.doi.org/10.5772/intechopen.75368

Abstract

Water soluble polymers are widely used as mobility control agents for enhanced oil recovery (EOR). Yet, in harsh reservoir environments (i.e., elevated temperatures and high ionic strength), the applicability of conventional polymers is limited. This issue has been somewhat resolved through the chemical synthesis of polymers having functional moieties such as sulfonic acid groups and/or n-vinylpyrrolidone. Another approach to circumvent expensive chemical syntheses, it is the formulation of supramolecular polymers built via non-covalent and β-cyclodextrin (β-CD) host-guest interactions. In this study, an advanced polymer-surfactant (SAP-AP1) system formulated via the self-assembling of an associative polymer with an anionic surfactant and β-CD was evaluated as a mobility control agent to displace and recover heavy oil (i.e., 2560 cP at 25°C). Displacement tests employing unconsolidated sand-pack systems were carried out at simulated heavy oil reservoir conditions. The experimental results demonstrate that the SAP-AP1 produces a stable viscous displacement front that results in more efficient volumetric sweep, faster reduction of the water/oil ratio (WOR), and incremental oil recovery (e.g., 19% higher incremental oil recovery relative to the baseline polymer). The SAP-AP1 system shows potential for EOR applications at economically favorable conditions.

Keywords: β-CD host-guest complexations, noncovalent bonding, associating polymers, supramolecular systems, enhanced oil recovery, mobility control, resistance factor, residual resistance factor, water-oil ratio

1. Introduction

Worldwide, polymer flooding is extensively applied as a mobility control agent to increase the sweep efficiency of the displacing fluid during enhanced oil recovery (EOR). As stated by

Perttamo, "[Compared] to conventional waterfloods on a timescale, polymer floods will accelerate the recovery process due to rapid viscosity build-up.... [that] will contribute to a faster and higher oil production. An incremental recovery factor of 5% [of the] original oil in place (OOIP) or more is regarded as a successful polymer application" [1]. Polymer flooding has been historically applied in light and medium gravity oil reservoirs. More recently, it has also been applied successfully in heavy oil reservoirs with oil viscosities ≥1200 cP, which expands the practical applicability of this EOR technique [2–11].

Incremental oil recovery by polymer flooding is induced by the following mechanisms: reduction of the water-oil mobility ratio by means of the increased viscosity of the displacing phase (i.e., injected water) and reduction of viscous fingering, decrease of the water relative permeability due to polymer retention within the formation rock, diversion of the injected water to unswept reservoir regions, improvement of the water-injection profile (i.e., preventing cross-flow between vertical heterogeneous layers), and the increase of oil fractional flow [4, 5, 12–21].

Field polymer-flood projects carry high chemical operating expenditures [5, 22]. Therefore, it is vital to carefully select the appropriate polymer chemistry for the specific reservoir characteristics. For instance, the practical applicability of polymer flooding is limited to reservoirs with moderate temperatures (i.e., <90°C) and formation brines containing low concentrations of divalent cations (e.g., Ca^{2+}, Mg^{2+}) to avoid the chemical degradation of the polymer guaranteeing the technical success of the process in a cost-effective fashion [12–17, 19, 23].

Divalent cations (i.e., Ca^{2+}, Mg^{2+}) significantly affect the viscosity of polymer solutions. The bridging effect of divalent cations with the negatively charged moieties (i.e., carboxyl groups) causes viscosity loss due to polymer coiling [12, 13, 15, 16, 23, 24]. Furthermore, the reaction of the carboxyl group with divalent cations causes polymer precipitation and phase separation [14]. Therefore, to compensate for the loss of viscosity, it is necessary to add higher polymer concentrations to the brine solution [12]. Alternatively, the reservoir could be conditioned before polymer flooding by injecting low-salinity water to prevent the mixing of the polymer slug with the high-salinity reservoir brine [16].

Several approaches have been taken to improve the chemistry of polymers to ensure their stability and functionality at elevated temperatures and in reservoir brine containing high salinity and hardness concentration. These polymeric systems have been customized by incorporating specific functional moieties that are covalently grafted onto the polymer structure. The attachment of sulfonic acid groups like allyl sulfonic acid, 2-acrylamido-2-methylpropane sulfonate (AMPS) and/or n-vinylpyrrolidone (n-VP) monomers increases the polymer resistance to hydrolysis and tolerance to high salinity and hardness. Shear stability and viscosifying power of polymers have been advanced by the introduction of hydrophobic groups like n-alkyl (i.e., $\geq C_6$ carbon numbers) acrylamide, styrene, ring structures, large and rigid side groups such as styrene sulfonic acid, n-alkyl maleimide, acrylamide-base long-chain alkyl acid, and 3-acrylamide-3-methyl butyric acid, among others [3, 13, 14, 17, 23, 24].

EOR polymers are shear sensitive, which is a downside for EOR applications. According to Zaitoun et al. [23] and Sheng et al. [17], shearing occurs within several devices during the different phases of polymer handling and injection process in the high flow rate region close to the injection well such as in shearing devices during polymer dissolution, during recirculation of the polymer solution through centrifugal pumps, polymer flow through chokes and downhole valves under high differential pressure, and during the flow of polymers at high flow rates through the perforations of the reservoir rock and sand face [25]. The shear degradation of the polymer structure consists of the breakage of the macromolecule chain reducing its molecular weight, size, and viscosifying power. Thus, shear degradation is irreversible [17, 23, 26].

The shear degradability of EOR polymers is directly related to the polymer molecular structure, molecular weight, and chain flexibility. The physics of polymer mechanical degradation is reported in [26]. As indicated by Jouenne et al., "flexible polymer chains have the ability to be extended under elongational flow fields [and the] … stretching of the polymer chains can lead to chain rupture" [27]. For example, xanthan gum, which is a rigid rod-like biopolymer with a double-strand helical structure that aligns in the direction of the flow [26], displays a very high shear resistance because it does not stretch under shearing/elongations forces, which reduces the friction forces on the carbon/carbon backbone. On the contrary, the linear polyacrylamide homopolymer is highly flexible and therefore very sensitive to shear degradation. The shear stability of polyacrylamide is commonly improved by introducing negative acrylate groups to the backbone, since it provides rigidity by means of electrostatic repulsion. Nevertheless, the presence of electrolytes (e.g., Na^+, Ca^{2+}, Mg^{2+}) shelters the negative charges of the acrylate groups inducing the coiling and folding of the polymer chain, which becomes less rigid and more flexible [17, 23]. Then, the stretching of the coiled (i.e., coiled-stretch transition) polymer chain under the influence of shear and elongational forces makes it vulnerable to chain breakage and irreversible shear degradation. Thus, the shear sensitivity of EOR polymers increases with brine salinity [23].

The shear stability of acrylamide copolymers can also be increased by introducing the polymer chain large functional hydrophobic groups such as the acrylamide tert-butyl sulfonate (ATBS) and the n-vinylpyrrolidone (n-VP) as they impart rigidity to the polymer structure [23]. The attachment of hydrophobic groups to the macromolecular backbone of EOR polymers to improve the shear and thermal stability, as well as the tolerance to brines with high salinity and hardness concentration, has been widely recognized. The main benefit of the incorporation of hydrophobic groups is as explained by Perttamo [1]: "the reorientation of the macromolecules due to polar and non-polar, results in [the] formation of hydrophobic associations between de incorporated hydrophobic groups," generating intramolecular and intermolecular associations forming supramolecular aggregates. These polymers are called associating hydrophobic polymers or hydrophobically modified polymers or for short associating polymers [1, 17, 19, 28, 29]. Under shear, these supramolecular aggregates can disassemble due to the reversible disruption of the hydrophobic bonds; therefore, at high shear rates, these systems show a remarkable shear-thinning behavior. As indicated by Dupuis et al. [19], these systems offer several benefits for field applications: "…reduced polymer concentration to achieve a required mobility ratio, extend the range of suitable reservoirs in terms of salinity, and facilitate the mixing, pumping, and injection procedures."

In this chapter, we evaluated the effectiveness of a supramolecular polymer-surfactant (SA-AP1) system as a mobility control agent for displacing heavy oil in high salinity and hardness concentration. The SAP-AP1 system was formulated via self-assembling driven by β-CD host-guest complexation, divalent cation bridges (i.e., Ca^{2+} or Mg^{2+}), hydrophobic interactions, and hydrogen bonding, among others. The SAP-AP1 system contains 0.75 wt% of an associating polymer (AP1), 0.007 wt% (70 ppm) of an anionic surfactant, and 0.007 wt% (70 ppm) of β-CD prepared in saline solution. Detailed information on the formulation and properties of the SAP-AP1 system is provided in the preceding chapter of this book.

In this chapter, we begin by describing the sand-pack core-flooding displacement test and the properties of the unconsolidated porous media and fluids (i.e., heavy oil and brine) employed. Next, we discuss the viscosifying power of the baseline AP1 polymer and the SAP-AP1 system during flow through porous media by means of the resistance factor (RF). Polymer retention in porous media is also analyzed through the residual resistance factor (RRF). The effectiveness of both polymer AP1 and the SAP-AP1 system in recovering heavy oil is analyzed next. Finally, we discuss the effect of AP1 and SAP-AP1 on the water to oil production ratio (WOR).

2. Sand-pack flooding displacement tests

The performance of the polymer AP1 and the SAP-AP1 system as a mobility control agent for heavy oil recovery was determined through routine oil sand-pack displacement tests at simulated reservoir conditions. The heavy oil used in these flooding tests was provided by Husky Energy Inc. (Calgary, AB, Canada) with a viscosity of 68,728 cP at 25°C that was adjusted to a viscosity of 2560 cP at 25°C by dilution with natural condensate produced from the McCully field, Corridor Resources Inc. (Sussex, NB, Canada). The density of the diluted crude oil was 0.954 g/ml at 25°C, the API corrected to 60°F was 15.27, and the interfacial tension (IFT) between the crude oil and the SAP-AP1 system was 0.032 dynes/cm at 25°C. The IFT was determined using a M6500 Spinning Drop Tensiometer manufactured by Grace Instrument (Houston, TX, USA). QUIKRETE® Premium Play Sand® (No. 1113), which is 100% quartz [30], was employed to prepare the unconsolidated sand packs. The sand-grain size distribution was determined by sieve analysis following the procedure described in [31], which conforms to ASTM C136/C136M-14. The sieve analysis indicated that the effective size of the sand, D10, and the uniformity coefficients, D60/D10, were 240 and 2.02 μm, respectively.

A total of four displacement tests were conducted at a temperature of 25°C using a brine concentration of 8.4 wt%. The synthetic brine composition was 6.9 wt% of NaCl, 0.18 wt% of $MgCl_2$, 1.3 wt% of $CaCl_2$, and 0.04 wt% of Na_2SO_4. Two displacement tests were conducted using the baseline polymer AP1 at the optimum concentration of 0.75 wt% (control tests) and two displacement tests were carried out using the optimum SAP-AP1 system which also contained a polymer concentration of 0.75 wt%. Displacement tests were carried out using a DCHH series core holder (pressure-tapped, biaxial-type loading) manufactured by Temco, Inc. (Tulsa, OK, USA). Two CFR series transfer vessels (Temco, Inc., Tulsa, OK, USA) were employed to displace brine, polymer, and crude oil through the sand pack. A Teledyne ISCO Syringe pump, model 100DX manufactured by Teledyne Isco, Inc. (Lincoln, NE, USA), was used to pump the fluids through the transfer vessels. Several PGT-30 series/stainless steel pressure gauges manufactured by Omega (Laval, Quebec, Canada) with an accuracy of 0.5% as a percent of full scale

(FS) were installed at the inlet of the core holder (P1), at the inlet cap to monitor the overburden pressure (P_{OP}), and two pressure gauges along the core holder (P2 and P3). **Figure 1** shows a simplified schematic of the experimental set-up used during the sand-pack displacement tests.

The sand-pack properties, such as pore volume (PV), porosity (φ), and permeability to brine (k) were determined following routine procedures as outlined in [32, 33]. **Table 1** presents the sand-pack properties for each of the displacement tests.

Heavy oil sand-pack displacement tests were carried out following a fluid injection scheme of four stages: heavy oil injection, waterflooding, polymer flooding, and post-polymer waterflooding. All the fluids were injected at a flow rate of 0.98 cm³/min, which is equivalent to a flow velocity of 0.0116 m/s (0.91 ft./day). During the heavy oil injection stage, oil was continuously injected until the production of brine stopped, which corresponds to a volume of oil equivalent to three pore volumes (3 PV).

Afterward, the oil-saturated sand packs were waterflooding to displace oil by injecting brine (8.4 wt%) at a flow rate of 0.98 cm³/min until no more oil was produced, which corresponds to a volume of brine equivalent to 6 pore volumes (6 PV). Right after the waterflooding stage was completed, 1 pore volume of AP1 or SAP-AP1 polymer solution was injected at a flow rate of 0.98 cm³/min to displace the unrecovered or "remaining" oil that was bypassed during the previous waterflooding stage [34]. Finally, a post-polymer waterflooding step was immediately initiated at the end of the polymer flooding stage. A total volume of 6.5 pore volumes of brine (8.4 wt%) was injected at a flow rate of 0.98 cm³/min. In each of the injection stages, the injection time, pressure readings, and the corresponding volumes of the fluids (brine and oil) produced

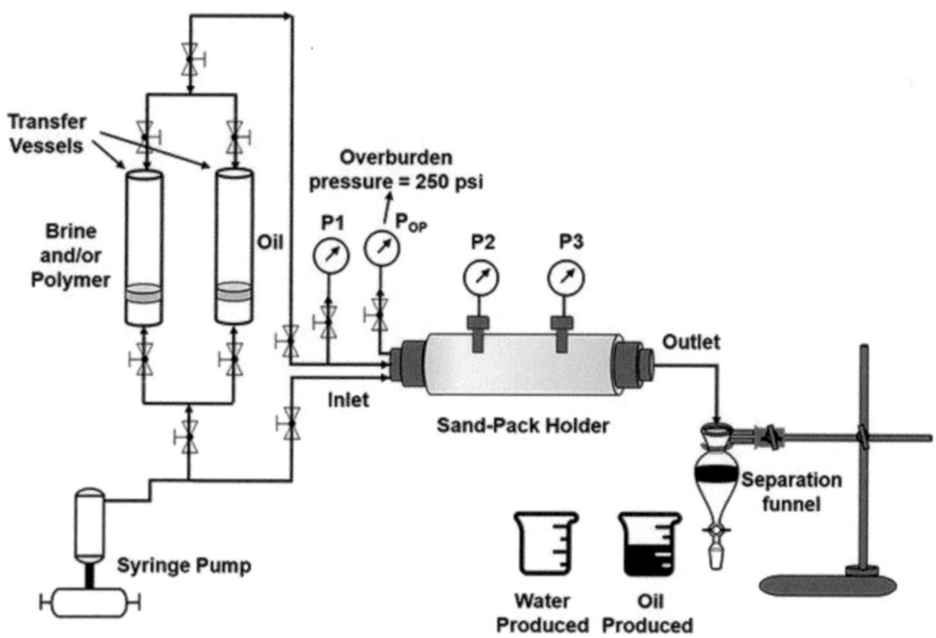

Figure 1. Experimental set-up for sand-pack displacement tests.

Displacement test #	Pore volume [cm³]	Porosity [%]	Permeability [mD]
Baseline # 1.1	173	24	3085
SAP AP1 # 1.2	177	25	2812
Baseline # 2.1	169	24	2746
SAP AP1 # 2.2	185	26	1758

Table 1. Sand-pack properties.

were monitored. Material balance was applied to determine oil and water saturations, as well as the percentage of oil recovered from each injection stage. More details of the experimental procedure employed during the sand-pack displacement tests are provided in [35].

The results obtained from the displacement tests were analyzed by plotting the resistance factor (RF), the residual resistance factor (RRF), percentage of cumulative oil recovery, ratio of remaining oil saturation over initial oil saturation (S_{ro}/S_{oi}), and water oil ratio (WOR) as a function of volume of fluid injected. The volume of fluid injected was expressed as a fraction of pore volume normalized by porosity and permeability using the capillary bundle parameter [36–39] to compare the displacement tests at the same porosity and permeability reference.

3. Resistance factor and residual resistance factor

Resistance factor (RF) provides information on the effective viscosity of the polymer solution during flow in porous media relative to water before polymer flooding [38]; therefore, RF indicates the effectiveness of the polymer system as a mobility control agent during enhanced oil recovery (EOR) [6, 19, 29, 40, 41]. While, the residual resistance factor (RRF) measures the increased pressure drop across the porous media due to polymer retention (mechanical entrapment and polymer adsorption) [5, 6, 19, 29, 37–44].

Figure 2(a) and **(b)** plots RF as a function of volume of fluid injected expressed as a fraction of pore volume (PV) normalized for permeability and porosity for tests—Baseline # 1.1 and SAP-AP1 # 1.2 and Baseline # 2.1 and SAP-AP1 # 2.2 respectively.

Figure 2 demonstrates that the RF curves of the AP1 polymer and the SAP-AP1 network have similar flow behavior. After the injection of 0.013 PV, the RF values increase continuously with increasing volume of fluid injected until a maximum RF value was reached at about 0.02 PV. Then, the RF values decreased with increased throughput in the sand packs until the RF values leveled off. The RF value plateaued out at an average value of 5.6 for the SAP-AP1 system, while for the baseline AP1 polymer, the RF value plateaued out at an average value of 2.0. Therefore, the SAP-AP1 system offered on average 3.6 times higher effective viscosity within the porous media relative to the effective viscosity achieved by polymer AP1.

The RF-curves eventually reached plateau values for both systems: the SAP-AP1 and the baseline. This may occur due to the dynamic disassembling and reassembling of the non-covalent

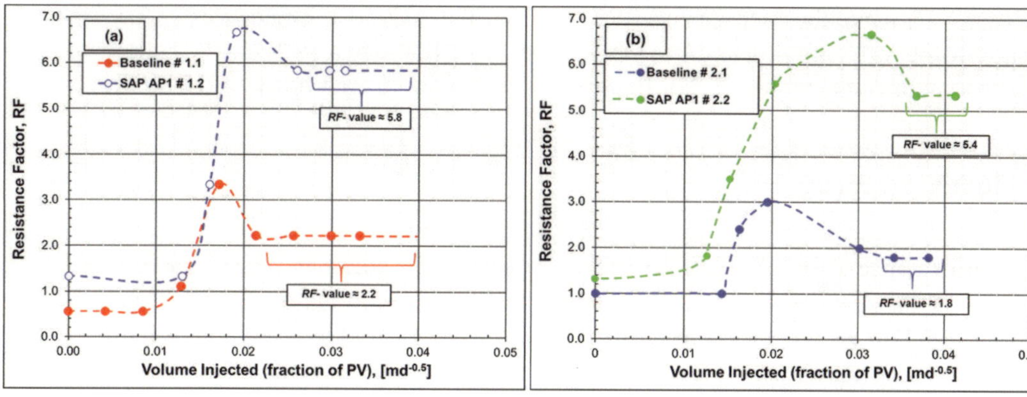

Figure 2. RF versus volume of fluid injected: (a) Baseline # 1.1 and SAP-AP1 # 1.2 and (b) Baseline # 2.1 and SAP-AP1 # 2.2.

intra- and intermolecular interactions (i.e., decomplexation and complexation of host-guest interactions, breaking and reforming of hydrophobic interactions, and hydrogen bonds, among others) under the influence of the shear forces imposed during the flow process. Eventually, equilibrium of the shear forces and the flow resistance forces of the network structures are reached, and the RF curves leveled off. The stabilization of the RF curves also suggests that propagation of the SAP-AP1 systems and the baseline polymers took place along the sand-pack systems. These observations agree with previous research on flow behavior of associating polymers through porous media [28, 38, 44]. Overall, the optimum SAP-AP1 formulation consistently provides higher resistance factors and consequently a better mobility control than the baseline polymer AP1.

The performance of both polymers AP1 and SAP-AP1 in terms of the residual resistance factor, RRF, is presented in **Figure 3(a)** and **3(b)**, which plots RRF values as a function of volume of brine injected for tests Baseline # 1.1 and SAP-AP1 # 1.2 and Baseline # 2.1 and SAP-AP1 # 2.2, respectively.

Figure 3 indicates that the RRF curves for both systems decrease continuously as the volume of brine injected increases that eventually stabilize. The average end RRF value for the

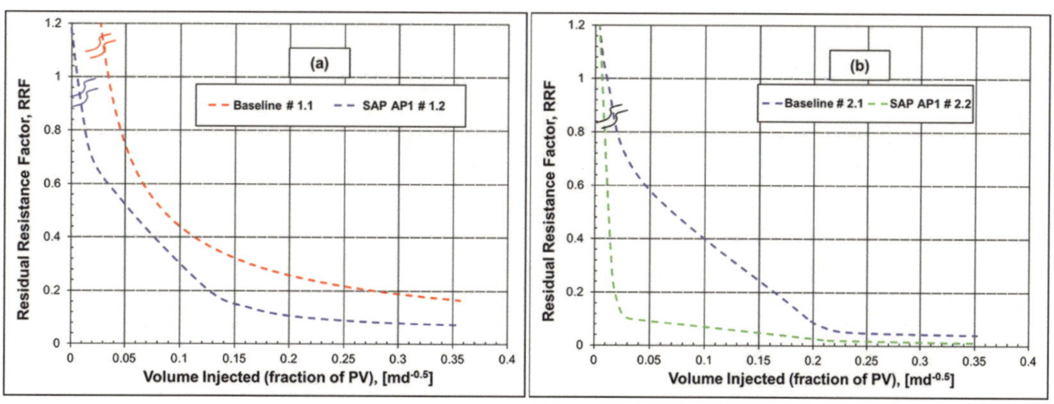

Figure 3. RRF versus volume of fluid injected: (a) Baseline # 1.1 and SAP-AP1 # 1.2 and (b) Baseline # 2.1 and SAP-AP1 # 2.2.

AP1 was 0.5, while the average end RRF value for the SAP-AP1 system was 0.02, suggesting insignificant pore plugging and/or permeability reduction due to polymer retention, which is expected in unconsolidated and/or high permeability porous media [37, 42].

4. Heavy oil recovery

The performance of the SAP-AP1 system and baseline AP1 polymer as mobility control agents for the displacement and recovery of heavy oil is displayed in **Figure 4**, which plots the percent of cumulative oil recovery as a function of volume of fluid injected and flooding stage.

Figure 4 reveals that the average oil recovery during the waterflooding stage was about 30% for all the displacement tests. The combined average cumulative oil recovery produced by polymer flooding and post-polymer waterflooding for the Baseline tests # 1.1 and # 2.1 was 37%, respectively, after subtracting the average oil recovery attributed to the initial waterflooding stage. Whereas, the combined average cumulative oil recovery produced by flooding and post-polymer waterflooding for the SAP-AP1 tests # 1.1 and # 2.1 was 56%, respectively, after subtracting the average oil recovery attributed to the initial waterflooding stage. These experimental observations demonstrate that the SAP-AP1 system produced an additional incremental oil recovery of 19% relative to the baseline AP1 polymer. In this analysis, average values of cumulative oil recovery were used as an alternative to the individual results from each of the displacement tests to provide a conservative assessment of the experimental results in terms of heavy oil recovery. This approach was necessary to avoid the overestimation of oil recovery from polymer flooding and the post-polymer waterflooding stage, because the oil and water separation process after polymer flooding was found to be a difficult and lengthy process, even though several experimental steps were carried out to achieve the most effective separation of water from the produced oil.

The ratio of remaining to initial oil saturation as a function of volume of fluid injected and injection step is presented in **Figure 5** for the baseline polymer and SAP-AP1 system.

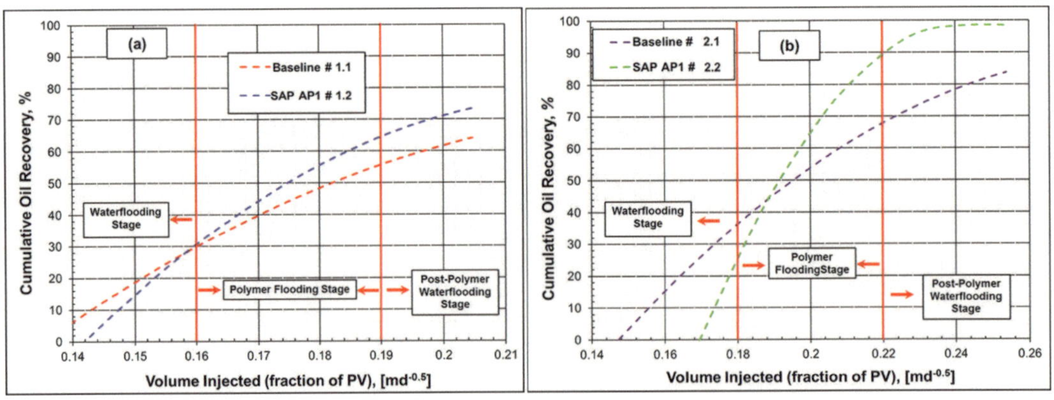

Figure 4. Percentage of cumulative oil recovery versus volume of fluid injected and flooding stage: (a) Baseline # 1.1 and SAP-AP1 # 1.2 and (b) Baseline # 2.1 and SAP-AP1 # 2.2.

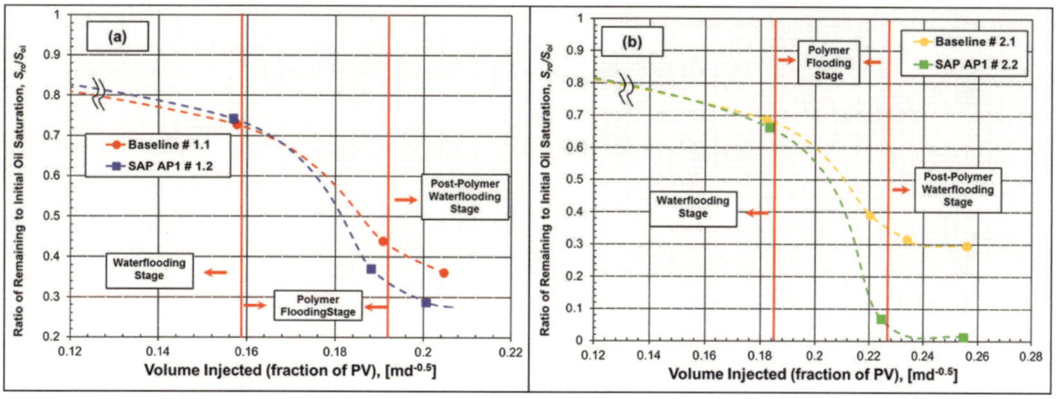

Figure 5. S_{ro}/S_{oi} versus volume of fluid injected and flooding stage: (a) Baseline # 1.1 and SAP-AP1 # 1.2 and (b) Baseline # 2.1 and SAP-AP1 # 2.2.

Figure 5 clearly shows the effect of each of the flooding stages on oil recovery. The polymer flooding stage produced a dramatic decrease of the S_{ro}/S_{oi} ratio that rapidly stabilized during the post-polymer waterflooding stage. The average S_{ro}/S_{oi} ratio obtained from the Baseline tests # 1.1 and # 2.1 was 0.33, while the average S_{ro}/S_{oi} ratio attained from the SAP-AP1 tests # 1.2 and # 2.2 was 0.14.

These experimental results demonstrate that the optimum SAP-AP1 system provided a more efficient mobility control compared to the baseline AP1, which resulted in a more stable viscous displacement and accelerated heavy oil recovery. Furthermore, the low concentration of anionic surfactant contained in the SAP-AP1 formulation reduces the interfacial tension (IFT) of the oil-brine system from 30 [45] to 0.032 dynes/cm. This remarkable reduction in IFT decreases capillary forces, which facilitates the detachment and mobilization of oil during SAP-AP1 flooding [46, 47]. Therefore, the SAP-AP1 system produces incremental oil recovery by the synergistic effect of greater mobility control functionality and by decreasing the IFT of the oil-brine system.

5. Water-to-oil ratio

During waterflooding of heavy oil, "the adverse mobility ratio between the viscous oil and the water induces high-water-cut production and poor sweep efficiency" [2]. Polymer flooding decreases the mobility of the injected water (i.e., augmented water viscosity) reducing the water-cut production levels. **Figure 6** presents the water/oil ratio (WOR) as a function of volume of fluid injected and the flooding stage for the baseline polymer and the SAP-AP1 system tests.

Figure 6 indicates that the average WOR at the end of the initial waterflooding stage for the displacement tests was about 10. As soon as the polymer flooding stage (i.e., baseline polymer AP1 and/or SAP-AP1 system) was initiated, WOR continuously decreased as the volume of polymer injection increased, reaching a minimum WOR value at the end of the polymer flooding stage. The WOR curves in **Figure 6** show that the SAP-AP1 system was more efficient in reducing and controlling the water-oil ratio by providing a faster response and lower average

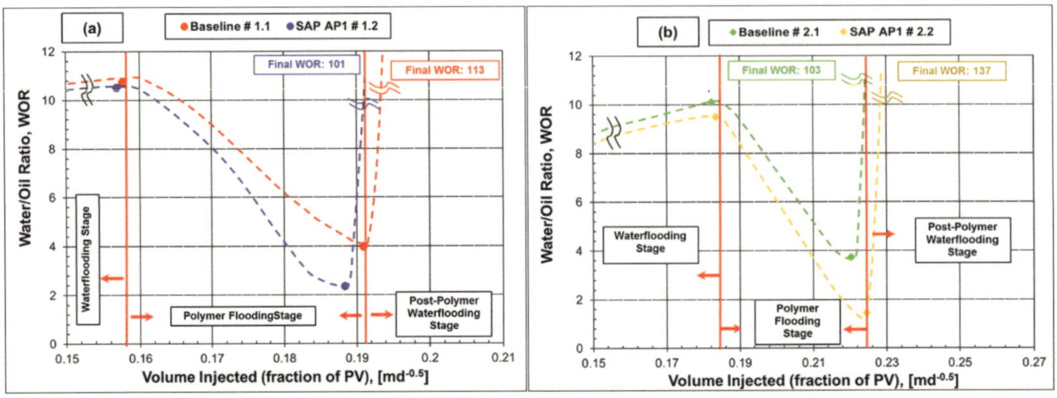

Figure 6. WOR versus volume of fluid injected and flooding stage: (a) Baseline # 1.1 and SAP-AP1 # 1.2 and (b) Baseline # 2.1 and SAP-AP1 # 2.2.

WOR value of 1.9. Whereas, the baseline AP1 polymer rendered a minimum average WOR value of 3.9 at the end of the polymer flooding stage. The WOR curves also show that immediately after the post-polymer waterflooding stage was initiated, an abrupt increase of the water to oil ratio took place that continued until the end of the post-polymer waterflooding stage. As explained by Seright "once brine injection [post-polymer flooding] begins, viscous fingering and [porous media] heterogeneities will quickly lead to severe channeling" [of the water to the production end] [6]. These results demonstrate that the optimum SAP-AP1 system offers superior mobility control functionality relative to the baseline AP1 polymer. The structural strength of the SAP-AP1 system is more effective in generating a stable viscous displacement that promotes a more efficient volumetric heavy oil sweep, a faster WOR reduction, and accelerated heavy oil recovery.

6. Conclusions

We discussed the performance of an advanced supramolecular polymer-surfactant system, SAP-AP1, driven by β-cyclodextrin host-guest complexations as mobility control agent to displace heavy oil (i.e., 2560 cP at 25°C).

Heavy oil recovery displacement tests demonstrated that the SAP-AP1 system shows suitable propagation and low retention in unconsolidated sand-pack systems. The SAP-AP1 system displays superior mobility control efficiency when compared to the baseline AP1 polymer. The higher structural strength of the SAP-AP1 system makes it more effective in generating a stable viscous displacement front that results in a more efficient volumetric sweep, a faster WOR reduction, and accelerated heavy oil recovery. An average additional incremental oil recovery of 19% was achieved relative to the baseline AP1 polymer.

The important incremental oil recovery achieved by the supramolecular polymer-surfactant system is also attributed to the synergistic effect of greater mobility control functionality and decreased interfacial tension (IFT) between the oil-brine system offered by the SAP-AP1 system.

Overall, the SAP-AP1 system offers the potential for increasing heavy oil recovery at economically favorable conditions.

Acknowledgements

We thank Nayef Alayad (Chemical Engineering Department, University of New Brunswick) for helping with the sand-pack displacement tests. We gratefully acknowledge Sasol Chemicals, SNF Floerger, Husky Energy Inc., and Corridor Resources Inc. for providing samples of surfactant, polymer, heavy oil, and natural condensate employed in this study. Financial support from the University of New Brunswick through the Sabbatical Research Grant and the Canada Foundation for Innovation, CFI, research grant is greatly appreciated.

Author details

Laura Romero-Zerón* and Xingzhi Jiang

*Address all correspondence to: laurarz@unb.ca

Chemical Engineering Department, University of New Brunswick, Fredericton, New Brunswick, Canada

References

[1] Perttamo EK. Characterization of Associating Polymer (AP) Solutions. Influences on flow behavior by the degree of hydrophobicity and salinity [thesis]. The University of Bergen; 2013

[2] Delamaide E, Zaitoun A, Renard G, Tabary R. Pelican Lake field: First successful application of polymer flooding in a heavy-oil reservoir. SPE Reservoir Evaluation & Engineering. 2014;**17**(03):340-354

[3] Kang PS, Lim JS, Huh C. A novel approach in estimating shear-thinning rheology of HPAM and AMPS polymers for enhanced oil recovery using artificial neural network. In: The Twenty-third International Offshore and Polar Engineering Conference; Jun 30-Jul 5, 2013; Anchorage, Alaska: ISOPE-I-13-171; 2013. pp. 81-85

[4] Koh H, Lee VB, Pope GA. Experimental investigation of the effect of polymers on residual oil saturation. Society of Petroleum Engineers. February 1, 2018. pp. 1-17. DOI: 10.2118/179683-PA

[5] Li K, Sun W, Li F, Qu Y, Yang Y. Novel method for characterizing single-phase polymer flooding. SPE Journal. 2014;**19**(04):695-702

[6] Seright RS. How much polymer should be injected during a polymer flood? Review of previous and current practices. SPE Journal. 2017;**22**(01):1-8

[7] Saleh LD, Wei M, Bai B. Data analysis and updated screening criteria for polymer flooding based on oilfield data. SPE Reservoir Evaluation & Engineering. 2014;**17**(01):15-21

[8] Morel DC, Vert M, Jouenne S, Gauchet R, Bouger Y. First polymer injection in deep offshore field Angola: Recent advances in the Dalia/Camelia field case. Oil Gas Facil. 2012;**1**(02):43-52

[9] Saleh L, Wei M, Zhang Y, Bai B. Data analysis for polymer flooding that is based on a comprehensive database. SPE Reservoir Evaluation & Engineering. 2016;**17**(01):1-18

[10] Li Z, Delshad M. Development of an analytical injectivity model for non-Newtonian polymer solutions. SPE Journal. 2014;**19**(03):381-389

[11] Pantus PV. Experimental investigation of viscosity ratio effect on displacement performance of polymer systems during heavy oil recovery [doctoral dissertation]. Regina, Canada: University of Regina; 2012

[12] Ayirala SC, Uehara-Nagamine E, Matzakos AN, Chin RW, Doe PH, van den Hoek PJ. A designer water process for offshore low salinity and polymer flooding applications. In: SPE Improved Oil Recovery Symposium; 24-28 April, 2010; Tulsa, Oklahoma; 2010. pp. 1-12

[13] Vermolen E, Van Haasterecht MJ, Masalmeh SK, Faber MJ, Boersma DM, Gruenenfelder MA. Pushing the envelope for polymer flooding towards high-temperature and high-salinity reservoirs with polyacrylamide based ter-polymers. In: SPE Middle East Oil and Gas Show; 25-28 September, 2011; Manama, Bahrain; 2011. pp. 1-9

[14] Wu Y, Mahmoudkhani A, Watson P, Fenderson TR, Nair M. Development of new polymers with better performance under conditions of high temperature and high salinity. In: SPE EOR Conference at Oil and Gas West Asia; 16-18 April, 2012; Muscat, Oman. 2012. pp. 1-11

[15] Han B, Lee J. Sensitivity analysis on the design parameters of enhanced oil recovery by polymer flooding with low salinity waterflooding. In: The Twenty-fourth International Ocean and Polar Engineering Conference; 15-20 June, 2014; Busan, Korea. 2014. pp. 147-151

[16] Khorsandi S, Qiao C, Johns RT. Displacement efficiency for low-salinity polymer flooding including wettability alteration. SPE Journal. 2017;**22**(02):417-430

[17] Sheng JJ, Leonhardt B, Azri N. Status of polymer-flooding technology. Journal of Canadian Petroleum Technology. 2015;**54**(02):116-126

[18] Rivas C, Gathier F. C-EOR projects–offshore challenges. In: The Twenty-Third International Offshore and Polar Engineering Conference; June 30–July 5, 2013; Anchorage, Alaska: ISOPE; 2013

[19] Dupuis G, Rousseau D, Tabary R, Grassl B. Flow of hydrophobically modified water-soluble polymers in porous media: Controlled resistance factors vs. flow-induced gelation in the semidilute regime. SPE Journal. 2012;**17**(04):1196-1206

[20] Choi SK, Sharma MM, Bryant S, Huh C. pH-sensitive polymers for novel conformance-control and polymer-flood applications. SPE Reservoir Evaluation & Engineering. 2010; **13**(06):926-939

[21] Zechner M, Clemens T, Suri A, Sharma MM. Simulation of polymer injection under fracturing conditions—An injectivity pilot in the Matzen field, Austria. SPE Reservoir Evaluation & Engineering. 2015;**18**(02):236-249

[22] Choudhuri B, Thakuria C, Belushi AA, Nurzaman Z, Al Hashmi K, Batycky R. Optimization of a large polymer flood with full-field streamline simulation. SPE Reservoir Evaluation & Engineering. 2015;**18**(03):318-328

[23] Zaitoun A, Makakou P, Blin N, Al-Maamari RS, Al-Hashmi AA, Abdel-Goad M. Shear stability of EOR polymers. SPE Journal. 2012;**17**(02):335-339

[24] Levitt D, Pope GA. Selection and screening of polymers for enhanced-oil recovery. In: SPE Symposium on Improved Oil Recovery; 19-23 April 2008; Tulsa, Oklahoma; 2008. pp. 1-18

[25] Lüftenegger M, Kadnar R, Puls C, Clemens T. Operational challenges and monitoring of a polymer pilot, Matzen field, Austria. SPE Production & Operations. 2016;**31**(03):228-237

[26] Jouenne S, Chakibi H, Levitt D. Polymer stability after successive mechanical-degradation events. Society of Petroleum Engineers. February 1, 2018. pp. 1-16. DOI: 10.2118/186103-PA

[27] Jouenne S, Anfray J, Levitt D, Souilem I, Marchal P, Lemaitre C, Choplin L, Nesvik J, Waldman T. Degradation (or lack thereof) and drag reduction of HPAM solutions during transport in turbulent flow in pipelines. Oil and Gas Facilities. 2015;**4**(01):80-92

[28] Guo Y, Hu J, Zhang X, Feng R, Li H. Flow behavior through porous media and microdisplacement performances of hydrophobically modified partially hydrolyzed polyacrylamide. SPE Journal. 2016;**03**:688-705

[29] Pancharoen M, Thiele MR, Kovscek AR. Inaccessible pore volume of associative polymer floods. In: SPE Improved Oil Recovery Symposium; 24-28 April, 2010; Tulsa, Oklahoma; 2010. pp. 1-15

[30] Products TM, QUIKRETE Cement & Concrete. GHS Safety Data Sheet. B4: Packaged Aggregates [Internet]. 2015. Available from https://www.quikrete.com/pdfs/sds-b4-playsand.pdf [Accessed: 2017-08-12]

[31] The Ohio State University. Sand Size Analysis for Omsite Wastewater Treatment Systems. Determination of Sand Effective Size and Uniformity Coefficient. [Internet]. 2017. Available from: https://ohioline.osu.edu/factsheet/aex-757. [Accessed: 2017-08-18]

[32] McPhee C, Reed J, Zubizarreta I. Core Analysis: A Best Practice Guide. Waltham, MA: Elsevier; 2015

[33] Satter A, Iqbal GM, Buchwalter JL. Practical Enhanced Reservoir Engineering: Assisted with Simulation Software. Tulsa, OK: Pennwell Books; 2007

[34] Qi P, Ehrenfried DH, Koh H, Balhoff MT. Reduction of residual oil saturation in sandstone cores by use of viscoelastic polymers. SPE Journal. 2017;**22**(02):447-458

[35] Riahinezhad M, Romero-Zerón L, McManus N, Penlidis A. Evaluating the performance of tailor-made water-soluble copolymers for enhanced oil recovery polymer flooding applications. Fuel. 2017;**203**:269-278

[36] Civan F. Porous Media Transport Phenomena. New Jersey: John Wiley & Sons; 2011. 488 p

[37] Seright RS. Use of Polymers to Recover Viscous Oil from Unconventional Reservoirs. Final Report. Contract No. DE-NT0006555. US Department of Energy. 2011

[38] Seright RS, Fan T, Wavrik K, Wan H, Gaillard N, Favéro C. Rheology of a new sulfonic associative polymer in porous media. SPE Reservoir Evaluation & Engineering. 2011;**14**(06):726-734

[39] Taylor KC, Nasr-El-Din HA. Water-soluble hydrophobically associating polymers for improved oil recovery: A literature review. Journal of Petroleum Science and Engineering. 1998;**19**(03):265-280

[40] Kazempour M, Sundstrom EA, Alvarado V. Effect of alkalinity on oil recovery during polymer floods in sandstone. SPE Reservoir Evaluation & Engineering. 2012;**15**(02):195-209

[41] Manichand RN, Seright R. Field vs. laboratory polymer-retention values for a polymer flood in the Tambaredjo field. SPE Reservoir Evaluation & Engineering. 2014;**17**(03):314-325

[42] Lotfollahi M, Farajzadeh R, Delshad M, Al-Abri AK, Wassing BM, Al-Mjeni R, Awan K, Bedrikovetsky P. Mechanistic simulation of polymer injectivity in field tests. SPE Journal. 2016;**21**(04):1178-1191

[43] Quadri SM, Shoaib M, AlSumaiti AM, Alhassan SM. Screening of polymers for EOR in high temperature, high salinity and carbonate reservoir conditions. In: International Petroleum Technology Conference; 6-9 December, 2015; Doha, Qatar; 2015. pp. 1-30

[44] Levitt DB. The optimal use of enhanced oil recovery polymers under hostile conditions [doctoral dissertation]. Austin, The University of Texas; 2009

[45] Strausz OP, Lown EM. The Chemistry of Alberta Oil Sands, Bitumens and Heavy Oils. Alberta Energy Research Institute: Calgary; 2003

[46] Watson A, Trahan GA, Sorensen W. An interim case study of an alkaline-surfactant-polymer flood in the mooney field, Alberta, Canada. In: SPE Improved Oil Recovery Symposium; 12-16 April, 2014; Tulsa, Oklahoma; 2014. pp. 1-16

[47] Wan W. Chemical flood under high total dissolved solids (TDS) conditions [doctoral dissertation]. Norman, Oklahoma: Mewbourne School of Petroleum and Geological Engineering, Graduate College, University of Oklahoma; 2014

Chapter 10

Promising Use of Cyclodextrin-Based Non-Viral Vectors for Gene and Oligonucleotide Drugs

Ahmed F.A. Mohammed, Keiichi Motoyama,
Taishi Higashi and Hidetoshi Arima

Additional information is available at the end of the chapter

http://dx.doi.org/10.5772/intechopen.74614

Abstract

Genes, short-hairpin RNA (shRNA), small-interfering RNA (siRNA), and decoy DNA can be principally used as tools for the treatment and prevention of many disorders, including but not limited to cancers, genetic disorders, and inherited diseases. This is accomplished by introducing exogenous nucleic acids into mammalian cells to modulate gene expression. However, direct use of such oligonucleotide drugs is hampered by several barriers, including their degradation by nucleases present in the blood and extracellular fluid, cell-membrane impermeability, and their retention in endosomes. To address this issue, the development of safe and effective delivery vectors has emerged as the main fundamental challenge for successful gene and oligonucleotide therapy. Due to the intrinsic risks associated with viral vectors, non-viral vectors have attracted increasing attention as gene and oligonucleotide carriers. We originally developed various cyclodextrin (CyD) conjugates with polyamidoamine (PAMAM) dendrimers as novel CyD-based polymers for the delivery of plasmid DNA, siRNA, shRNA, and decoy DNA. In this review, we describe the recent findings on PAMAM dendrimer conjugates using CyDs as carriers for gene, shRNA, siRNA, and decoy DNA delivery.

Keywords: cyclodextrin, polyamidoamine dendrimer, conjugate, DNA delivery, shRNA delivery, siRNA delivery

1. Introduction

The principle of somatic gene therapy is to introduce certain genes into selected cells to treat a genetic or acquired disease by interfering with the expression of specific proteins or potentially fixing a genetic mutation. Since the discovery of gene therapy, new perspectives for diagnosis, prevention, and treatment of incurable diseases have evolved, including those related to various types of cancers, cardiovascular diseases, dermatological diseases, and even vision loss [1–3]. Successful gene therapy requires a delivery system capable of efficiently delivering the genetic cargo to all target cells. There are two different technologies that have been developed for this purpose: (1) viral and two types of non-viral vectors, and (2) lipofection and polyfection. As of 2012, >1800 clinical trials have been performed in >31 countries [4], with lipofection and naked DNA representing only 6.2 and 18.6% of the methods studied, respectively. However, polyfection has never been applied to such clinical trials. Notably, in July 2012, the European Medicines Agency recommended Glybera, which is a gene-therapy product based on an engineered adeno-associated viral vector for the treatment of severe lipoprotein-lipase deficiency in muscles, for approval as the first gene-therapy drug in the European Union. More recently, Strimvelis was approved by the European Commission in 2016 as the first *ex vivo* stem cell gene therapy for the treatment of patients with a very rare disorder called severe combined immunodeficiency due to adenosine deaminase deficiency. In some cases, viral vectors have the advantage of producing higher gene expression, although there are potential safety risks, such as immunogenicity that causes inflammatory reactions, toxin production, oncogenesis, and insertional mutagenesis [5]. The advantages of non-viral vectors in this context include higher safety levels, lower immunogenicity, cost-effectiveness, and their ability to attach a targeting ligand [6]. However, the efficiency of non-viral systems for gene delivery is markedly low in regard to gene transduction as compared with viral vectors, likely due to the lack of recognition receptors and lower endosomal-escape capability and nuclear-pore targeting [7]. Therefore, non-viral vectors have been continuously researched and developed in order to overcome their disadvantages and successfully achieve the desired effects of gene therapy.

The discovery of RNA interference (RNAi) as an endogenous tool capable of fine-tuning gene expression is considered as one of the most important discoveries in the recent years [8]. Since then, many efforts have been made to exploit this natural mechanism to experimentally silence target genes, as well as to study functional genomics by inhibiting gene expression via the degradation of specific mRNA. Selective gene silencing by RNAi can be achieved either by cytoplasmic delivery of double-stranded small-interfering RNA (siRNA; 21–27 bp) or by nuclear delivery of gene-expression cassettes that express short-hairpin RNA (shRNA) [9, 10].

There are many types of nucleic acid drugs, including siRNA, microRNA (miRNA), antisense oligonucleotides, decoy DNA, aptamers, and ribozymes. siRNA is a double-stranded RNA that targets specific mRNA harboring a complementary sequence and causes its degradation; whereas, miRNA is a small noncoding-RNA molecule found in plants and animals and exhibiting functions related to RNA silencing and posttranscriptional regulation of gene expression. Furthermore, antisense oligonucleotides are single-stranded DNA harboring a complementary sequence with that of RNA molecules used to inhibit its expression. Decoy DNAs are short double-stranded oligodeoxynucleotides containing the binding sequence of a transcription factor, which prevents the transcription factor from binding to the genomic promoter, thereby resulting in transcription inhibition. Additionally, an aptamer is a nucleic acid molecule (DNA or RNA) that binds to targets with high selectivity and sensitivity, whereas ribozymes are RNA enzymes capable of binding to specific RNA, resulting in sequence-specific cleavage. Considering the marketed nucleic acid drugs, we can find only two antisense products and one aptamer product. Vitravene (fomivirsen) is the first marketed antisense product approved by the Food and Drug Administration, and represents an antiviral antisense drug used for the treatment of patients with cytomegalovirus retinitis or those with acquired immune deficiency syndrome. The second antisense product is Kynamro (mipomersen sodium), which is used to treat patients with homozygous familial hypercholesterolemia. The aptamer product is Macugen (pegaptanib sodium), which is an anti-angiogenic drug used for the treatment of age-related macular degeneration and that binds specifically to vascular endothelial growth factor protein to block its activity. However, there are currently no commercially marketed or approved decoy DNA or siRNA drugs, although several siRNA and miRNA drugs have been in clinical trials [11].

The direct use of such nucleic acid drugs has been hindered by multiple barriers that prevent these genes and nucleic acids from reaching their targets. These barriers include degradation of nucleic acids by nucleases in the blood and extracellular fluid, elimination by the reticuloendothelial system, cell-membrane impermeability, and their retention and in endosomes and subsequent degradation [12]. Consequently, the development of a delivery system capable of overcoming these barriers is extremely important to achieving the required effects of these nucleic acid drugs. Non-viral vectors have been developed for this purpose due to their higher safety levels, non-immunogenicity, and non-pathogenicity. Methods using non-viral carriers include lipofection, which depends upon liposomes to deliver the nucleic acid cargo, polyfection, which is based on the use of a polycation complex with nucleic acids, and lipopolyfection.

Polycation-based nucleic acid drug-delivery methods offer adequate efficiency in protecting the therapeutic genes against the aforementioned barriers and also help in overcoming intracellular barriers [13]. One of the most promising approaches to non-viral delivery of nucleic acids and genes involves cationic polymers due to their ease of manufacture, flexible properties, and robustness [14]. Polyamidoamine (PAMAM) Starburst dendrimers (dendrimers), which were developed by Tomalia et al., represent the first polycations exhibiting intrinsic endosomal-release activity, thereby removing the need to add external endosomolytic agents as required by other polycations, such as poly(L-lysine) [15–17]. The dendrimers are biocompatible, non-immunogenic, and water-soluble polymers possessing unique, spherical, and highly ordered structures with a narrow molecular-weight distribution, very low polydipersity, and specific size [18, 19].

These characteristics attracted many nanotechnology, pharmaceutical, and medicinal chemistry scientists interested in their use for various applications. Because, dendrimers have terminal modifiable amino-functional groups positively charged at physiological pH [20], and they are capable of forming complexes with genes [15, 21] and oligonucleotides [22] through electrostatic interactions, as well as with glycosaminoglycans on cell surfaces [23]. Furthermore, their proton-sponge effect, as well their defined shapes, explains their high degree of transfection efficiency. In this case, the proton-sponge effect is produced by the large buffering capacity of cationic polymers that stimulate endosome swelling and eventual endosomal-membrane disruption and release of nucleic acid drugs into the cytosol [24]. The ability of dendrimers to act as non-viral vectors depends greatly upon their number of generations, with the gene-transfer activity of dendrimers with a high number of generations exceeding that of those with a low number of generations, although their cytotoxicity increases as the number of generations increases [25, 26]. Consequently, there has been growing interest in

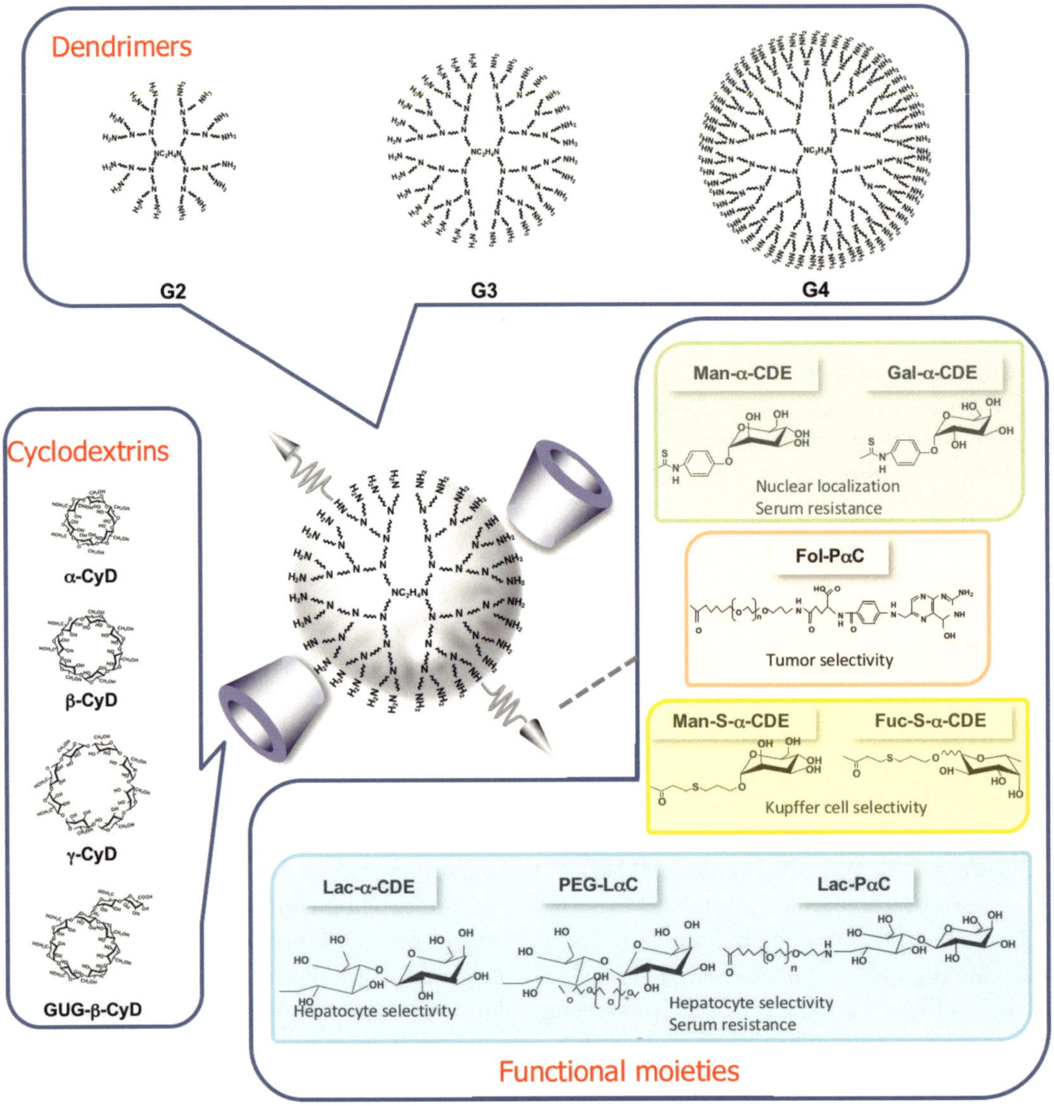

Figure 1. PAMAM dendrimer conjugates with various CyD (CDEs) and targeting ligands. GUG-β-CyD, glucuronylglucosyl-β-CyD; Man-α-CDE, mannosylated-CDE; Gal-α-CDE, galactosylated-CDE; Fol-PαC, folate-PEG-appended α-CDE; Man-S-α-CDE, mannosyloxypropylthioalkylated-α-CDE; Fuc-S-α-CDE, fucosyloxypropylthioalkylated-α-CDE; Lac-α-CDE, lactosylated-CDE; PEG-LαC, PEGylated Lac-α-CDE; Lac-PαC, lactose-PEG-appended α-CDE; PAMAM, polyamidoamine.

developing modified dendrimers possessing high levels of safety along with low numbers of generations due to their extremely low cytotoxicity [27–29].

The potential use of cyclodextrins (CyDs) as carriers of nucleic acids based of their direct interaction would not be expected, given that they exhibit very weak interactions with nucleic acids [30]. Therefore, combining CyDs with cell-penetrating nucleic acid carriers (cationic polymers) or modifying the CyD structure was necessary for their internalization. Various methods have been adopted to enhance the interactions between CyD polymers and conjugates with nucleic acids [31]. Davis et al. modified the β-CyD structure to form a bifunctionalized β-CyD with two amine groups, allowing its incorporation into the backbone of other linear polymers [32]. The new polymer, CALAA-01, was ultimately established as the first targeted delivery mechanism of synthetic siRNA in humans for the treatment of solid tumors [33]. However, various CyD-containing polymers used as nucleic acid vectors have also been reported [34, 35]. In this regard, Arima and Motoyama originally developed various CyD conjugates with dendrimers as CyD-based polymers for the delivery of various nucleic acid drugs (**Figure 1**).

2. Polyamidoamine (PAMAM) dendrimer conjugates with α-CyDs (α-CDEs) as nucleic acid carriers

2.1. α-CDEs as plasmid (p) DNA carriers

Arima et al. prepared a variety of CyD conjugates with PAMAM dendrimers (CDE) and utilized them as gene and nucleic acid drug carriers. Originally, Arima et al. prepared dendrimer [generation 2 (G2)] conjugates with α-, β-, and γ-CyDs and named them α-, β-, and γ-CDEs (G2), respectively [36]. Among these CDEs, α-CDE (G2) exhibited the most prominent gene-transfer activity, showing 100-fold higher luciferase gene-transfer activity as compared with that of dendrimer (G2) alone or its physical mixture with α-CyD in NIH3T3 cells, a mouse embryo fibroblast cell line, and RAW264.7 cells, a mouse macrophage-like cell line. This was attributed not only to increased levels of cellular association, but also to the augmented endosomal-escape ability of the pDNA complex due to the synergy of both the proton-sponge effect and the ability of α-CyD to disrupt the endosomal membrane. Afterward, Kihara et al. examined the optimal dendrimer generation (G2, G3, or G4) and degree of substitution (DS) for α-CyD in the α-CDE molecule [37]. Consequently, α-CDE (G3, DS2) showed the highest transfection efficiency along with low cytotoxicity, which was superior to that of TransFact and Lipofectin when tested in NIH3T3 cells. Furthermore, to elucidate the reason behind the superior gene-transfer activity of α-CDE (G3, DS2), Arima et al. investigated the cellular uptake, intracellular distribution, and the physicochemical properties of pDNA complexes involving both α-CDE (G3, DS2) and the dendrimer (G3). The particle sizes, as well as the ζ-potential values, were nearly the same for both complexes. Furthermore, the enhanced gene-transfer activity could not be explained based on cellular uptake, as the values of the complexes with α-CDEs (G2, G3, and G4) were equivalent to those observed with their parent dendrimers, suggesting that factors other than cellular uptake or the physicochemical properties of the α-CDE (G3, DS2)/pDNA complexes might be strongly associated with improving gene-transfer activity. To elucidate the mechanism of cellular uptake, Arima et al. studied the effect of different endocytosis inhibitors on the cellular uptake of fluorescein isothiocyanate-labeled pDNA [(FITC)-pDNA] complexes with tetramethylrhodamine-5-(and 6)-isothiocyanate-labeled α-CDE [TRITC-α-CDE (G3)] transfection in A549 cells, ultimately using flow cytometry and confocal laser-scanning microscopy (CLSM) to observe the colocalization of TRITC-α-CDE (G3), FITC-endocytosis markers, and FITC-pDNA after transfection.

Consequently, after transfection of pDNA complexes, the complexes with TRITC-α-CDE (G3, DS2) colocalized with the endocytosis markers FITC-transferrin and FITC-cholera toxin B. Similarly, the gene-transfer activity of α-CDE (G3, DS2) was markedly lowered by the addition of clathrin-dependent endocytosis inhibitors (i.e., chlorpromazine and sucrose) and raft-dependent endocytosis inhibitors (i.e., nystatin and filipin), but not by amiloride, the macropinocytosis inhibitor. These results suggested that the main mechanism of α-CDE (G3, DS2) uptake involved clathrin- and raft-dependent endocytosis. To confirm the release of the complex from endosomes, we observed the intracellular distribution of the α-CDE (G3, DS2)/FITC-pDNA complex by CLSM, finding that the green fluorescence originating from FITC-pDNA in the case of the α-CDE (G3, DS2) complex was predominantly localized in the cytoplasm to a much greater degree than that of the dendrimer system, confirming the improved capability for endosomal disruption conferred by the synergy between α-CDE and the proton-sponge effect of the dendrimer.

Moreover, *in vivo* studies of α-CDE (G3, DS2), as well as dendrimer complexes with pDNA, were evaluated after intravenous administration of 50 μg pDNA/mouse at a charge ratio of 10. After 12 h, organs were collected, and pDNA levels in various organs were determined. The results showed that α-CDE (G3, DS2) delivered pDNA more efficiently in the liver and kidney; however, the highest gene-expression levels were observed in the spleen. Blood-chemistry data related to α-CDE administration, including aspartate aminotransferase (AST), lactate dehydrogenase (LDH), blood urea nitrogen (BUN), alanine aminotransferase (ALT), and creatinine (CRE) concentrations, showed minor changes as compared with those associated with the dendrimer [38]. These results suggested potential use of α-CDE (G3, DS2) as a safe and promising non-viral pDNA vector, although additional modifications of the α-CDE (G3, DS2) chemical structure were required to improve its nuclear translocation enable improved gene-expression results [39].

2.2. α-CDE (G3) as siRNA carriers

Assessment of the siRNA-carrier ability of α-CDE (G3, DS2) was performed by Tsutsumi et al. using a luciferase-reporter system [40, 41]. Their results showed significant reductions in the luciferase activity of the α-CDE (G3, DS2) system as compared with that observed in the control and accompanied by negligible cytotoxicity after transfection with the ternary complex pDNA/siRNA/α-CDE (G3, DS2). In this study, α-CDE (G3, DS2) showed superior transfection efficiency relative to that of the dendrimer and other commercially available transfection reagents. Additionally, they observed that the complex localized exclusively to the cytoplasm, where RNAi-related activity occurs, and as a result of the lack of a nuclear-translocation moiety in α-CDE (G3, DS2). Tsutsumi et al. also reported efficient knockdown of the firefly luciferase gene using a α-CDE (G3, DS2)/siRNA binary system accompanied by negligible cytotoxicity as compared with the use of siRNA complexes with commercially available transfection reagent (Lipofectamine 2000 and RNAiFect). Furthermore, the physicochemical properties, local irritation, cytotoxicity, interferon response, cellular uptake, and intracellular distribution of the siRNA complexes, as well as the RNAi activity associated with the α-CDE (G3, DS2)/siRNA complex, were evaluated on endogenous gene-expression in Colon-26-luc and NIH3T3-luc cells stably expressing the pGL3 luciferase gene. The results revealed potent RNAi activity against Lamin A/C and Fas expression along with minor cytotoxicity as compared with commercial transfection agents [40]. Additionally, siRNA complexed with α-CDE (G3, DS2) was protected from degradation by serum nucleases. Intriguingly and somewhat similar to results observed with the α-CDE (G3, DS2)/pDNA complex, α-CDE (G3, DS2)/FITC-siRNA was delivered exclusively to the cytoplasm in NIH3T3-luc cells. Additionally, when this system was applied *in vivo* in mice bearing Colon-26-luc tumors, α-CDE (G3, DS2)/siGL3 (siRNA against pGL3 gene) showed potent RNAi activity against pGL3 expression after intravenous, as well as intratumoral, injection. Moreover, the siRNA complex neither triggered the immune response nor changed blood-chemistry data, indicating its safety. These results suggested the potential of α-CDE (G3, DS2) as a novel siRNA-carrier candidate for both *in vivo* and *in vitro* applications.

2.3. α-CDE (G3) as shRNA carriers

To improve upon and prolong the duration of RNAi-mediated gene knockdown, vector-based shRNA-expression systems were developed [42]. Upon shRNA transfection into mammalian cells, the insert containing the vector is transferred to the nucleus, integrated into the host genome, expressed, and quickly processed by Dicer-dependent cleavage and loaded into the RNA-induced silencing complex, which is then directed to the target mRNA, resulting in its degradation. A previous study demonstrated the potential of α-CDE (G3, DS2) as a novel carrier of pDNA expressing shRNA [43]. In this study, the authors used pDNA expressing shRNA against the pGL3 firefly luciferase gene (shGL3). Even at a low charge ratio, α-CDE (G3, DS2) capably formed a stable condensed complex with shGL3 and induced the conformational transition of shRNA in solution from B-form to the more compact C-form DNA. Furthermore, α-CDE (G3, DS2) markedly inhibited shGL3 degradation by DNase I, and the α-CDE (G3, DS2)/shGL3 complex showed the most potent RNAi activity at a charge ratio of 20 along with negligible cytotoxicity and without off-target effects in A549 cells, a human alveolar adenocarcinoma cell line, while also transiently expressing the luciferase gene in NIH3T3-luc cells. Moreover, the addition of sufficient amounts of siGL3 along with α-CDE (G3, DS2)/shGL3 dramatically enhanced the RNAi activity, which was ascribed to the stabilizing effect of α-CDE (G3, DS2) against DNase I degradation of the shRNA accompanied by negligible cytotoxicity. These results suggested that α-CDE (G3, DS2) possessed the potential to be a novel shRNA carrier.

2.4. Functionalized α-CDEs as cell-specific DNA and siRNA carriers

In the early 1900s, the German Noble laureate Paul Ehrlich first introduced the concept of targeted drug delivery [44, 45]. At that time, he called it the "magic bullet", as it was able to deliver the drug specifically to microbes (such as bacteria) without harming the body. His continuous research eventually led to the development of the first effective drug against syphilis (Salvarsan). The primary aim of targeted drug-delivery systems is to increase the concentration of the medication in specific areas of the body relative to others, thereby improving its therapeutic index and reducing cytotoxicity. Various approaches have been adopted to target medications to the disease site [46, 47] with many compounds internalized inside of cells via receptor-mediated endocytosis. Receptor-mediated techniques use ligands attached to polyplexes to transfect cells with selected genes. Endocytosis is then mediated by various receptors, such as asialoglycoprotein receptor (ASGPR), mannose receptor (ManR), fucose receptor (FucR), and folate receptor (FR). However, despite the advantages offered by such systems, some drawbacks exist, including immune reactions against the carriers and rapid disposition of the carriers, as well as redistribution of the drugs after their release from the carriers [48]. Therefore, Arima et al. focused on how to improve CDEs for efficient delivery of gene and nucleic acid drugs to various organs through the attachment of various ligands to existing CDEs to aid the process of receptor-mediated endocytosis.

2.4.1. *Galactosylated α-CDE as a hepatocyte-selective pDNA carrier*

The liver consists mainly of hepatocytes (nearly 70%) and parenchymal cells. Gene- or drug-targeting systems designed to target the liver are usually directed to hepatocytes, which overexpress ASGPRs on their cell surface. These ASGPRs mediate the removal of potentially hazardous glycoconjugates from the blood; therefore, ASGPRs are usually targeted using galactose residues coupled with a core molecule to enhance binding [49, 50]. Many approaches, including but not limited to (1) intravenous injection of pDNA within liposomes [51, 52] or via ASGPR targeting [53] and (2) intra-portal injection of recombinant adenovirus [54] and retroviral vectors [55] have been adopted to deliver foreign genes *in vivo* to hepatocytes. Moreover, an *in vitro* system that takes advantage of ASGPR-mediated endocytosis to transfect hepatocytes with an exogenous DNA using a soluble DNA carrier was developed [56]. Various ASGPR-mediated gene-delivery systems using different polymers have also been described, including galactose-polyethylene glycol (PEG)-poly(L-lysine) [57], galactosylated PEG-graft-polyethylenimine (PEI) [58, 59], and galactosylated chitosan-grafted-PEI [60]. In order to attain hepatocyte-specificity and/or improve the efficacy of α-CDE as a gene-delivery carrier, Arima et al. attached a galactose residue to form Gal-α-CDE (G2) as a novel non-viral carrier [61]. The galactose moieties were attached to the primary amino groups of α-CDE (G2) through a spacer consisting of α-D-galctopyranosylphenyl isothiocyanate, achieving various degrees of substitutions of galactose (DSGs; 1, 4, 5, 8, and 15). Evaluation of the complexation ability of the pDNA complexes by electrophoresis showed that the Gal-α-CDEs (G2) could form complexes with pDNA; however, complexation ability decreased along with increasing DSG values, possibly due to decreases in free positive-charged primary amino groups. Moreover, the ability of these carriers to protect the pDNA from degradation by serum nucleases also decreased along with increasing DSGs, likely due to the attenuated interactions and loss of pDNA-condensation ability in the presence of high DSGs in the conjugates. Additionally, they observed that Gal-α-CDEs (G2, DSG4) elicited the most prominent gene-transfer activity relative to that of the dendrimer (G2), and α-CDE (G2, DS2) in HepG2 cells, a human hepatoma cell line, NIH3T3 cells, and A549 cells showed no cytotoxicity up to a charge ratio of 200. Surprisingly, this was independent of ASGPR expression, possibly due to the inability of the spacer to properly present the galactose residue for receptor recognition. Therefore, these results suggested the potential of Gal-α-CDE (G2, DSG4) as a novel non-viral vector independent of cell-surface ASGPR expression. It is worth mentioning that in this study, the authors used a cancer-cell line (HepG2) and not normal hepatocytes, given that they both express ASGPR at similar levels, and HepG2 cells are widely used by scientists engaged in targeting studies for genes using non-viral carriers. Surprisingly, the addition of 10% fetal bovine serum (FBS) did not alter the gene transfer activity of Gal-α-CDE (G2, DSG4); however, this activity on the part of the dendrimer (G2), as well as that of α-CDE (G2, DSG2),

http://dx.doi.org/10.5772/intechopen.74614

as pDNA vectors was decreased by the addition of FBS. Additionally, after co-incubation of Gal-α-CDE (G2, DSG4), dendrimer (G2), or Gal-α-CDE (G2, DS2) pDNA complex with asialofetuin and galactose, only a slight decrease in gene-transfer activity was observed in HepG2 cells, with no competitive effects. Consequently, these latter results confirmed that the mechanism associated with the enhanced gene-transfer activity of Gal-α-CDE (G2, DSG4) was not ASGPR-specific, but rather possibly due to other factors, such as increased stability of the pDNA complex or changes in intracellular trafficking. Collectively, these results suggested that Gal-α-CDE (G2, DSG4) exhibited enhanced pDNA-transport activity along with low cytotoxicity and considerable resistance to serum-associated degradation and could, therefore, represent an excellent non-viral gene-delivery carrier.

2.4.2. Lactosylated α-CDEs as hepatocyte-selective pDNA and siRNA carriers

As previously mentioned, Gal-α-CDE (G2, DSG4) did not exhibit hepatocyte-specific gene-delivery activity. Therefore, Arima et al. prepared lactose-appended α-CDEs (Lac-α-CDEs) containing a glucose moiety as a spacer between the dendrimer and the lactose moiety [62, 63]. Of the various Lac-α-CDEs (G2) harboring different degrees of substitution of lactose (DSLs; 1.2, 2.6, 4.6, 6.2, and 10.2), Lac-α-CDE (G2, DSL2.6) exhibited the highest gene-transfer activity in HepG2 cells along with negligible cytotoxicity. To verify whether Lac-α-CDE (G2, DSL 2.6) could bind to galactose-binding lectins, the association constant of Lac-α-CDE (G2, DSL2.6) with peanut lectin was determined and compared with that of α-CDE (G2, DS2) using surface plasmon resonance. The results showed that the association constant of Lac-α-CDE (G2, DSL2.6) was 100-fold greater than that of α-CDE (G2, DS2). It was previously reported that the dissociation constant of asialofetuin for ASGPR located on HepG2 cells is ~3.61×10^{-9} M [64]. These results indicated that the specific galactose-binding ability of Lac-α-CDE (G2, DSL2.6) was maintained, although the magnitude was not as strong as that of asialofetuin. To confirm the ASGPR-mediated gene-transfer activity of Lac-α-CDE (G2, DSL2.6), the effects of asialofetuin, as a competitor on this activity, was examined in HepG2 cells. The results revealed that the gene-transfer activity of Lac-α-CDE (G2, DSL2.6) was significantly suppressed by asialofetuin, and that cellular association of the Lac-α-CDE (G2, DSL2.6)/Alexa-pDNA complex was markedly reduced in the presence of asialofetuin. However, no inhibitory effect of asialofetuin was observed on the activity of α-CDE (G2, DS2)/Alexa-pDNA in HepG2 cells. These results indicated that the gene-transfer activity of Lac-α-CDE (G2, DSL2.6) was mediated by ASGPR endocytosis. Arima et al. then evaluated the ability of Lac-α-CDE (G2, DSL2.6) to deliver pDNA *in vivo* using the pGL3 luciferase system in mice. The complexes were intravenously injected into mice, and after 12 h, luciferase activity was determined. They observed that luciferase activity in the Lac-α-CDE (G2, DSL2.6) system was significantly higher than that observed in the α-CDE (G2) system in the liver. Furthermore, to estimate the safety profile of Lac-α-CDE (G2, DSL2.6), the effect of its pDNA complex on blood-chemistry data, such as CRE, BUN, AST, ALT, and LDH concentrations, was analyzed following intravenous administration to mice. The ALT concentration in the Lac-α-CDE (G2, DSL2.6) system, as well as in the jetPEI-hepatocyte system, was slightly elevated (no significant difference) as compared with that observed in the control. By contrast, all other parameters in the Lac-α-CDE (G2, DSL2.6) system were almost equivalent to those of controls. These results strongly suggested that Lac-α-CDE (G2, DSL2.6) exhibited hepatocyte-specific gene-transfer activity along with a good safety profile *in vivo*.

Recently, Hayashi et al. prepared a PEGylated Lac-α-CDE [PEG-LαC (G3)] to improve *in vivo* gene-transfer activity by enhancing complex stability, as well as prolonging the half-life in circulation [62]. Of the various PEG-LαCs (G3), those with degrees of substitution of the PEG moiety (DSPs) of 2.1 [PEG-LαC (G3, DSP2.1] showed higher luciferase gene-transfer activity than other PEG-LαCs (G3, DSP4.0 and DSP6.2) in HepG2 cells along with negligible cytotoxicity up to a charge ratio of 50. Additionally, its gene-transfer activity decreased in the presence of asialofetuin, whereas it retained significantly higher gene-transfer activity, even in the presence of 50% serum. Additionally, PEG-LαC (G3, DSP2.1) showed selective gene-transfer activity into hepatic parenchymal cells rather than hepatic non-parenchymal cells *in vivo*. Furthermore, blood-chemistry values, such as CRE, BUN, AST, ALT, and LDH concentrations, following administration of the PEG-LαC (G3, DSP2.1)/pDNA complex system were almost equivalent with those in controls, suggesting that PEG-LαC (G3, DSP2.1) showed potential as a hepatocyte-selective gene carrier both *in vitro* and *in vivo*.

The potential of PEG-LαC (G3) as a siRNA carrier was also evaluated (unpublished data), with an siRNA against transthyretin (TTR) mRNA (siTTR) used to treat TTR-related familial amyloidotic polyneuropathy (TTR-FAP). The results indicated that the PEG-LαC (G3)/siTTR complex significantly reduced TTR mRNA expression in the liver as compared with the Lac-α-CDE (G3)/siTTR complex, suggesting the potential use of PEG-LαC (G3) as a hepatocyte-selective siRNA carrier (**Figure 2**).

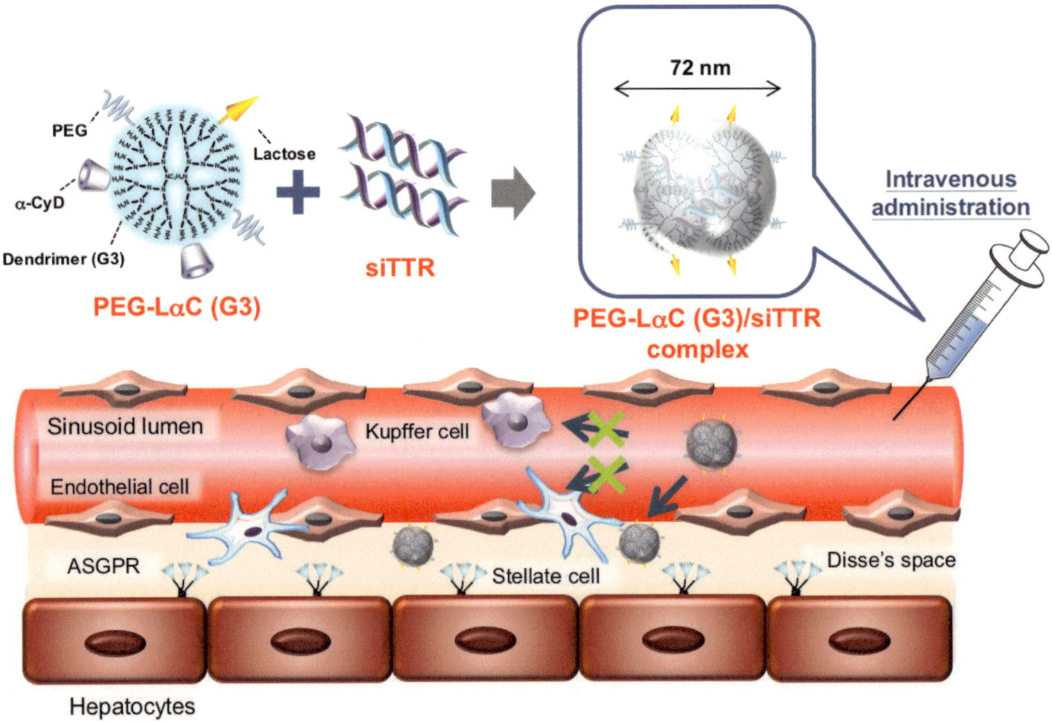

Figure 2. PEG-LαC (G3) as a targeting carrier for siTTR delivery to hepatocytes.

http://dx.doi.org/10.5772/intechopen.74614

2.4.3. Mannose and fucose-appended α-CDEs as Kupffer cell (KC)-selective pDNA and siRNA carriers

KCs are reticuloendothelial cells that reside within the lumen of the liver sinusoid and adhere to endothelial cells that form blood-vessel walls. These non-parenchymal cells represent ~15% of the total liver cells in the human body [65] and are the first macrophages that come into contact with bacteria and bacterial toxins derived from the gastrointestinal tract [66]. KCs also play a critical role in removing harmful materials circulating in the blood. Moreover, KCs are considered an essential part of innate immunity and play an important role in the rapid response to threatening stimuli. They are also involved in the pathogenesis of different liver diseases, including viral hepatitis, alcoholic liver disease, non-alcoholic fatty liver, development of liver fibrosis, and portal hypertension [67]. Importantly, KCs express ManR and FucR; therefore, both fucose and mannose can be used as targeting ligands on KCs used for nucleic acid delivery [68].

Arima et al. prepared mannose-appended α-CDEs (Man-α-CDE, G2) [39] to develop ManR-targeted non-viral carriers by attaching mannose residues to primary amino acid residues of α-CDE (G2) using an α-D-mannopyranosylphenyl isothiocyanate spacer. Of the various Man-α-CDEs (G2) with degrees of substitution of mannose (DSMs) of 1.1, 3.3, 4.9, and 8.3, Man-α-CDEs (G2, DSM3.3 and DSM4.9) showed higher gene-transfer activity as compared with that of dendrimer (G2) and α-CDE (G2) in NR8383 cells, a rat lung macrophage cell line, with no cytotoxicity observed up to charge ratio of 200 (carrier/pDNA). However, Man-α-CDE (G2) also showed high gene-transfer activity in A549 cells [ManR (−)], suggesting its low selectivity for ManR, possibly due to the rigidity of the spacer.

More recently, Arima et al. prepared a new Man-α-CDE with a α-D-mannopyranosylprop ylthiopropionylated α-CDE (G3) spacer [Man-S-α-CDEs (G3)], which was longer and more flexible than that in Man-α-CDE [69, 70]. In this study, nuclear factor (NF)-κB was targeted due to its important role in the inflammatory response, and because it is found in almost all animal cell types. Therefore, to suppress NF-κB activation, NF-κB siRNA and NF-κB-decoy DNA were employed, with both strategies potentially attractive for the treatment of cytokine-related liver diseases, such as fulminant hepatitis. Of the various Man-S-α-CDEs with different DSMs, the NF-κB p65-specific siRNA (sip65) complex with Man-S-α-CDE (G3, DSM4) showed significantly lower *NF-κB p65* mRNA levels and nitric oxide levels in lipopolysaccharide (LPS)-stimulated NR8383 cells following ManR-mediated endocytosis (**Figure 3**). Intravenous administration of the Man-S-α-CDE (G3, DSM4)/sip65 complex increased the survival rate of the LPS-induced fulminant hepatitis mouse model via significant *in vivo* RNAi activity. These results suggested that Man-S-α-CDE (G3, DSM4) represented a potential novel KC-selective siRNA carrier.

Several reports demonstrated that NF-κB-decoy complexes harboring a liposome-functionalized fucose moiety showed higher gene-transfer efficiency as compared with mannose-appended liposomes specific for KCs [71, 72]. Therefore, Akao et al. prepared thioalkylated fucose-appended α-CDEs [Fuc-S-α-CDE (G2)] and assessed their potential as KC-selective carriers of decoy DNA (**Figure 4**) [73]. The NF-κB-decoy in complex with Fuc-S-α-CDE (G2) with an average degree of substitution of fucose (DSFuc) of two suppressed the production of nitric oxide, as well as tumor necrosis factor-α (TNF-α) expression, in LPS-simulated NR8383 cells through FucR-mediated cellular uptake and successful endosomal escape. This complex also improved survival rates following intravenous injection into a fulminant hepatitis mouse model. Moreover, Fuc-S-α-CDE (G2, DSFuc2)/NF-κB decoy complexes showed marked accumulation in the liver relative to that observed in other organs. Furthermore, serum ALT and AST levels, as well as TNF-α levels, significantly decreased after intravenous administration of the complex in mice with fulminant hepatitis. These results suggested the potential of the Fuc-S-α-CDE (G2)/NF-κB decoy complex as an oligonucleotide therapy for fulminant hepatitis. There are many other receptors available for KC-specific drug targeting, including galactose receptors, scavenger receptors, CD36, lol-density lipoprotein receptor, and complement receptors [74]; therefore, future studies on the utility of fucose in this context should focus on its efficacy with different ligands.

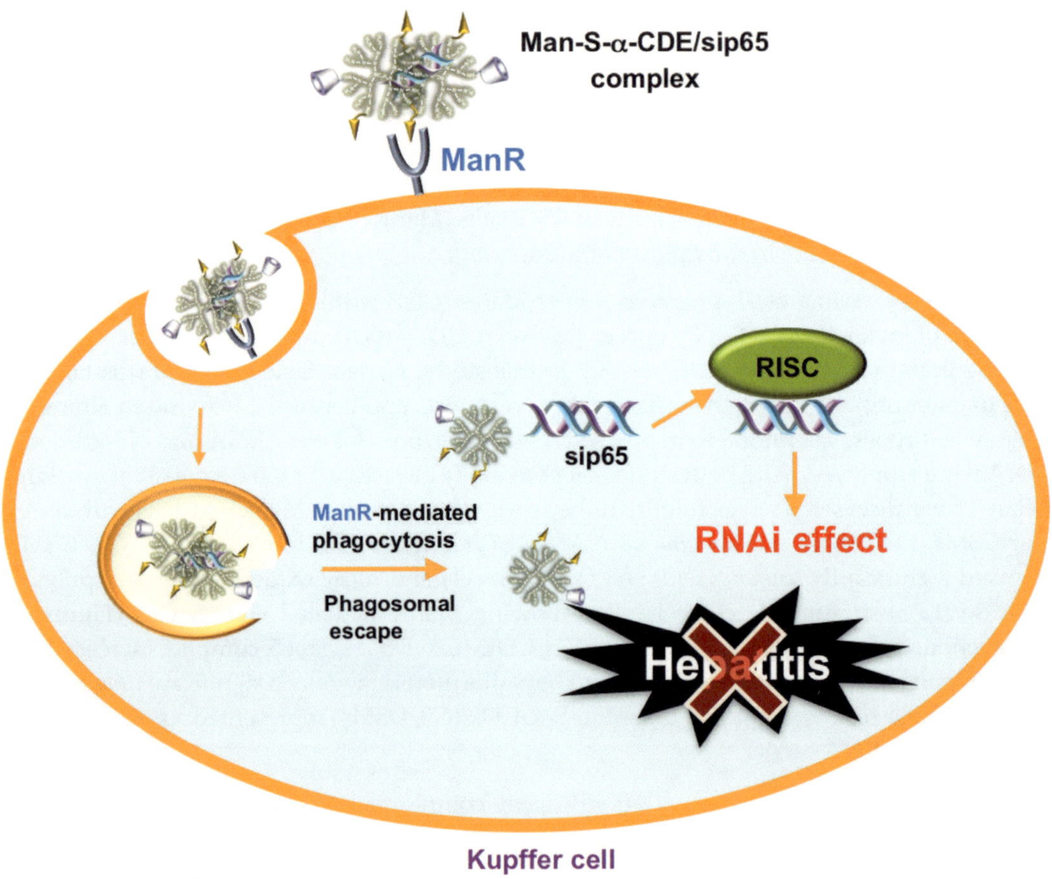

Figure 3. Man-S-α-CDE (G3) as a KC-specific targeting carrier via ManR.

http://dx.doi.org/10.5772/intechopen.74614

2.4.4. *Folate PEG-appended α-CDEs as cancer-cell-selective pDNA and siRNA carriers*

To achieve maximum effective therapeutic effects against cancer using siRNAs, the design of tumor-selective delivery systems is extremely crucial. Folic acid has often been used as a tumor-specific ligand [75, 76], because it is relatively affordable as compared with other cancer-targeting ligands and is capable of high-affinity interactions with the FR-α receptor

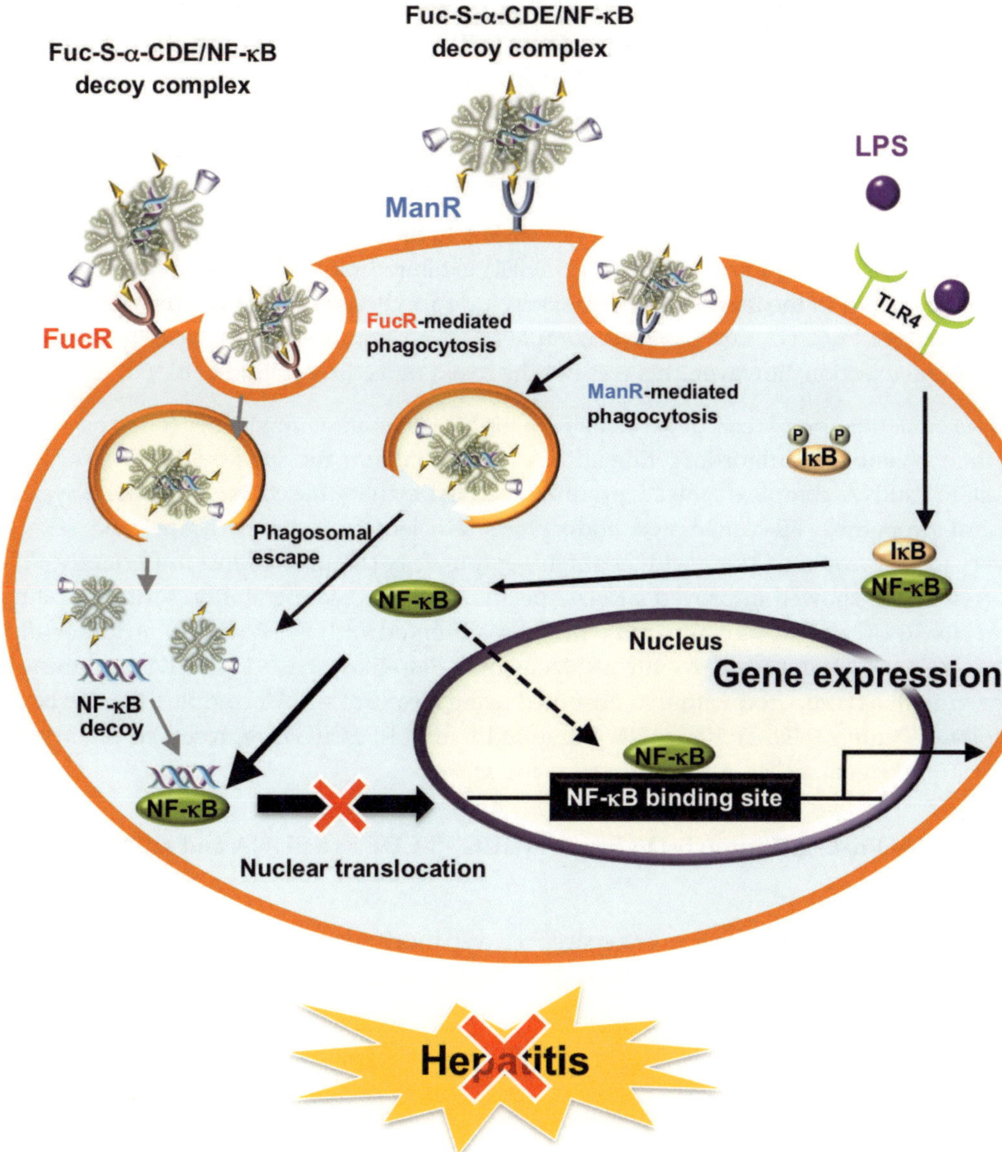

Figure 4. Fuc-S-α-CDE (G2) as targeting decoy DNA carrier to KCs.

expressed on the surface of many cancer cells ($k_d > 10^{-9}$–10^{-10} M). FR-α is highly expressed in several tumor cells, including those associated with lung, ovary, breast, kidney, and brain cancers, and is negligibly expressed in normal tissues. Additionally, as the cancer progresses in stage, the FR-α expression increases substantially. Therefore, folic acid is considered an ideal candidate cancer-cell-selective ligand.

Arima et al. prepared a folic acid-appended α-CDE (G3) with a PEG spacer [Fol-PαC (G3)] to fabricate a cancer-selective gene and siRNA carrier. Fol-PαC (G3) showed selective FR-α-overexpressing tumor-cell gene-transfer activity [77]. Specifically, Fol-PαC (G3) with an average degree of substitution of folate (DSF) of five showed significantly higher gene-transfer activity as compared with that of α-CDE (G3) in KB cells [FR-α (+)], but not in A549 [FR-α (−)] cells along with negligible cytotoxicity. Moreover, Fol-PαC (G3, DSF5) showed higher gene-transfer activity than α-CDE (G3) after intratumoral injection in mice bearing tumors.

The potential of Fol-PαC (G3) for delivery of siRNA to FR-α-overexpressing cancer cells was evaluated [78], showing that Fol-PαC (G3, DSF4) exhibited high siRNA-transfer activity in KB cells [FR-α (+)] in the absence of cytotoxicity up to a charge ratio of 100 (carrier/siRNA). Notably, the Fol-PαC (G3, DSF4)/siRNA complex showed significant RNAi activity following intratumoral injection; however, this was not the result of its dissociation in blood.

Ohyama et al. then prepared Fol-PαCs using a higher-generation dendrimer (G4) and evaluated their potential as tumor-targeting siRNA carriers *in vitro* and *in vivo* [79]. The Fol-PαC (G4, DSF2)/siRNA complex showed prominent RNAi activity based on adequate physicochemical properties, FR-α-mediated endocytosis, efficient endosomal escape, and siRNA delivery to the cytoplasm along with negligible cytotoxicity (**Figure 5**). Most importantly, Fol-PαC (G4, DSF2) showed improved siRNA-specific blood-circulating ability, serum stability, and *in vivo* RNAi activity as compared with those observed with Fol-PαC (G3). Additionally, Fol-PαC (G4, DSF2) in complex with siRNA against Polo-like kinase 1 (siPLK1) suppressed tumor growth as compared with that observed using a control siRNA complex in mice bearing colon-26 tumor cells. These results suggested that Fol-PαC (G4) represented a potential novel tumor-targeting siRNA carrier *in vitro* and *in vivo*.

2.5. 6-*O*-α-(4-*O*-α-D-glucuronyl)-D-glucosyl (GUG)-β-CDE as a pDNA and siRNA carrier

Arima et al. clarified the importance of a spacer between the dendrimers and the targeting ligands for providing cell-specific pDNA delivery [78, 80]. However, the effect of a spacer between the CyD and the dendrimer on the gene-transfer activity of the CDE remained unknown. Consequently, a new CDE was prepared (GUG-β-CDE) utilizing a glucuronyl-glucosyl group as a spacer between the CyD and the dendrimer. Additionally, GUG-β-CyD has many advantages over the parent β-CyD, including higher water solubility, lower hemolytic activity, and increased bioadaptability [81]. Moreover, it contains a carboxyl group capable of interacting with primary amino groups present in dendrimers. Of the various GUG-β-CDEs (G2) having different DS values, GUG-β-CDE (G2, DS1.8) showed higher gene-transfer activity *in vitro* as compared with other GUG-β-CDEs (DS1.2, DS2.5, and DS4.5) [82]. Additionally, GUG-β-CDE (G2, DS1.8) showed higher gene-transfer activity relative to that of α-CDE (G2, DS1.2) and β-CDE

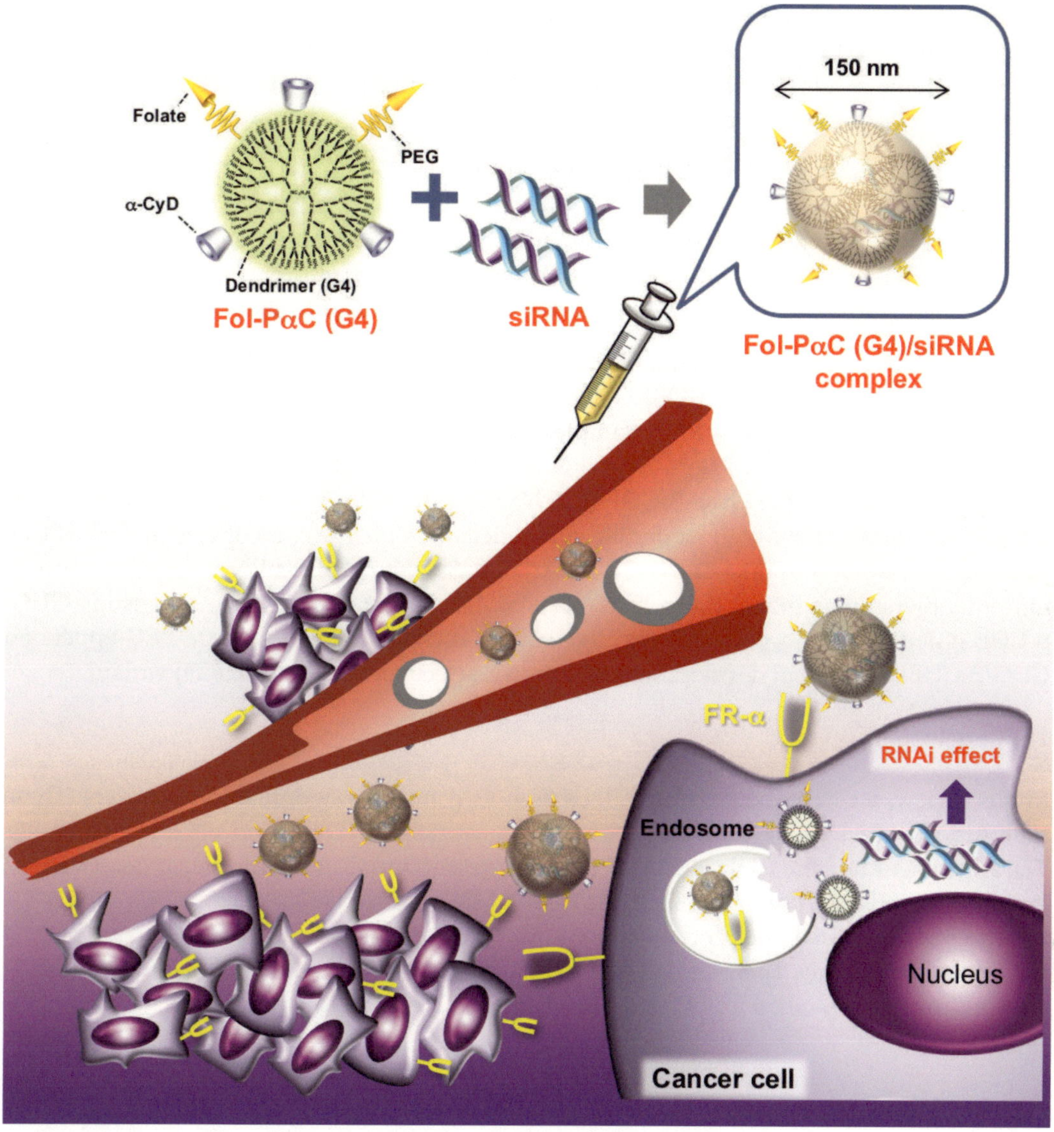

Figure 5. Fol-PαC (G4) as a targeting-siRNA carrier to folate-expressing tumor cells.

(G2, DS1.3) in A549 cells and RAW264.7 cells, respectively, possibly due to the pDNA complex exhibiting an increased ability to escape endosomes and a high degree of nuclear localization [39, 83, 84]. Moreover, *in vivo* GUG-β-CDE (G2, DS1.8) gene-transfer activity was much higher than that of α-CDE (G2, DS1.2) or β-CDE (G2, DS1.3) in kidney at 12 h after intravenous injection of the complexes in mice [85]. Therefore, GUG-β-CDE (DS1.8) might have potential as a pDNA carrier for gene-transfer targeting the kidney. Furthermore, no cytotoxicity was observed in A549 cells or RAW264.7 cells up to a charge ratio of 200 (carrier/pDNA). The hemolytic activity of GUG-β-CDE (G2, DS1.8) in rabbit red blood cells was also substantially lower than that associated with the dendrimer, and negligible changes in the blood-chemistry data were observed 12 h after intravenous administration of the GUG-β-CDE (G2, DS1.8)/pDNA complex in mice. These results strongly suggested that this complex showed a good safety profile *in vivo* and *in vitro*, and that it might constitute an adequate carrier for gene therapy targeting kidney diseases, such as polycystic kidney disease, Alport syndrome, renal cancers, glomerulonephritis, and renal fibrosis.

Additionally, Anno et al. evaluated the potential of GUG-β-CDE (G2) as a siRNA carrier. GUG-β-CDE (G2, DS1.8) in complex with siTTR showed high RNAi activity with no cytotoxicity in HepG2 cells. Moreover, *TTR* mRNA-expression levels were reduced after intravenous administration of the complex to BALB/c mice, with only minor changes in blood-chemistry parameters, suggesting the potential of GUG-β-CDE (G2, DS1.8) as a siRNA carrier for the treatment of TTR-FAP [86].

Moreover, Ahmed et al. prepared a GUG-β-CDE using a higher-generation dendrimer (G3) [87]. Various GUG-β-CDEs (G3, DS1.6, DS3.0, DS3.7, DS5.0, and DS8.6) were prepared, with the GUG-β-CDE (G3, DS3.7)/siRNA complex showing the highest RNAi activity in KB cells transiently expressing the luciferase gene and colon 26-luc cells stably expressing the luciferase gene. Moreover, the GUG-β-CDE (G3, DS3.7)/FITC-siRNA complex showed the highest cellular uptake along with negligible cytotoxicity at a charge ratio of 20 (carrier/siRNA). Additionally, cellular uptake of the GUG-β-CDE (G3, DS3.7)/FITC-siRNA complex was significantly higher than that of the α-CDE (G3, DS2.4)/FITC-siRNA complex, suggesting GUG-β-CDE (G3, DS3.7) as a potential effective siRNA carrier. Currently, folate-PEG-appended GUG-β-CDEs (G3) are in development as a cancer-selective GUG-β-CDE (G3) variant.

2.6. Conclusion

In this review, we described various CDEs used as gene and oligonucleotide carriers. These multifunctional CDEs showed great potential as carriers for DNA and nucleic acid drugs. The advantages of these CDEs included (1) low cytotoxicity; (2) facile modification of various targeting ligands and polymers, such as PEG; and (3) enhanced endosomal-escape ability via the synergistic action of the proton-sponge effect in the dendrimer and the interaction of CyD with membrane lipids. Therefore, these CyD-based carriers have the potential for utilization as multifunctional carriers for pDNA, siRNA, decoy DNA, and shRNA.

Acknowledgements

We would like to thank Editage (www.editage.jp) for English language editing.

Author details

Ahmed F.A. Mohammed[1,2], Keiichi Motoyama[1], Taishi Higashi[1] and Hidetoshi Arima[1,2*]

*Address all correspondence to: arimah@gpo.kumamoto-u.ac.jp

1 Graduate School of Pharmaceutical Sciences, Kumamoto University, Japan

2 Program for Leading Graduate Schools "HIGO (Health Life Science: Interdisciplinary and Glocal Oriented) Program", Kumamoto University, Japan

http://dx.doi.org/10.5772/intechopen.74614

References

[1] Pfeifer A, Verma IM. Gene therapy: Promises and problems. Annual Review of Genomics & Human Genetics. 2001;**2**(1):177

[2] Elsabahy M, Nazarali A, Foldvari M. Non-viral nucleic acid delivery: Key challenges and future directions. Current Drug Delivery. 2011;**8**(3):235-244

[3] Feng Q, Yu M-Z, Wang J-C, Hou W-J, Gao L-Y, Ma X-F, et al. Synergistic inhibition of breast cancer by co-delivery of VEGF siRNA and paclitaxel via vapreotide-modified core–shell nanoparticles. Biomaterials. 2014;**35**(18):5028-5038

[4] Ginn SL, Alexander IE, Edelstein ML, Abedi MR, Wixon J. Gene therapy clinical trials worldwide to 2012 – An update. The Journal of Gene Medicine. 2013;**15**(2):65-77

[5] Nayerossadat N, Maedeh T, Ali PA. Viral and nonviral delivery systems for gene delivery. Advanced Biomedical Research. 2012;**1**:27

[6] Luo D, Saltzman WM. Synthetic DNA delivery systems. Nature Biotechnology. 2000;**18**(1):33-37

[7] Subramanian A, Ranganathan P, Diamond SL. Nuclear targeting peptide scaffolds for lipofection of nondividing mammalian cells. Nature Biotechnology. 1999;**17**(9):873-877

[8] Fire A, Xu S, Montgomery MK, Kostas SA, Driver SE, Mello CC. Potent and specific genetic interference by double-stranded RNA in *Caenorhabditis elegans*. Nature. 1998;**391**(6669):806-811

[9] Oliveira S, Storm G, Schiffelers RM. Targeted delivery of siRNA. Journal of Biomedicine and Biotechnology. 2006;**2006**:63675

[10] Sliva K, Schnierle BS. Selective gene silencing by viral delivery of short hairpin RNA. Virology Journal. 2010;**7**:248

[11] Kanasty R, Dorkin JR, Vegas A, Anderson D. Delivery materials for siRNA therapeutics. Nature Materials. 2013;**12**(11):967-977

[12] Li SD, Huang L. Gene therapy progress and prospects: Non-viral gene therapy by systemic delivery. Gene Therapy. 2006;**13**(18):1313-1319

[13] Tiera MJ, Shi Q, Winnik FM, Fernandes JC. Polycation-based gene therapy: Current knowledge and new perspectives. Current Gene Therapy. 2011;**11**(4):288-306

[14] Liu Z, Zhang Z, Zhou C, Jiao Y. Hydrophobic modifications of cationic polymers for gene delivery. Progress in Polymer Science. 2010;**35**(9):1144-1162

[15] Resnier P, Montier T, Mathieu V, Benoit J-P, Passirani C. A review of the current status of siRNA nanomedicines in the treatment of cancer. Biomaterials. 2013;**34**(27):6429-6443

[16] Creixell M, Peppas NA. Co-delivery of siRNA and therapeutic agents using nanocarriers to overcome cancer resistance. Nano Today. 2012;**7**(4):367-379

[17] Zheng C, Zheng M, Gong P, Deng J, Yi H, Zhang P, et al. Polypeptide cationic micelles mediated co-delivery of docetaxel and siRNA for synergistic tumor therapy. Biomaterials. 2013;**34**(13):3431-3438

[18] Challa T, Goud BA, Baskar S, Chandra Mouli G, Jukuri R. Dendrimers: A novel polymer for drug delivery. International Journal of Pharmaceutical Sciences Review & Research. 2011;**9**(1):88-99

[19] Tomalia DA, Baker H, Dewald J, Hall M, Kallos G, Martin S, et al. A new class of polymers: Starburst-dendritic macromolecules. Polymer Journal. 1985;**17**(1):117-132

[20] Esfand R, Tomalia DA. Poly(amidoamine) (PAMAM) dendrimers: From biomimicry to drug delivery and biomedical applications. Drug Discovery Today. 2001;**6**(8):427-436

[21] Wei W, Lv P-P, Chen X-M, Yue Z-G, Fu Q, Liu S-Y, et al. Codelivery of mTERT siRNA and paclitaxel by chitosan-based nanoparticles promoted synergistic tumor suppression. Biomaterials. 2013;**34**(15):3912-3923

[22] Wu J, Zhou J, Qu F, Bao P, Zhang Y, Peng L. Polycationic dendrimers interact with RNA molecules: Polyamine dendrimers inhibit the catalytic activity of Candida ribozymes. Chemical Communications. 2005;(3):313-315

[23] Wang Z, Liu G, Zheng H, Chen X. Rigid nanoparticle-based delivery of anti-cancer siRNA: Challenges and opportunities. Biotechnology Advances. 2014;**32**(4):831-843

[24] Zeng L, Li J, Wang Y, Qian C, Chen Y, Zhang Q, et al. Combination of siRNA-directed Kras oncogene silencing and arsenic-induced apoptosis using a nanomedicine strategy for the effective treatment of pancreatic cancer. Nanomedicine: Nanotechnology, Biology and Medicine. 2014;**10**(2):463-472

[25] Kukowska-Latallo JF, Bielinska AU, Johnson J, Spindler R, Tomalia DA, Baker JR. Efficient transfer of genetic material into mammalian cells using starburst polyamidoamine dendrimers. Proceedings of the National Academy of Sciences. 1996;**93**(10):4897-4902

[26] Eichman JD, Bielinska AU, Kukowska-Latallo JF, Baker Jr JR. The use of PAMAM dendrimers in the efficient transfer of genetic material into cells. Pharmaceutical Science & Technology Today. 2000;**3**(7):232-245

[27] Fatemian T, Othman I, Chowdhury EH. Strategies and validation for siRNA-based therapeutics for the reversal of multi-drug resistance in cancer. Drug Discovery Today. 2014;**19**(1):71-78

[28] Nam HY, Hahn HJ, Nam K, Choi WH, Jeong Y, Kim DE, et al. Evaluation of generations 2, 3 and 4 arginine modified PAMAM dendrimers for gene delivery. International Journal of Pharmaceutics. 2008;**363**(1-2):199-205

[29] Navarro G, Maiwald G, Haase R, Rogach AL, Wagner E, de Ilarduya CT, et al. Low generation PAMAM dendrimer and CpG free plasmids allow targeted and extended transgene expression in tumors after systemic delivery. Journal of Controlled Release. 2010;**146**(1):99-105

[30] Arima H, Motoyama K. Recent findings concerning PAMAM dendrimer conjugates with cyclodextrins as carriers of DNA and RNA. Sensors. 2009;**9**(8):6346

[31] Ortiz Mellet C, Garcia Fernandez JM, Benito JM. Cyclodextrin-based gene delivery systems. Chemical Society Reviews. 2011;**40**(3):1586-1608

[32] Gonzalez H, Hwang SJ, Davis ME. New class of polymers for the delivery of macromolecular therapeutics. Bioconjugate Chemistry. 1999;**10**(6):1068-1074

[33] Davis ME. The first targeted delivery of siRNA in humans via a self-assembling, cyclodextrin polymer-based nanoparticle: From concept to clinic. Molecular Pharmaceutics. 2009;**6**(3):659-668

[34] Pun SH, Bellocq NC, Liu A, Jensen G, Machemer T, Quijano E, et al. Cyclodextrin-modified Polyethylenimine polymers for gene delivery. Bioconjugate Chemistry. 2004;**15**(4): 831-840

[35] Forrest ML, Gabrielson N, Pack DW. Cyclodextrin–polyethylenimine conjugates for targeted *in vitro* gene delivery. Biotechnology and Bioengineering. 2005;**89**(4):416-423

[36] Arima H, Kihara F, Hirayama F, Uekama K. Enhancement of gene expression by polyamidoamine dendrimer conjugates with α-, β-, and γ-cyclodextrins. Bioconjugate Chemistry. 2001;**12**(4):476-484

[37] Kihara F, Arima H, Tsutsumi T, Hirayama F, Uekama K. Effects of structure of polyamidoamine dendrimer on gene transfer efficiency of the dendrimer conjugate with α-cyclodextrin. Bioconjugate Chemistry. 2002;**13**(6):1211-1219

[38] Kihara F, Arima H, Tsutsumi T, Hirayama F, Uekama K. *In vitro* and *in vivo* gene transfer by an optimized α-cyclodextrin conjugate with polyamidoamine dendrimer. Bioconjugate Chemistry. 2003;**14**(2):342-350

[39] Wada K, Arima H, Tsutsumi T, Chihara Y, Hattori K, Hirayama F, et al. Improvement of gene delivery mediated by mannosylated dendrimer/α-cyclodextrin conjugates. Journal of Controlled Release. 2005;**104**(2):397-413

[40] Arima H, Tsutsumi T, Yoshimatsu A, Ikeda H, Motoyama K, Higashi T, et al. Inhibitory effect of siRNA complexes with polyamidoamine dendrimer/α-cyclodextrin conjugate (generation 3, G3) on endogenous gene expression. European Journal of Pharmaceutical Sciences. 2011;**44**(3):375-384

[41] Tsutsumi T, Hirayama F, Uekama K, Arima H. Evaluation of polyamidoamine dendrimer/α-cyclodextrin conjugate (generation 3, G3) as a novel carrier for small interfering RNA (siRNA). Journal of Controlled Release. 2007;**119**(3):349-359

[42] Moore CB, Guthrie EH, Huang MT-H, Taxman DJ. Short hairpin RNA (shRNA): Design, delivery, and assessment of gene knockdown. Methods in Molecular Biology. 2010;**629**: 141-158

[43] Tsutsumi T, Hirayama F, Uekama K, Arima H. Potential use of polyamidoamine dendrimer/α-cyclodextrin conjugate (generation 3, G3) as a novel carrier for short hairpin RNA-expressing plasmid DNA. Journal of Pharmaceutical Sciences. 2008;**97**(8):3022-3034

[44] Garcia-Sanchez JE, Garcia E, Merino ML. 100 years of Dr. Ehrlich's magic bullet (1909-2009). Enfermedades Infecciosas y Microbiologia Clinica. 2010;**28**(8):521-533

[45] Strebhardt K, Ullrich A. Paul Ehrlich's magic bullet concept: 100 years of progress. Nature Review Cancer. 2008;**8**(6):473-480

[46] Tavano L, Muzzalupo R. Multi-functional vesicles for cancer therapy: The ultimate magic bullet. Colloids and Surfaces B, Biointerfaces. 2016;**147**:161-171

[47] Al-Jamal KT. Active drug targeting: Lessons learned and new things to consider. International Journal of Pharmaceutics. 2013;**454**(1):525-526

[48] Fahmy TM, Fong PM, Goyal A, Saltzman WM. Targeted for drug delivery. Materials Today. 2005;**8**(8):18-26

[49] Wu GY, Wu CH. Delivery systems for gene therapy. Biotherapy. 1991;**3**(1):87-95

[50] Ahmed M, Narain R. Carbohydrate-based materials for targeted delivery of drugs and genes to the liver. Nanomedicine. 2015;**10**(14):2263-2288

[51] Kaneda Y, Iwai K, Uchida T. Introduction and expression of the human insulin gene in adult rat liver. The Journal of Biological Chemistry. 1989;**264**(21):12126-12129

[52] Tomita N, Morishita R, Higaki J, Tomita S, Aoki M, Ogihara T, et al. *In vivo* gene transfer of insulin gene into neonatal rats by the HVJ-liposome method resulted in sustained transgene expression. Gene Therapy. 1996;**3**(6):477-482

[53] Wu GY, Wu CH. Receptor-mediated gene delivery and expression *in vivo*. The Journal of Biological Chemistry. 1988;**263**(29):14621-14624

[54] Nakatani T, Kuriyama S, Tominaga K, Tsujimoto T, Mitoro A, Yamazaki M, et al. Assessment of efficiency and safety of adenovirus mediated gene transfer into normal and damaged murine livers. Gut. 2000;**47**(4):563-570

[55] Rettinger SD, Kennedy SC, Wu X, Saylors RL, Hafenrichter DG, Flye MW, et al. Liver-directed gene therapy: Quantitative evaluation of promoter elements by using *in vivo* retroviral transduction. Proceedings of the National Academy of Sciences of the United States of America. 1994;**91**(4):1460-1464

[56] Wu GY, Wu CH. Evidence for targeted gene delivery to Hep G2 hepatoma cells in vitro. Biochemistry. 1988;**27**(3):887-892

[57] Hu HM, Zhang X, Zhong NQ, Pan SR. Study on galactose-poly(ethylene glycol)-poly(L-lysine) as novel gene vector for targeting hepatocytes *in vitro*. Journal of Biomaterials Science Polymer Edition. 2012;**23**(5):677-695

[58] Nie C, Liu C, Chen G, Dai J, Li H, Shuai X. Hepatocyte-targeted psiRNA delivery mediated by galactosylated poly(ethylene glycol)-graft-polyethylenimine *in vitro*. Journal of Biomaterials Applications. 2011;**26**(3):255-275

[59] Wang Y, Su J, Cai W, Lu P, Yuan L, Jin T, et al. Hepatocyte-targeting gene transfer mediated by galactosylated poly(ethylene glycol)-graft-polyethylenimine derivative. Drug Design, Development and Therapy. 2013;**7**:211-221

[60] Jiang HL, Kwon JT, Kim YK, Kim EM, Arote R, Jeong HJ, et al. Galactosylated chitosan-graft-polyethylenimine as a gene carrier for hepatocyte targeting. Gene Therapy. 2007;**14**(19):1389-1398

[61] Wada K, Arima H, Tsutsumi T, Hirayama F, Uekama K. Enhancing effects of galactosylated dendrimer/α-cyclodextrin conjugates on gene transfer efficiency. Biological & Pharmaceutical Bulletin. 2005;**28**(3):500-505

[62] Hayashi Y, Higashi T, Motoyama K, Mori Y, Jono H, Ando Y, et al. Design and evaluation of polyamidoamine dendrimer conjugate with PEG, α-cyclodextrin and lactose as a novel hepatocyte-selective gene carrier *in vitro* and *in vivo*. Journal of Drug Targeting. 2013;**21**(5):487-496

[63] Hayashi Y, Mori Y, Yamashita S, Motoyama K, Higashi T, Jono H, et al. Potential use of lactosylated dendrimer (G3)/α-cyclodextrin conjugates as hepatocyte-specific siRNA carriers for the treatment of familial amyloidotic polyneuropathy. Molecular Pharmaceutics. 2012;**9**(6):1645-1653

[64] Ishihara H, Hayashi Y, Hara T, Aramaki Y, Tsuchiya S, Koike K. Specific uptake of asialofetuin-tacked liposomes encapsulating interferon-γ by human hepatoma cells and its inhibitory effect on hepatitis B virus replication. Biochemical and Biophysical Research Communications. 1991;**174**(2):839-845

[65] Sitia G, Iannacone M, Aiolfi R, Isogawa M, van Rooijen N, Scozzesi C, et al. Kupffer cells hasten resolution of liver immunopathology in mouse models of viral hepatitis. PLoS Pathogens. 2011;**7**(6):e1002061

[66] Fox ES, Thomas P, Broitman SA. Comparative studies of endotoxin uptake by isolated rat Kupffer and peritoneal cells. Infection and Immunity. 1987;**55**(12):2962-2966

[67] Bilzer M, Roggel F, Gerbes AL. Role of Kupffer cells in host defense and liver disease. Liver International. 2006;**26**(10):1175-1186

[68] Haltiwanger RS, Lehrman MA, Eckhardt AE, Hill RL. The distribution and localization of the fucose-binding lectin in rat tissues and the identification of a high affinity form of the mannose/N-acetylglucosamine-binding lectin in rat liver. The Journal of Biological Chemistry. 1986;**261**(16):7433-7439

[69] Motoyama K, Mitsuyasu R, Akao C, Abu H II, Sato N, Tanaka T, et al. Potential use of thioalkylated mannose-modified dendrimer (G3)/α-cyclodextrin conjugate as an NF-κB siRNA carrier for the treatment of fulminant hepatitis. Molecular Pharmaceutics. 2015;**12**(9):3129-3136

[70] Motoyama K, Mitsuyasu R, Akao C, Tanaka T, Ohyama A, Sato N, et al. Design and evaluation of thioalkylated mannose-modified dendrimer (G3)/α-cyclodextrin conjugates as antigen-presenting cell-selective siRNA carriers. The AAPS Journal. 2014;**16**(6):1298-1308

[71] Yeeprae W, Kawakami S, Higuchi Y, Yamashita F, Hashida M. Biodistribution characteristics of mannosylated and fucosylated O/W emulsions in mice. Journal of Drug Targeting. 2005;**13**(8-9):479-487

[72] Higuchi Y, Kawakami S, Yamashita F, Hashida M. The potential role of fucosylated cationic liposome/NFκB decoy complexes in the treatment of cytokine-related liver disease. Biomaterials. 2007;**28**(3):532-539

[73] Akao C, Tanaka T, Onodera R, Ohyama A, Sato N, Motoyama K, et al. Potential use of fucose-appended dendrimer/α-cyclodextrin conjugates as NF-κB decoy carriers for the treatment of lipopolysaccharide-induced fulminant hepatitis in mice. Journal of Controlled Release. 2014;**193**:35-41

[74] Mishra N, Yadav NP, Rai VK, Sinha P, Yadav KS, Jain S, et al. Efficient hepatic delivery of drugs: Novel strategies and their significance. BioMed Research International. 2013;**2013**:20

[75] Chen H, Ahn R, Van den Bossche J, Thompson DH, O'Halloran TV. Folate-mediated intracellular drug delivery increases the anticancer efficacy of nanoparticulate formulation of arsenic trioxide. Molecular Cancer Therapeutics. 2009;**8**(7):1955-1963

[76] Lu Y, Low PS. Folate-mediated delivery of macromolecular anticancer therapeutic agents. Advanced Drug Delivery Reviews. 2002;**54**(5):675-693

[77] Arima H, Arizono M, Higashi T, Yoshimatsu A, Ikeda H, Motoyama K, et al. Potential use of folate-polyethylene glycol (PEG)-appended dendrimer (G3) conjugate with α-cyclodextrin as DNA carriers to tumor cells. Cancer Gene Therapy. 2012;**19**(5):358-366

[78] Arima H, Yoshimatsu A, Ikeda H, Ohyama A, Motoyama K, Higashi T, et al. Folate-PEG-appended dendrimer conjugate with α-cyclodextrin as a novel cancer cell-selective siRNA delivery carrier. Molecular Pharmaceutics. 2012;**9**(9):2591-2604

[79] Ohyama A, Higashi T, Motoyama K, Arima H. *In vitro* and *in vivo* tumor-targeting siRNA delivery using folate-PEG-appended dendrimer (G4)/α-cyclodextrin conjugates. Bioconjugate Chemistry. 2016;**27**(3):521-532

[80] Arima H, Yamashita S, Mori Y, Hayashi Y, Motoyama K, Hattori K, et al. *In vitro* and *in vivo* gene delivery mediated by lactosylated dendrimer/α-cyclodextrin conjugates (G2) into hepatocytes. Journal of Controlled Release. 2010;**146**(1):106-117

[81] Tavornvipas S, Hirayama F, Arima H, Uekama K, Ishiguro T, Oka M, et al. 6-*O*-α-(4-*O*-α-D-glucuronyl)-D-glucosyl-β-cyclodextrin: Solubilizing ability and some cellular effects. International Journal of Pharmaceutics. 2002;**249**(1-2):199-209

[82] Anno T, Motoyama K, Higashi T, Hirayama F, Uekama K, Arima H. Preparation and evaluation of polyamidoamine dendrimer (G2)/branched-β-cyclodextrin conjugate as a novel gene transfer carrier. Journal of Inclusion Phenomena and Macrocyclic Chemistry. 2011;**70**(3-4):339-344

[83] Schroder HC, Ushijima H, Theis C, Seve AP, Hubert J, Muller WE. Expression of nuclear lectin carbohydrate-binding protein 35 in human immunodeficiency virus type 1-infected Molt-3 cells. Journal of Acquired Immune Deficiency Syndromes and Human Retrovirology. 1995;**9**(4):340-348

[84] Feinberg H, Park-Snyder S, Kolatkar AR, Heise CT, Taylor ME, Weis WI. Structure of a C-type carbohydrate recognition domain from the macrophage mannose receptor. The Journal of Biological Chemistry. 2000;**275**(28):21539-21548

[85] Anno T, Higashi T, Motoyama K, Hirayama F, Uekama K, Arima H. Potential use of glucuronylglucosyl-β-cyclodextrin/dendrimer conjugate (G2) as a DNA carrier *in vitro* and *in vivo*. Journal of Drug Targeting. 2011;**20**(3):272-280

[86] Anno T, Higashi T, Hayashi Y, Motoyama K, Jono H, Ando Y, et al. Potential use of glucuronylglucosyl-β-cyclodextrin/dendrimer conjugate (G2) as a siRNA carrier for the treatment of familial amyloidotic polyneuropathy. Journal of Drug Targeting. 2014;**22**(10):883-890

[87] Abdelwahab AF, Ohyama A, Higashi T, Motoyama K, Khaled KA, Sarhan HA, et al. Preparation and evaluation of polyamidoamine dendrimer conjugate with glucuronylglucosyl-β-cyclodextrin (G3) as a novel carrier for siRNA. Journal of Drug Targeting. 2014;**22**(10):927-934

Chapter 11

Encapsulation of Essential Oils by Cyclodextrins: Characterization and Evaluation

Jaruporn Rakmai, Benjamas Cheirsilp, Antonio Cid,
Ana Torrado-Agrasar, Juan Carlos Mejuto and
Jesus Simal-Gandara

Additional information is available at the end of the chapter

http://dx.doi.org/10.5772/intechopen.73589

Abstract

The essential oils normally had low physicochemical stability and low solubility in water. These facts limit their industrial applications in general and in food formulations particularly. This chapter characterizes the physicochemical properties and the antioxidant and antimicrobial activities of three encapsulated essential oils – guava leaf, yarrow and black pepper essential oils – in hydroxypropyl-β-cyclodextrin (HPβCD).

Keywords: essential oils, cyclodextrins, food technology applications, pharmacological applications, antioxidant activity, antimicrobial activity

1. CDs in food science and food technology

There is much interest in manipulating the complex-forming ability of cyclodextrins (CDs) with a view to developing applications [1–10]. In the last years, several reviews describing the use of CDs in food and flavor applications have been published [5, 6, 11–16]). CDs have been recommended for applications in food processing and as food additives with a variety of aims: (i) to protect lipophilic food components that are sensitive to oxygen and light- or heat-induced degradation; (ii) to solubilize food colorings and vitamins; (iii) to stabilize fragrances, flavors, vitamins, and essential oils against unwanted changes; (iv) to suppress unpleasant odors or tastes and (v) to achieve a controlled release of certain food constituents.

Indeed, CDs form inclusion complexes with a variety of molecules including fats, flavors and colors. For instance, they are used for the removal and masking of undesirable components and controlled release of desired food constituents [17]. Moreover, CDs are used in food formulations for flavor protection or flavor delivery [18]. Most natural and artificial flavors are volatile oils or liquids, and complexation with CDs provides a promising alternative to the conventional encapsulation technologies for flavor protection. CDs act as molecular encapsulants, protecting the flavor throughout many rigorous food-processing methods such as freezing, thawing and microwaving. β-CD as a molecular encapsulant allows the flavor quality and quantity to be preserved to a greater extent and longer period compared to other encapsulants and provides longevity to the food item [19]. In Japan, CDs have been approved as "modified starch" for food applications for more than two decades, serving to mask odors in fresh food and to stabilize fish oils. One or two European countries—for example, Hungary—have approved γ-CD for use in certain applications because of its low toxicity. It was proved that CDs may alter the sensory profile of a food and the flavor release depends of the CD type [20], the temperature [21] and may depend the solvent nature that is, water, water/alcohol mixtures, etc. [22]. Their beneficial effects essentially derive from the ability to form stable inclusion complexes with sensitive lipophilic nutrients and constituents of flavor and taste, making easy to prepare powdered flavor materials [23–25] and even to release such flavors during cooking [26]. Toxicological data are examined and an assessment of CDs from the standpoint of safety for human consumption is made [27]. Regulations are covered, showing a general trend toward a wider acceptance of CDs as food additives. The growing health consciousness of consumers and expanding market for functional foods and nutraceutical products are opening up to CDs a promising future in food industry [11].

The complexation of CDs with sweetening agents such as aspartame stabilizes and improves the taste. It also eliminates the bitter aftertaste of other sweeteners such as stevioside, glycyrrhizin and rubusoside. CD itself is a promising new sweetener. Enhancement of flavor by CDs has been also claimed for alcoholic beverages such as whisky and beer [28]. The bitterness of citrus fruit juices is a major problem in the industry caused by the presence of limonoids (mainly limonin) and flavonoids (mainly naringin). Cross-linked CD polymers are useful to remove these bitter components by inclusion complexation [29]. CDs are also used to control bitterness in tannins, plant and fungal extracts; skim milk hydrolyses and overcooked tea and coffee [30]. They can also be used to keep the profile of oil volatiles in paste samples that were vacuum- or spray-dried [31, 32], due to their high encapsulation efficiency. The most prevalent use of CD in process aids is the removal of cholesterol from animal products such as eggs or dairy products, like cheese [33]. CD-treated material shows 80% removal of cholesterol. Free fatty acids can also be removed from fats using CDs, thus improving the frying property of fat (e.g., reduced smoke formation, less foaming, less browning and deposition of oil residues on surfaces) [30]. Fruits and vegetable juices are also treated with CD to remove phenolic compounds, which cause enzymatic browning. In juices, polyphenol oxidase converts the colorless polyphenols to colored compounds and addition of CDs removes polyphenoloxidase from juices by complexation. Sojo et al. [34] studied the effect of CDs on the oxidation of *o*-diphenol by banana polyphenoloxidase and found that CDs act as activator as well as inhibitor. By combining 1–4% CD with chopped ginger root, Sung [35] established that it could be stored in vacuum at cold temperature for 4 weeks or longer without browning or rotting.

Other studies describes the development of a gas chromatography-mass spectrometry (GC-MS) library to identify optically active compounds in the flavor and fragrance field using enantioselective GC with CD derivatives (CDs) as chiral selectors in combination with MS

[36, 37], but also olfactometry can be used for detection to have extra information about flavors [38]. The ability to separate and quantitate enantiomers at low levels should be useful for detecting adulterated products, for evaluating fermentation processes and for the accurate characterization of enantiomeric flavor components, growth regulators, pesticides, and herbicides as well as their chiral environmental degradation products and metabolites [39].

Flavonoids and terpenoids are good for human health because of their antioxidative and antimicrobial properties but they cannot be utilized as foodstuff owing to their very low aqueous solubility and bitter taste. Sumiyoshi [40] discussed the improvement of the properties of these plant components (flavonoids and terpenoids) with CD complexation. CDs are used in preparation of foodstuffs in different ways. For example, highly branched CDs are used in flour-based items like noodles, pie dough, pizza sheets and rice cakes to impart elasticity and flexibility to dough [41]. They are also used in the preparation of antimicrobial food preservatives containing trans-2-hexanalin in apple juice preparation [42] and in the processing of medicinal mushrooms for the preparation of crude drugs and health foods. CDs are used in the preparation of controlled release powdered flavors and confectionery items and are also used in chewing gum to retain its flavor for longer duration, a property highly valued by customers [43]. CDs are also used in the detection of aflatoxin in food samples [44].

A large variety of commercial encapsulation practices are currently followed, however, those involving the formation of flavor/CD molecular-inclusion complexes offer great potential for protection of volatile and labile flavoring materials present in a multicomponent food system throughout many rigorous food-processing methods (cooking, pasteurization, etc.) [14, 45–47]. In the same way, CDs can eliminate some taste. In fact, a bitter taste is the main reason for the rejection of various food products although exceptions to this rule are rooted in many cultures: in some foods and beverages, such as coffee, beer, and wine, a certain degree of bitterness is expected [2, 48–51]. Bitterness, however, has proved a major limitation in the acceptance of commercial citrus juices. A commercial process is needed that removes bitter components without adding anything to the juice, while still maintaining the expected flavor and nutritional value of the product. CDs can be used for the removal or masking of undesirable components. Some foods have a peculiar smell, but, when CDs are added in their manufacture, these components form CD-inclusion complexes deodorizing the result product. For instance, this process is used for deodorizing soybean milk and soy protein, and also for removing the peculiar fish odors, seafood and meat products [52–54]. On the other hand, the formation of inclusion complexes with CDs can protect some lipophilic food components that are sensitive to oxygen and heat- or light-induced degradation [55]. In addition, CDs protected phenolic compounds from enzymatic oxidation by forming inclusion complexes [56–59].

2. Essential oils

Both *in vitro* and *in vivo* studies have demonstrated the important applications of essential oils, such an antioxidant or antibacterial activity, even antitumor or anti-inflammatory, with important technological applications in food science and pharmacology [60–64]. Indeed the presence of eugenol, carvacrol or thymol as main component of these oils guarantee their properties both antioxidants and antibiotics (**Figure 1**).

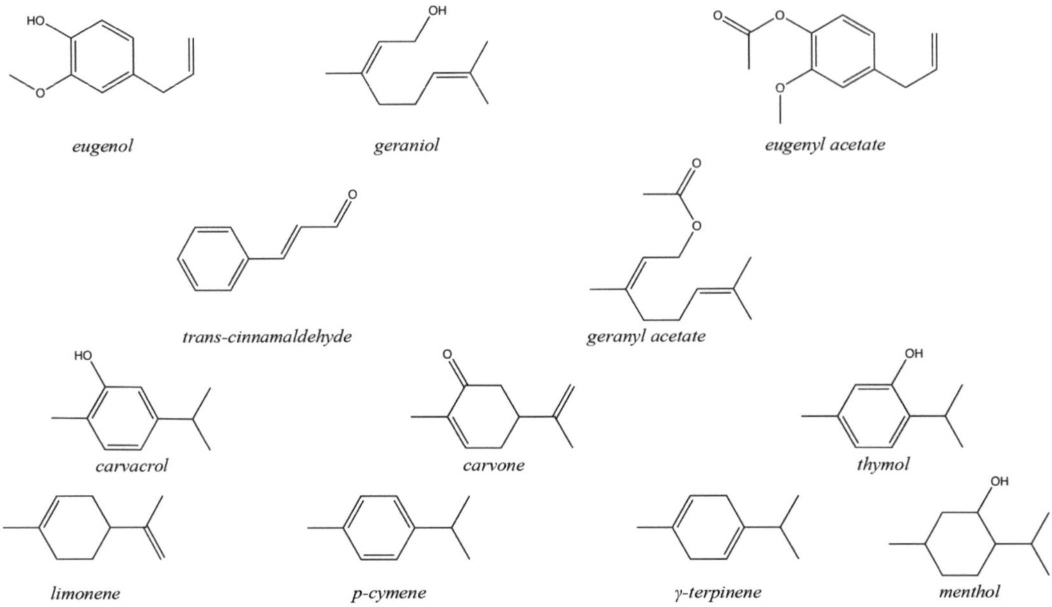

Figure 1. Main components of essential oils [60].

As example, in the present chapter we have selected some essential oils to characterize their inclusion complex in (hydroxypropyl-β-CD) (**Figure 2**). They were black pepper essential oil, guava essential oil and yarrow essential oil.

Black pepper (*Piper nigrum* L.) is considered the king of spices because of its pungent of piperine [65]. It can be used for different purposes such as medicine, human dietaries, preservatives and bio control agents [65–67]. It has been already reported that essential oil from black pepper possess antioxidant [68] and antimicrobial activities [69]. Black pepper oil is basically composed of terpenes, which have been found to be β-caryophyllene, limonene, δ-3-carene and pinene (**Figure 3**) [68, 70]. The major composition of black pepper oil was found to be β-caryophyllene [68, 70]. Nevertheless, some active compounds in essential oils are sensitive toward the chemical modification under effect of some external factors such as temperature, light, oxygen, etc. [71]. Besides, to apply in foods, an extremely low flavor threshold of essential oils can drastically change the sensory properties of foods, and highly water insoluble may have limited contact with pathogens [72].

Guava (*Psidium guajava* L.) has been used as a traditional medicine because of its biological properties [73–75]. Essential oil from guava leaves contains several bioactive compounds, which are responsible for anti-proliferation, antioxidant and antimicrobial activities [76, 77]. Limonene, β-caryophyllene, 1,8-cineole and α-pinene are the major constituents (**Figure 4**) [78, 79]. However, essential oils have some limitations for food applications. Their low solubility in water limits contact with food pathogens in aqueous matrices [72]. Besides, some active compounds in essential oils are sensitive to chemical modifications under the effect of external factors such as temperature, light or oxygen [71].

http://dx.doi.org/10.5772/intechopen.73589

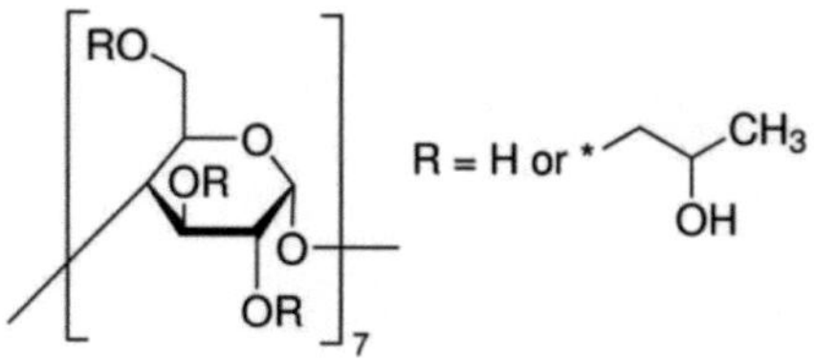

Figure 2. (2-Hydroxypropyl)-β-CD.

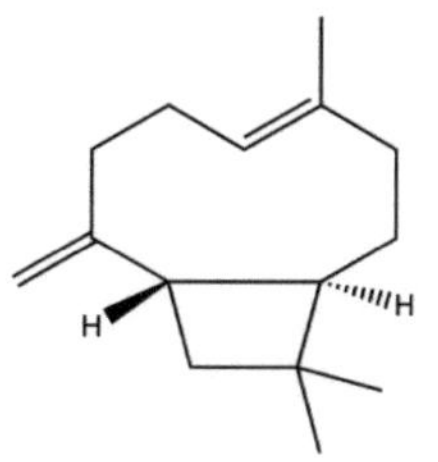

caryophyllene
4,11,11-trimethyl-8-methylene-bicyclo[7.2.0]undec-4-ene

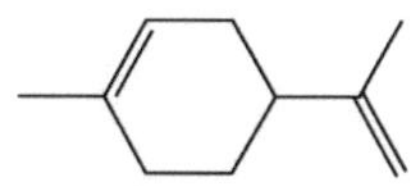

limonene

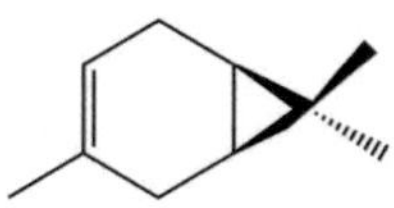

carene
3,7,7-Trimethylbicyclo[4.1.0]hept-3-ene

pinene
(1S,5S)-6,6-dimethyl-2-methylenebicyclo[3.1.1]heptane

Figure 3. Main components of black pepper [68, 70].

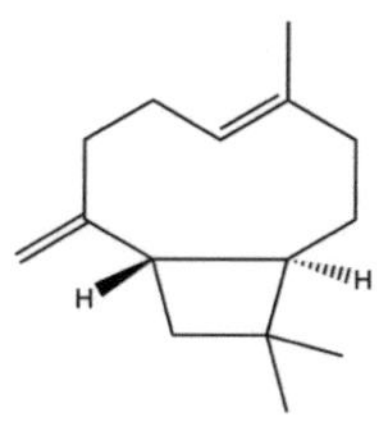

caryophyllene
4,11,11-trimethyl-8-methylene-bicyclo[7.2.0]undec-4-ene

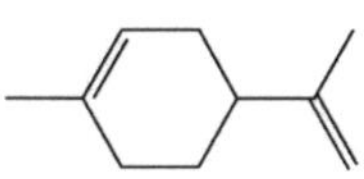

limonene

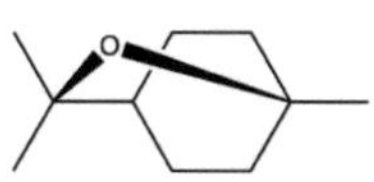

eucalyptol
1,8-cineole

pinene
(1S,5S)-6,6-dimethyl-2-methylenebicyclo[3.1.1]heptane

Figure 4. Main components of guava [78, 79].

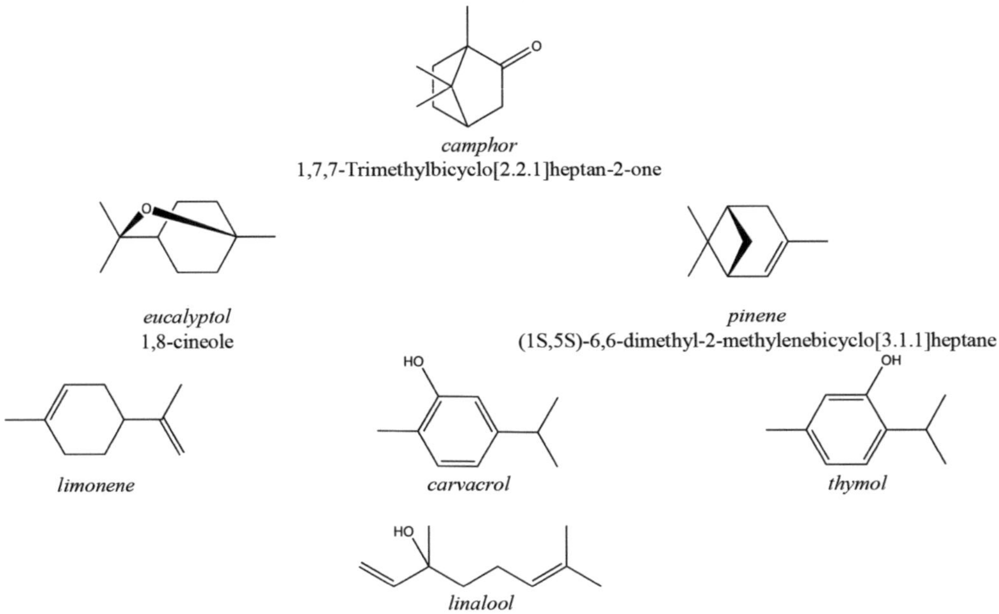

Figure 5. Main components of yarrow [87].

Yarrow (*Achillea millefolium* L. s. l.) has a broad spectrum of pharmacological activities. It is widely used in folk medicine [80]. In Europe, it has been used as a remedy to treat digestive problems, diabetes, hepatic-biliary diseases, amenorrhea, and consumed for its antitumor and anti-inflammatory properties [81–83]. In addition, antimicrobial and antioxidant properties of yarrow have also been reported [84–86]. Chemical components of yarrow essential oil have been found to be carvacrol, linalool, 1,8-cineole, camphor and thymol was mostly found as a major component (**Figure 5**) [87]. However, some active chemical components of yarrow oil (such as carvacrol and thymol) are sensitive to environmental factors such as, light, oxygen and temperature. Encapsulation of yarrow essential oil could offer possible solutions for the limitation.

3. CDs and essential oils

The use of CDs for the essential oils encapsulation can protect the active compounds of essential oils from environmental conditions [13, 14] and improve the aqueous solubility of essential oils for increasing their capacity to functionalize the products in which it is used as additive [88]. As quote above, CDs are cyclic oligosaccharides consisting of glucopyranosyl units linked by α-(1,4) bonds [89]. The widely used natural CDs are α-, β- and γ-CD consisting of 6, 7 and 8 glucopyranose units, respectively [90, 91]. These molecules have a unique structure with a hydrophobic cavity and a hydrophilic surface, which can form inclusion complex with a wide variety of guests. They can be used to enhance the solubility of insoluble compounds, stabilize labile guests against oxidation, control volatility and sublimation, modify taste by masking off flavors, entrap odors and control the releasing of drugs and flavors [92]. Among those CDs, β-CD is the most widely applicable kind because of its suitable cavity size for common guests with molecular weights between 200 and 800 g/mol and its availability and reasonable price [93]. Although β-CD can be used with many guests, its solubility in water is low (1.8 g in 100 mL water at 25°C). In some cases, there is a need to enhance water solubility of β-CD by adding the hydroxyl-alkyl groups on the β-CD surface. A hydroxyl-alkylated or hydroxypropyl-β-CD derivative (HPβCD) is relatively high aqueous solubility (above 60 g in 100 mL water at 25°C) with low toxicity and satisfactory inclusion ability [94].

On the other hand, encapsulation of essential oils or their chemical components with CDs or CD derivatives for improvement of biological properties have been observed [5, 95–98] or their antimicrobial activity [99].

Indeed, a large amount of contributions about technologic applications of CD-inclusion complex of essential oils and their main components has been published in the last 10 years, some of them are included in **Table 1**.

Essential oil	Guest	References	Essential oil component	Guest	References
Black pepper essential oil	Hydroxypropyl-β-CD	[100]	Allyl isothiocyanate	α-CD	[101]
Cinnamon essential oil	β-CD	[99, 102, 103]	Allyl isothiocyanate	β-CD	[101, 104]
Citronella oil	β-CD	[105]	Barbigerone	Hydroxypropyl-β-CD	[106]
Clove bod oil	β-CD	[99, 107]	Carvacrol	β-CD	[108, 109]
Coriander essential oil	β-CD	[71]	Carvacrol	Hydroxypropyl-β-CD	[110]
Garlic oil	β-CD	[102, 111]	Cinnamaldehyde	β-CD	[99, 103]
Guava leaf oil	Hydroxypropyl-β-CD	[112]	Citronellal	β-CD	[105]
Lemon oil	β-CD	[113]	Citronellol	β-CD	[105]
Olive leaf oil	β-CD	[114]	Eugenol	β-CD	[99, 115–118]
Oregano essential oil	β-CD	[107, 119]	Limonene	β-CD	[120]
Thyme essential oil	β-CD	[121, 122]	2-Nonanone	β-CD	[123]
Sweet basil essential oils	β-CD	[124]	Thymol	β-CD	[103, 109, 121]
Yarrow essential oil	Hydroxypropyl-β-CD	[125]	Vanillin	β-CD	[126, 127]

Table 1. Contributions abut host-guest complex formation between CDs and CDs derivatives and essential oils.

3.1. Encapsulation efficiency

As quoted above, we present the encapsulation efficiency of three essential oils (guava oil, yarrow oil and black pepper oil) in hydroxypropyl-β-CD (HPβCD).

In the case of yarrow oil and carvacrol (yarrow oil major component), there efficiency were 45.05 and 86.59%, respectively [125] see **Table 2**. Black pepper [100] exhibit similar behavior with efficiency of 50.55 and 85.30, respectively, for essential oil and its main component (β-caryophyllene). Finally, guava leaf oil encapsulation efficiency was 52.5%, while it reached 91.8% for limonene, the major pure compound of guava leaf oil [112].

This difference in encapsulation efficiency of the pure compound and the essential oil results from the presence of other minority components. In the case of yarrow oil and carvacrol [125], the other components like 1,8 cineole, thymol, camphor and linalool have also high affinities for CD [6, 121, 128–132] that compete for inclusion complex formations. Kamimura et al. [110] reported that the encapsulation efficiency values of pure carvacrol in HPβCD prepared by kneading and freeze-drying methods were around 78 and 84%, respectively.

Similar explanation would justify the diferences in encapsulation efficiency of the pure compound and the black pepper oil [100] because the presence of other components in the black pepper oil such as limonene, δ-3-carene and pinene [68] that also have high affinities for HPβCD. In the case of guava leaf oil [112], the large values found are due to minority components, such as β-caryophyllene, 1,8-cineole and α-pinene, exhibit low affinity for the β-CD that are not easily encapsulated and the competition between the other host for the guest in not so important.

Similar observation has been reported for other authors in the literature [99] showing that encapsulation efficiencies of cinnamon oil and clove oil were 41.72 and 77.74%, respectively. The encapsulation efficiencies of major components including trans-cinnamaldehyde in cinnamol oil and eugenol in clove oil were also examined and showed higher encapsulation efficiency of 84.70 and 90.15%, respectively. In addition, comparable values of encapsulation efficiency were found in other carriers such as alginate-chitosan system. In this case, the yarrow oil components exhibited 82.4% efficiency of polyphenol encapsulation [133, 134].

Compound	Encapsulation efficiency (%)	Compound	Encapsulation efficiency (%)
Black pepper oil	50.55	β-caryophyllene	85.30
Yarrow oil	45.05	Carvacrol	86.59
Guava leaf oil	52.50	Limonene	91.80
Cinnamon oil	41.72	Cinnamaldehyde	84.70
Clove oil	77.74	Eugenol	90.15

Table 2. Encapsulation efficiency value in HPβCD.

3.2. Characterization of host-guest complex

3.2.1. Morphological examinations

It is well known that the inclusion complex formation would change the morphology of CDs [135]. **Figure 6** presents the morphology of the encapsulated oils studied by SEM.

The particle shape and morphology of the encapsulated oil were similar to those of free HPβCD in the cases evaluated – guava, yarrow and black pepper – see **Figure 7**. It indicates the hydrogen bonding of the free HPβCD molecules interact with each other in solution producing the cluster of HPβCD [136, 137]. This case not occurs in inclusion complex because inclusion complex formation also induces the conformation change of CD and obstructs the agglomeration among them. Similar observations have been previously reported that the distribution of inclusion complex of carvacrol and β-CD, and the gathering of free β-CD were also found [135].

By contrast, the free HPβCD particle sizes are much larger than those of the encapsulated products. These results are in agreement with Guimaraes et al. [135], who analyze carvacrol encapsulation with β-CD. Considering that HPβCD form clusters in solution through intermolecular hydrogen bonds [136, 137], it seems that the incorporation of different essential oils interferes in these interactions and reduces particle size.

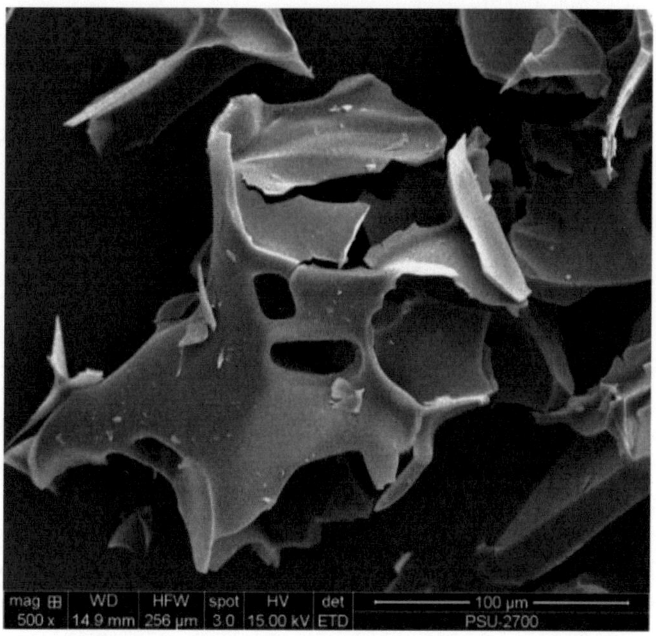

Figure 6. SEM micrographs of free HPβCD at 500 times magnification.

Guava essential oil - HPβCD host-guest complex | *Black pepper essential oil - HPβCD host-guest complex* | *Yarrow essential oil - HPβCD host-guest complex*

Figure 7. SEM micrographs of encapsulated essential oils at 500 times magnification.

3.2.2. Fourier-transform infrared spectroscopy (FT-IR)

FT-IR technique can be used to investigate the variation of shape, intensity and position of peaks [138].

FT-IR spectrum of black pepper oil consisted of the prominent absorption bands at 2954, 2923 and 2865 cm^{-1} for C—H stretching vibration of methylene group, 1638 cm^{-1} for H—O—H bending, 1447 cm^{-1} for C—H scissoring vibration, 1369 cm^{-1} for symmetrical deformation vibration of CH_3, 886 cm^{-1} for C—H deformation vibration and 789 cm^{-1} for S—C absorption. However, FT-IR spectrum of the encapsulated black pepper oil showed that no character similar to the free black pepper oil. All bands of black pepper oil spectrum were totally obscured by HPβCD bands it was possible that black pepper oil entered the cavity of HPβCD and inclusion complex was formed.

In the case of yarrow oil, its FT-IR spectrum of yarrow oil shows prominent absorption bands at 2956 cm^{-1} for $=CH_2$ symmetrical stretching vibration, 2926 cm^{-1} for C—H stretching vibration of methylene group, 2869 cm^{-1} for —CH stretching, 1652 cm^{-1} for H—O—H bending, 1626 cm^{-1} for C=C stretching vibration of the allyl group, 1446 cm^{-1} for C—H scissoring vibration, 1380 cm^{-1} for symmetrical deformation vibration of $-CH_3$, 1366 cm^{-1} for symmetrical deformation vibration of $-CH_3$, 1240 cm^{-1} and 1103 cm^{-1} for P—O and P=O, 1022 cm^{-1} for C—O—C stretching vibration, 916 cm^{-1} for C—S—C stretching vibration, 875 cm^{-1} and 865 cm^{-1} for C—H bending of aromatic ring. The spectrum of HPβCD shows prominent absorption bands at 3406 cm^{-1} for O—H stretching vibration, 2970 cm^{-1} for $=CH_2$ symmetrical stretching vibration, 2930 cm^{-1} for C—H stretching vibration, 1646 cm^{-1} for H—O—H bending vibration, 1157 cm^{-1} for C—O—C asymmetrical stretching vibration, 1083 cm^{-1} and 1033 cm^{-1} for symmetric C—O—C stretching vibration [139]. FT-IR spectrum of inclusion complex was identical to HPβCD and no feature similar to yarrow oil. The results indicated that HPβCD covered all the absorption bands of yarrow oil in the spectrum of inclusion complex indicating the entering to the cavity of HPβCD and the formation of inclusion complex.

Finally, FT-IR spectrum of guava leaf oil showed prominent absorption bands at 2921 cm^{-1} for C—H stretching vibration of methylene group, 1642 cm^{-1} for H—O—H bending, 1447 cm^{-1} for C—H scissoring vibration, 1376 cm^{-1} for symmetrical deformation vibration of CH3, 886 cm^{-1} for C—H deformation vibration and 789 cm^{-1} for S—C absorption. FT-IR spectrum of encapsulated guava leaf oil shows no feature similar to the free guava oil. The bands of guava leaf oil spectrum were almost completely concealed by very intense and broad bands of HPβCD. However, the absorption band at 608 cm^{-1} of HPβCD disappeared in encapsulated guava leaf oil. This change may be related to the interaction between guava leaf oil and HPβCD in the inclusion complex.

The inclusion complex formation of β-CD was also investigated by Liu et al. [140] using FT-IR analysis. The absorption bands of β-caryophyllene were not detected in the spectrum of inclusion complex. The changes were related to the inclusion complex formation of β-CD and the guests which whole of guest could be contained in the CD cavity. Wang et al. [139] have reported similar results. In their study, the inclusion complex formation of soybean lecithin and β-CD was determined by FT-IR. All the absorption bands of soybean lecithin encapsulated in β-CD were obscured by β-CD spectrum showing that inclusion complex of β-CD and soybean lecithin was formed. However, Gomes et al. [141] reported that the absorption band at 1738 cm^{-1} of the red bell pepper pigments was observed after encapsulation in β-CD indicating that some region of the encapsulated molecules was not contained in the cavity of β-CD.

3.2.3. Ultraviolet-visible spectrophotometry (UV-Vis)

Essential oils contain various bioactive chemicals, which adsorb ultraviolet (UV) or visible light (Vis) at different wavelengths. CD host-guest complex formation would alter UV-Vis absorption spectra [142]. Otherwise, the spectra of the guests appear in line of CD [140]. Therefore, UV-Vis spectrophotometry, evaluated the inclusion complex formation of HPβCD and the three essential oils. The UV absorption spectra of guava leaf oil, limonene and their inclusion complexes were compared. Indeed, maximum absorption value of guava leaf oil was at 214.5 nm, which was mainly attributed to limonene. The absorption peak at 205 nm corresponds to β-caryophyllene and/or pinene. The peak at 275 nm of guava leaf oil was ascribed to 1,8-cineole.

The spectra of the physical mixture of HPβCD with guava leaf oil and with limonene before complexation were consistent with that of guava leaf oil or pure compound. The absorption spectra of the physical mixture of HPβCD with guava leaf oil and with limonene were in accordance to with the spectra of guava leaf oil and pure limonene, respectively. When the active compounds in essential oil or the pure compound were encapsulated into the cavity of HPβCD, the absorption peaks of each compound disappeared in the spectra of the inclusion complexes. To recover active compounds encapsulated in the HPβCD cavity, the active compounds were extracted from HPβCD by dissolving the inclusion complexes in 95% acetonitrile. The encapsulated compounds were released from the cavity of HPβCD and HPβCD was separated from guava leaf essential oil or limonene in solution by centrifugation. The solution was diluted 100 times with acetonitrile and the absorbance was measured by UV spectrophotometer.

After extraction from the inclusion complexes, the absorption peaks of encapsulated compounds in guava leaf oil could be observed. In this line, besides limonene, the absorption peaks at 205 and 275 nm suggested the presence of β-caryophyllene and 1,8-cineole, respectively. The results indicated that the active compounds in guava leaf oil had formed inclusion complex with HPβCD. Therefore, the chemical components of guava leaf oil were successfully encapsulated in the HPβCD.

UV spectrum of yarrow oil shows peaks at 270–275 nm indicated the presence of carvacrol, 1,8-cineole, thymol and camphor. A minor peak at 243 nm attributed to linalool. The spectra of the physical mixture of HPβCD with yarrow oil and with pure compound (carvacrol) conformed to UV spectra of yarrow oil and pure carvacrol, respectively. When the active compounds in yarrow oil or carvacrol were entrapped with HPβCD, the absorption peaks of each compound also disappeared in the spectrum of the inclusion complexes.

After extraction from the inclusion complex, the absorption peaks of entrapped compounds in yarrow oil appeared at 270–275 nm implying carvacrol and also are 1,8-cineole, thymol, camphor and linalool. In this study, the chemical components of yarrow oil were successfully entrapped in the HPβCD, as in the previous case. However, the encapsulation efficiency of yarrow oil was much lower than those of its pure compound. This was likely because the competition of major active compound among other components in essential oil has occurred during inclusion complex formation.

Finally, the absorption spectrum of black pepper oil was recorded with absorption peaks at 200, 205 and 214.5 nm for δ-3-carene, β-caryophyllene and limonene, respectively [140]. The maximum absorption peak at 205 nm was ascribed to β-caryophyllene. The spectra of the physical mixture of HPβCD with black pepper oil and with β-caryophyllene accorded with UV spectra of black pepper oil and pure β-caryophyllene, respectively. When the active compounds in black pepper oil or the pure compound (β-caryophyllene) were entrapped into the cavity of HPβCD, the absorption peaks of the compounds also disappeared in the spectrum of the inclusion complex.

After extraction from the complex, the observable peaks of entrapped compounds in black pepper oil could be seen. The spectrum of encapsulated compounds from black pepper oil show absorption peaks at 205 and 214.5 nm indicating β-caryophyllene and limonene, respectively. The UV spectrum indicated that the chemical components of black pepper oil were successfully entrapped in the HPβCD. As in the previous cases, the encapsulation efficiency of active compounds of black pepper oil was much lower than those of its pure compound. This was likely because the competition of major active compound among other components in black pepper oil has occurred during inclusion complex formation.

3.2.4. Phase solubility

Phase solubility study is generally performed to evaluate the stability and to classify the inclusion complex when they are in the solution. The phase solubility profiles can be obtained from the interaction between the guests (encapsulated compounds) and the hosts (CDs or derivatives) in the solution. In solution, a fundamental parameter such as stability constant (K_s) of inclusion complex formation can be used to evaluate the stability of the inclusion complex [143] – see **Table 3**.

Inclusion complex	T/°C	K_s/M^{-1}	Inclusion complex	T/°C	K_s/M^{-1}
Black pepper oil-HPβCD	25	104.5	β-caryophyllene-HPβCD	25	132.8
Black pepper oil-HPβCD	35	100.0	β-caryophyllene-HPβCD	35	114.0
Guava leaf oil-HPβCD	25	25.0	Limonene-HPβCD	25	628.0
Guava leaf oil-HPβCD	35	33.8	Limonene-HPβCD	35	605.9
Yarrow oil-HPβCD	25	106.6	Carvacrol-HPβCD	25	360.9
Yarrow oil-HPβCD	35	92.0	Carvacrol-HPβCD	35	309.7

Table 3. Phase solubility parameters and stability constants (K_s) of encapsulated essential oil and their main component.

In the case of black pepper, A linear relationship between the amount of dissolved essential oil or β-caryophyllene and the concentrations of HPβCD in this study with slope <1 was classified as a typical A_L-type (type A reveals an inclusion complex formation where the amount of encapsulated compounds increase as the HPβCD concentration increases, subscript L indicates a 1:1 molecular ratio formation of soluble complexes) [144]. As the majority of encapsulated compounds are mono- and sesquiterpenoids and phenylpropane derivatives of an average molecular weight of 120–160 g/mol, a 1:1 complex formation is observed [16]. The molar ratio of host to guest molecules is usually 1:1 for inclusion complexes formed in solution, except for complexes with long-chain or bifunctional guest molecules (e.g. guest molecules having two aromatic rings on opposite sides of a small central molecule segment). In aqueous system, black pepper oil and β-caryophyllene show difference in stability of complex form with the K_s of 104.5 and 132.8 L/mol at 25°C, respectively. This might be because of the other components in black pepper oil might compete to HPβCD form complex with β-caryophyllene. The decreases in K_s values with increasing temperatures were expected for exothermic processes [99].

Equivalent results were observed for yarrow oil host-guest complex, as we can observe in **Table 3**. In agreement with the results reported in **Table 3** – for black pepper essential oil and yarrow essential oil – similar Hill et al. [99] and Kamimura et al. [110] have reported observations. The water solubility of trans-cinnamaldehyde, eugenol, cinnamon bark extracts and clove bud extract samples increased with increasing temperatures while the K_s value of the samples decreased with increasing temperature [99]. Kamimura et al. [110] reported that water solubility of the pure carvacrol increased and the K_s value decreased with increasing temperatures.

Regarding to guava leaf essential oil – see **Table 3**, low K_s value were obtained for guava leaf oil than for limonene. They were in the order of those for β-CD complexes according to Connors [145]. This might be due to the competence of the other components in guava leaf oil with limonene to form HPβCD complexes. In addition, the decrease in K_s values with increasing temperature reflects that complex formation is an exothermic process [99]. However, these results reflect that the aqueous solubility of guava leaf oil can be increased with increasing HPβCD concentration. Considering that very labile complexes (K_s < 100 L/mol) result in premature release of the guests because of the weak interaction between hosts and guests [92], the very labile encapsulated guava leaf oil could be useful for fast release systems such as pharmaceutical applications.

3.3. Evaluation of antioxidant activity of host-guest complex

Antioxidant activity was evaluated in terms of DPPH scavenging capacity (%) of free and encapsulated guava leaf oil compared to a synthetic chemical antioxidant (BHT) at concentrations ranged from 5 to 50 μg/mL.

It was established that the components responsible for the antioxidant activity of guava leaf oil are limonene, α-pinene and β-caryophyllene [146]. While limonene has a moderate antioxidant activity [147], β-caryophyllene and α-pinene show weak and moderate DPPH scavenging activity, respectively [146, 147]. Unfortunately, the encapsulated guava leaf oil gave slightly lower DPPH scavenging activity than that of the free guava leaf oil. This could be because HPβCD blocks the functional groups of the active compounds that react with DPPH radicals [110].

In the case of yarrow oil carvacrol as a major component shows strong antioxidant activity (72% DPPH scavenging at 50 μg/mL). The most effective antioxidants usually contain aromatic or phenolic rings, which interrupt the free radical chain reaction by donating H• to the free radicals [148]. The encapsulated yarrow oil gave slightly lower antioxidant activity than the activity of the free yarrow oil. It was a result of the HPβCD was blocking the functional groups of active compounds during reacting with DPPH radicals [110]. However, the encapsulation has been reported to increase the stability of the essential oils [13, 14].

Black pepper oil shows antioxidant activity with 54% DPPH scavenging (50 μg/mL black pepper oil) (**Figure 5**). It was established that the components responsible for the antioxidant activity are β-caryophyllene, limonene and α-pinene [146]. β-caryophyllene, a major component of black pepper oil, was found to give a weak DPPH scavenging activity [146]. Limonene, a minor composition, has been reported to give a moderate antioxidant activity and another component, α-pinene, also possesses a moderate antioxidant property [147]. It should be noted that free HPβCD did not show antioxidant activity.

However, the inclusion complexes have been reported to increase the stability of the essential oils [13, 14]. After exposure to sunlight, the DPPH scavenging of free guava leaf oil drastically decreased around 43–54% at all tested concentrations (5–50 μg/mL), which was likely due to limonene and pinene sensitive to sunlight [149]. Then, the inclusion complexation of guava leaf oil with HPβCD could protect the active components against the effect of light. In effect, after sunlight exposure, the DPPH radical scavenging capacity of the encapsulated guava leaf oil was more stable than the free guava leaf oil by 26–38%.

Similar results were found for yarrow essential oil, where DPPH radical scavenging (with concentration range from 5 to 35 μg/mL of essential oil) decreased around 41–51% after exposure to sunlight for 12 h. The yarrow oil with higher concentration range (40–50 μg/mL) exhibited lower loss of DPPH radical scavenging (36–37%). Obviously, as in the previous case, the encapsulation of yarrow oil in HPβCD could protect the active components against the effect of sunlight. The complexation with HPβCD improved the stability of yarrow oil by 27–30% in a similar range that guava leaf oil (26–38% *-vide supra-*).

The DPPH radical scavenging capacity of black pepper oil drastically decreased after 12 h exposure to sunlight (**Figure 4**). At the sample concentration range of 5–25 μg/mL, the DPPH

http://dx.doi.org/10.5772/intechopen.73589

scavenging capacity decreased around 42–48%, while the decreasing of 30–39% was found at higher concentration range (30–50 μg/mL). The stability of encapsulated black pepper oil was improved from the free black pepper oil by 18–24%. This effect is lower that observed for guava and yarrow essential oils (26–38 and 27–30%, respectively).

3.4. Evaluation of antibacterial activity of host-guest complex

Table 4 shows minimum inhibitory concentration (MIC) and minimum bactericidal concentration (MBC) values of essential oil for *Staphylococcus aureus* and *Escherichia coli.*

Guava leaf oil displayed antibacterial activity against both bacteria with MIC value of 500 μg/mL, that could be attributed to guava leaf oil monoterpenes (such as limonene) which have been found to play efficient role in antimicrobial activity via membrane structures increasing membrane fluidity and permeability [150]. Pure limonene was reported to give antimicrobial activity against *S. aureus* and *E. coli* with MIC values of 8.0 and 10.0 μg/mL, respectively [151].

The antibacterial activity of guava leaf oil was improved after encapsulation in HPβCD by 4 and 2 times against *S. aureus* and *E. coli,* respectively. It has been reported that inclusion complexes with HPβCD could increase aqueous solubility of the encapsulated guests, thus improving the antimicrobial efficiency of essential oils at lower concentrations [99] due to a better accessibility of the active compounds to cells [111].

Yarrow oil exhibited antibacterial activity against *S. aureus* and *E. coli* with the MIC values of 250 μg/mL and 500 μg/mL, respectively. The antimicrobial activity of yarrow essential oil might be because its oxygenated phenolic compounds, such as carvacrol and thymol, have been reported to give strong antimicrobial activity. These compounds were found to increase membrane permeability and membrane disruption of microbial cells (*Pseudomonas aeruginosa* and *S. aureus*) [152]. Antimicrobial potential of oxygenated phenolic compounds, were also reported in the literature [153–157]. In addition, *S. aureus,* a representative for Gram-positive bacteria, was more sensitive to tested samples than *E. coli*. This was because the external surface of outer membrane of *E. coli* that composes of lipopolysaccharides and proteins is more

Antimicrobial compound	***S. aureus***		***E. coli***	
	MIC (μg/mL)	**MBC (μg/mL)**	**MIC (μg/mL)**	**MBC (μg/mL)**
Free guava leaf oil	500	1000	500	1000
Encapsulated guava leaf oil*	125	250	250	500
Yarrow oil	250	500	500	1000
Encapsulated yarrow oil*	62.5	125	62.5	125
Black pepper oil	1000	2000	2000	>2000
Black pepper oil-HPβCD complex*	250	500	500	1000

*Values were based on the actual concentrations of essential oil encapsulated in the HPβCD (calculated from encapsulation efficiency).

Table 4. Minimum inhibitory and bactericidal concentration (MIC, MBC) against *Staphylococcus aureus* and *Escherichia coli* for both free and encapsulated essential oil.

tolerate to the tested samples, and the O-side chains of the lipopolysaccharides of *E. coli* has a hydrophilic surface protecting the hydrophobic molecules to enter the bilayer [146].

The antibacterial efficacy of yarrow oil was much improved after encapsulated in HPβCD by 4 and 8 times against *S. aureus* and *E. coli*, respectively, while antibacterial activity of black pepper oil was improved by 4 times against both *S. aureus* and *E. coli*. As quote above, inclusion complex formation with HPβCD could increase aqueous solubility and improve antimicrobial efficacy at lower concentrations of encapsulated compounds [99]. As the primitive sites for antimicrobial action were found at the cell membrane and inside the cytoplasm, HPβCD may enhance the accession of essential oils to these regions by improving water solubility of essential oils [111].

4. Conclusions

Microencapsulation of essential oils in HPβCD was achieved proving that the host-guest complex formation implies different physicochemical characteristics from free essential oil. As advantage, the DPPH radical scavenging capacity of the encapsulated oil was more stable than for the free oil indicating that the inclusion complex with HPβCD could protect the active components of oil against the effect of sunlight. As well, encapsulation also increased the antibacterial activity of essential oils against both *S. aureus* and *E. coli* the observed behavior implies an important increase.

Acknowledgements

The Graduate School of Prince of Songkla University and Thai Government under Grant No.AGR560387S financially supported this work.

Author details

Jaruporn Rakmai[1], Benjamas Cheirsilp[1], Antonio Cid[2,3], Ana Torrado-Agrasar[4], Juan Carlos Mejuto[3] and Jesus Simal-Gandara[4*]

*Address all correspondence to: jsimal@uvigo.es

1 Department of Industrial Biotechnology, Faculty of Agro-Industry, Prince of Songkla University, Hat Yai, Thailand

2 UCIBIO, REQUIMTE, Departamento de Química, Faculdade de Ciências e Tecnología, Universidade NOVA de Lisboa, Caparica, Portugal

3 Department of Physical Chemistry, Faculty of Science, University of Vigo, Ourense, Spain

4 Department of Analytical Chemistry and Food Science, Faculty of Science, University of Vigo, Ourense, Spain

References

[1] Szejtli J, Osa T. Comprehensive Supramolecular Chemistry. Vol. 3. Cyclodextrins. New York: Pergamont; 1996

[2] Singh M, Sharma R, Banerjee UC. Biotechnological applications of cyclodextrins. Biotechnology Advances. 2002;**20**:341-359

[3] Arias-Estevez M, García-Río L, Mejuto JC, Rodríguez-Dafonte P, Simal-Gandara J. Influence of micelles on the basic degradation of carbofuran. Journal of Agricultural and Food Chemistry. 2005;**53**(18):7172-7178

[4] Arias-Estevez M, García-Falcón MS, García-Río L, Mejuto JC, Rial-Otero R, Simal-Gandara J. Binding constants of oxytetracycline to animal feed divalent cations. Journal of Food Engineering. 2007;**78**(1):69-73

[5] Astray G, Gonzalez-Barreiro C, Mejuto JC, Rial-Otero R, Simal-Gandara J. A review on the use of cyclodextrins in foods. Food Hydrocolloids. 2009;**3**:1931-1640. DOI: 10.1016/j.foodhyd.2009.01.001

[6] Astray G, Mejuto JC, Morales J, Rial-Otero R, Simal-Gandara J. Factors controlling flavors binding constants to cyclodextrins and their applications in foods. Food Research International. 2010;**43**(4):1212-1218

[7] Gonzalez-Barreiro C, Rial-Otero R, Simal-Gandara J, Astray G, Cid A, Mejuto JC, Manso JA, Morales J. Chapter 8: In starch-based polymeric materials and nanocomposites. Chemistry, processing, and applications. In: Starch-Derived Cyclodextrins and Their Future in the Food Biopolymer Industry. Boca Ratón: CRC Press; 2012. pp. 167-182. ISBN: 978-1-4398-5117-3

[8] Cid A, Mejuto JC, Orellana PG, López-Fernández O, Rial-Otero R, Simal-Gandara J. Effects of ascorbic acid on the microstructure and properties of SDS micellar aggregates for potential food applications. Food Research International. 2013;**50**(1):143-148

[9] Cid A, Morales J, Mejuto JC, Briz-Cid N, Rial-Otero R, Simal-Gandara J. Thermodynamics of sodium dodecyl sulphate-salicylic acid based micellar systems and their potential use in fruits postharvest. Food Chemistry. 2014;**151**:358-363

[10] Cid A, Morales J, Dieguez-Perez M, Rial-Otero R, Mejuto JC, Simal-Gandara J. Chapter X: In cyclodextrines: propriétés, chimie et applications: Application des cyclodextrines comme catalyseurs dans les médias micro-hétérogènes. Besançon: Presses Universitaires de Franche-Comté; 2014. ISBN: 978-2-84867-520-6

[11] Cravotto G, Binello A, Baranelli E, Carraro P, Trotta F. Cyclodextrins as food additives and in food processing. Current Nutrition Food Science. 2006;**2**(4):343-350

[12] Hedges AR, McBride CU. Of b-cyclodextrin in food. Cereal Foods World. 1999;**44**(10):700-704

[13] Hedges AR, Shieh WJ, Sikorski CT. Use of cyclodextrins for encapsulation in the use and treatment of food products. In: Risch SJ, Reineccius GA, editors. Encapsulation and Controlled Release of Food Ingredients. Vol. 590. ACS Symposium Series. Washington DC: ACS; 1995. pp. 60-71

[14] Qi ZH, Hedges AR. Use of cyclodextrins for flavours. In: Ho CT, Tan CT, Tong CH, editors. Flavour Technology: Physical Chemistry, Modification and Process. Vol. 610. ACS Symposium Series. Washington DC: ACS; 1995. pp. 231-243

[15] Samant SK, Pai JSC. New versatile food additive. Indian Food Packer. 1991;**45**(3):55-65

[16] Szente L, Szejtli J. Cyclodextrins as food ingredients. Trends in Food Science and Technology. 2004;**15**:137-142

[17] Prasad N, Strauss D, Reichart G. Cyclodextrins inclusion for food, cosmetics and pharmaceuticals. European Patent 1084625. 1999

[18] Kant A, Linforth RST, Hort J, Taylor AJ. Effect of β-cyclodextrin on aroma release and flavor perception. Journal of Agricultural and Food Chemistry. 2004;**52**:2028-2035

[19] Muñoz-Botella S, del Castillo B, Martín MA. Cyclodextrin properties and applications of inclusion complex formation. Ars Pharmaceutica. 1995;**36**:187-198

[20] Reineccius TA, Reineccius GA, Peppard TL. Encapsulation of flavors using cyclodextrins: Comparison of flavor retention in alpha, beta, and gamma types. Journal of Food Science. 2002;**67**(9):3271-3279

[21] Reineccius TA, Reineccius GA, Peppard TL. Flavor release from cyclodextrin complexes: Comparison of alpha, beta, and gamma types. Journal of Food Science. 2003;**68**(4): 1234-1239

[22] Reineccius TA, Reineccius GA, Peppard TL. The effect of solvent interactions on α-, β-, and γ-cyclodextrin/flavor molecular inclusion complexes. Journal of Agricultural and Food Chemistry. 2005;**53**:388-392

[23] Liu X-D, Furuta T, Yoshii H, Linko P, Coumans WJ. Cyclodextrin encapsulation to prevent the loss of l-menthol and its retention during drying. Bioscience, Biotechnology, and Biochemistry. 2000;**64**(8):1608-1613

[24] Tobitsuka K, Miura M, Kobayashi S. Interaction of cyclodextrins with aliphatic acetate esters and aroma components of La France pear. Journal of Agricultural and Food Chemistry. 2005;**53**:5402-5406

[25] Tobitsuka K, Miura M, Kobayashi S. Retention of a European pear aroma model mixture using different types of saccharides. Journal of Agricultural and Food Chemistry. 2006;**54**:5069-5076

[26] Shiga H, Yoshii H, Taguchi R, Nishiyama T, Furuta T, Linko P. Release characteristics of flavor from spray-dried powder in boiling water and during rice cooking. Bioscience, Biotechnology, and Biochemistry. 2003;**67**(2):426-428

[27] Munro IC, Newberne PM, Young VR, Bär A. Safety assessment of γ-cyclodextrin. Regulatory Toxicology and Pharmacology. 2004;**39**:S3-S13

[28] Parrish MA. Cyclodextrins – A Review. Newcastle-Upon-Tyne, England: Sterling Organics; 1988

[29] Szejtli J, Szente L. Elimination of bitter, disgusting tastes of drugs and foods by cyclodextrins. European Journal of Pharmaceutics and Biopharmaceutics. 2005;**1**:115-125

[30] Hedges AR. Industrial applications of cyclodextrins. Chemical Reviews. 1998;**98**:2035-2044

[31] Bhandari BR, D'Arcy BR, Padukka I. Encapsulation of lemon oil by paste method using β-cyclodextrin: Encapsulation efficiency and profile of oil volatiles. Journal of Agricultural and Food Chemistry. 1999;**47**:5194-5197

[32] Shiga H, Yoshii H, Ohe H, Yasuda M, Furuta T, Kuwahara H, Ohkawara M, Linko P. Encapsulation of shiitake (Lenthinus edodes) flavors by spray drying. Bioscience, Biotechnology, and Biochemistry. 2004;**68**(1):66-71

[33] Kwak HS, Jung CS, Shim SY, Ahn J. Removal of cholesterol from cheddar cheese by β-cyclodextrin. Journal of Agricultural and Food Chemistry. 2002;**50**:7293-7298

[34] Sojo MM, Nuñez-Delicado E, Garcia-Carmona F, Sanchez-Ferrer A. Cyclodextrins as activator and inhibitor of latent banana pulp polyphenol oxidase. Journal of Agricultural and Food Chemistry. 1999;**47**:518-523

[35] Sung H. Composition for ginger preservation. Repub Korea KR 9707148. 1997

[36] Bicchi C, Liberto E, Cagliero C, Cordero C, Sgorbini B, Rubiolo P. Conventional and narrow bore short capillary columns with cyclodextrin derivatives as chiral selectors to speed-up enantioselective gas chromatography and enantioselective gas chromatography-mass spectrometry analyses. Journal of Chromatography A. 2008;**1212**:114-123

[37] Liberto E, Cagliero C, Sgorbini B, Bicchi C, Sciarrone D, D'Acampora-Zellner B, Mondello L, Rubiolo P. Enantiomer identification in the flavour and fragrance fields by "interactive" combination of linear retention indices from enantioselective gas chromatography and mass spectrometry. Journal of Chromatography A. 2008;**1195**:117-126

[38] Wüst M, Mosandl A. Important chiral monoterpenoid ethers in flavours and essential oils -enantioselective analysis and biogenesis. European Food Research and Technology. 1999;**209**:3-11

[39] Armstrong DW, Chang C-D, Li WY. Relevance of enantiomeric separations in food and beverage analyses. Journal of Agricultural and Food Chemistry. 1990;**38**:1674-1677

[40] Sumiyoshi H. Utilisation of inclusion complexes with plant components for foods. Nippon Shokuhin Shinsozai Kenkyukaishi. 1999;**2**:109-114

[41] Fujishima N, Kusaka K, Umino T, Urushinata T, Terumi K. Flour based foods containing highly branched cyclodextrins. Japanese Patent JP 136898. 2001

[42] Takeshita K, Urata T. Antimicrobial food preservatives containing cyclodextrin inclusion complexes. Japanese Patent JP 29054. 2001

[43] Mabuchi N, Ngoa M. Controlled release powdered flavour preparations and confectioneries containing preparations. Japanese Patent JP 128638. 2001

[44] Chiavaro E, Dallasta C, Galaverna G, Biancardi A, Gambarelli E, Dossena A, et al. New reversed-phase liquid chromatographic method to detect aflatoxins in food and feed with cyclodextrins as fluorescence enhancers added to the eluent. Journal of Chromatography. A. 2001;**937**:31-40

[45] Jouquand C, Ducruet V, Giampaoli P. Partition coefficients of aroma compounds in polysaccharide solutions by the phase ratio variation method. Food Chemistry. 2004;**85**: 467-474

[46] Pagington JS. Beta-cyclodextrin – The success of molecular inclusion. Chemistry in Britain. 1987;**23**:455-458

[47] Bhandari B, D'Arcy B, Young G. Flavour retention during high temperature short time extrusion cooking process: A review. International Journal of Food Science and Technology. 2001;**36**:453-461

[48] Suzuki J. Japan Kokai, JP 7569100. 1975

[49] Shaw PE, Tatum JH, Wilson CW Improved flavor of navel orange and grapefruit juices by removal of bitter components with β-cyclodextrin polymer. Journal of Agricultural and Food Chemistry. 1984;**32**:832-836

[50] Binello A, Cravotto G, Nano GM, Spagliardi P. Synthesis of chitosan-cyclodextrin adducts and evaluation of their bitter-masking properties. Flavour and Fragrance Journal. 2004;**19**(5):394-400

[51] Binello A, Robaldo B, Barge A, Cavalli R, Cravotto G. Synthesis of cyclodextrin-based polymers and their use as debittering agents. Journal Applied Polymer Science. 2008; **107**:2549-2557

[52] Sakakibara S, Sugisawa K, Matsui F, Sengoku K. Japan Patent JP 851248075. 1985

[53] Takeda Chem. Ind. Ltd. Japan Kokai JP 81127058. 1981

[54] Kuwabara N, Takaku H, Oku S, Kopure Y. Japan Kokai JP 88267246. 1988

[55] Del Valle EMM. Cyclodextrins and their uses. Process Biochemistry. 2004;**39**:1033-1046

[56] López-Nicolas JM, Nuñez-Delicado E, Sánchez-Ferrer A, García-Carmona F. Kinetic model of apple juice enzymatic browning in the presence of cyclodextrins: The use of maltosyl-beta-cyclodextrin as secondary antioxidant. Food Chemistry. 2007;**101**:1164-1171

[57] López-Nicolas JM, Pérez-López AJ, Carbonell-Barrachina A, García-Carmona F. Use of natural and modified cyclodextrins as inhibiting agents of peach juice enzymatic browning. Journal of Agricultural and Food Chemistry. 2007;**55**:5312-5319

[58] López-Nicolas JM, Pérez-López AJ, Carbonell-Barrachina A, García-Carmona F. Kinetic study of the activation of banana juice enzymatic browning by the addition of maltosyl-beta-cyclodextrin. Journal of Agricultural and Food Chemistry. 2007;**55**:9655-9662

[59] López-Nicolas JM, García-Carmona F. Use of cyclodextrins as secondary antioxidants to improve the color of fresh pear juice. Journal of Agricultural and Food Chemistry. 2007;**55**:6330-6338

[60] Burt S. Essential oils: Their antibacterial properties and potential applications in foods – A review. International Journal of Food Microbiology. 2004;**94**:223-253

[61] Valero M, Salmerón M. Antibacterial activity of 11 essential oils against *Bacillus cereus* in tyndallized carrot broth. International Journal of Food Microbiology. 2003;**85**:73-81

[62] Dorman HJD, Surain P, Deans SG. Vitro antioxidant activity of a number of plant essential oils and phytoconstituents. Journal of Essential Oil Research. 2000;**12**:241-248

[63] Helander IM, Alakomi HL, Latva-Kala K, Mattila-Sandholm T, Pol I, Smid EJ. Characterization of the action of essential oil components on Gram-negative bacteria. Journal of Agricultural and Food Chemistry. 1998;**46**:3590-3595

[64] Calo JR, Crandall PG, O'Brian CA, Ricke SC. Essential oils as antimicrobials in food systems – A review. Food Control. 2015;**54**:111-119

[65] Srinivasan K. Black pepper and its pungent principle-piperine. A review of diverse physiological effects. Critical Reviews in Food Science and Nutrition. 2007;**47**:735-748

[66] Awen BZ, Ganapati S, Chandu BR. Influence of sapindus mukorossi on the permeability of ethyl cellulose free film for transedermal use. Research Journal of Pharmaceutical, Biological and Chemical Sciences. 2010;**1**:35-38

[67] Hussain A, Naz S, Nazir H, Shinwari ZK. Tissue culture of black pepper (*Piper nigrum* L.) in Pakistan. Pakistan Journal of Botany. 2011;**43**:1069-1078

[68] Singh G, Marimuthu P, Catalan C, de Lampasona MP. Chemical, antioxidant and antifungal activities of volatile oil of black pepper and its acetone extract. Journal of the Science of Food and Agriculture. 2004;**84**:1878-1884

[69] Dorman HJD, Deans SG. Antimicrobial agents from plants: Antibacterial activity of plant volatile oils. Journal of Applied Microbiology. 2000;**88**:308-316

[70] Menon AN, Padmakumari KP, Jayalekshmy A. Essential oil composition of four major cultivars of black pepper (*Piper nigrum* L.) III. Journal of Essential Oil Research. 2003;**15**:155-157

[71] Dima C, Cotarlet M, Tiberius B, Bahrim G, Alexe P, Dima S. Encapsulation of coriander essential oil in beta-cyclodextrin: Antioxidant and antimicrobial properties evaluation. Romanian Biotechnological Letters. 2014;**19**:9128-9140

[72] Kalemba D, Kunicka A. Antibacterial and antifungal properties of essential oils. Current Medical Chemistry. 2003;**10**:813-829

[73] Jaiarj P, Khoohaswan P, Wongkrajang Y. Anticough and antimicrobial activities of *Psidium guajava* Linn. leaf extract. Journal of Ethnopharmacology. 1999;**67**:203-212

[74] Lozoya X, Reyes-Morales H, Chávez-Soto MA, Martínez-García MC, Soto-González Y, Doubova SV. Intestinal anti-spasmodic effect of a phytodrug of *Psidium guajava* folia in the treatment of acute diarrheic disease. Journal of Ethnopharmacology. 2002;**83**:19-24

[75] Oh WK, Lee CH, Lee MS, Bae EY, Sohn CB, Oh H. Antidiabetic effects of extracts from *Psidium guajava*. Journal of Ethnopharmacology. 2005;**96**:411-415

[76] Sacchetti G, Maietti S, Muzzoli MV, Scaglianti M, Manfredini S, Radice M, Bruni R. Comparative evaluation of 11 essential oils of different originas functional antioxidants, antiradicals and antimicrobials in foods. Food Chemistry. 2005;**91**:621-632

[77] Manosroi J, Dhumtanom P, Manosroi A. Anti-proliferative activity of essential oil extracted from Thai medicinal plants on KB and P388 cell lines. Cancer Letters. 2006;**235**:114-120

[78] Hsin-Chun C, Ming-Jen S, Li-Yun L, Chung-May W. Chemical composition of the leaf essential oil of *Psidium guajava* L. from Taiwan. Journal of Essential Oil Research. 2007;**19**:345-347

[79] Ogunwande IA, Olawore NO, Adeleke KA, Ekundayo O, Koenig WA. Chemical composition of the leaf volatile oil of *Psidium guajava* L. growing in Nigeria. Flavour and Fragrance Journal. 2003;**18**:136-138

[80] Wichtl, M. Teedrogen und phytopharmaka. Stuttgart: Wissenschaftl Verlagsges. mbH; 1997. pp. 395-399

[81] Cavalcanti AM, Baggio CH, Freitas CS, Rieck L, Sousa RS, Santos JES. Safety and antiulcer efficacy studies of *Achillea millefolium* L. after chronic treatment in Wistar rats. Journal of Ethnopharmacology. 2006;**107**:277-284. DOI: 10.1016/j.jep.2006.03.011

[82] Jonsdottir G, Omarsdottird S, Vikingssona A, Hardardottirc I, Freysdottir J. Aqueous extracts from *Menyanthes trifoliate* and *Achillea millefolium* affect maturation of human dendritic cells and their activation of allogeneic CD4+ T cells in vitro. Journal of Ethnopharmacology. 2011;**136**:88-93

[83] Baretta IP, Felizardo RA, Bimbato VF, Santos MGJ, Kassuya CAL, Junior AG. Anxiolytic-like effects of acute and chronic treatment with *Achillea millefolium* L. extract. Journal of Ethnopharmacology. 2012;**140**:46-54. DOI: 10.1016/j.jep.2011.11.047

[84] Candan F, Unlu M, Tepe B, Daferera D, Polissiou M, Sokmenc A, Akpulat HAA. Antimicrobial activity of the essential oil and methanol extracts of *Achillea millefolium* ssp. millefolium Afan (Asteraceae). Journal of Ethnopharmacology. 2003;**87**:215-220. DOI: 10.1016/S0378-8741(03)00149-1

[85] Potrich FB, Allemand A, Silva LM, Santos AC, Baggio CH, Freitas CS, Mendes DAGB, Andre E, Werner MFP, Marques MCA. Antiulcerogenic activity of hydroalcoholic extract of *Achillea millefolium* L.: Involvement of the antioxidant system. Journal of Ethnopharmacology. 2010;**130**:85-92. DOI: 10.1016/j.jep.2010.04.014

[86] Trumbeckaite S, Benetis R, Bumblauskiene L, Burdulis D, Janulis V, Toleikis A. *Achillea millefolium* L. s. l. herb extract: Antioxidant activity and effect on the rat heart mitochondrial functions. Food Chemistry. 2011;**127**:1540-1548

[87] Alfatemi SMH, Rad JS, Rad MS, Mohsenzadeh S, da Silva JAT. Chemical composition, antioxidant activity and in vitro antibacterial activity of *Achilleab wilhelmsii* C. Koch essential oil on methicillin-susceptible and methicillin-resistant *Staphylococcus aureus* spp. 3. Biotechnology. 2014;**5**:39-44. DOI: 10.1007/s13205-014-0197-x

[88] Helena M, Cabral M. A review on cyclodextrin encapsulation of essential oils and volatiles. Flavour and Fragrance Journal. 2010;**25**:313-326

[89] Schmann HJ, Schollmeyer E. Applications of cyclodextrins in cosmetic products: A review. Journal of Cosmetic Science. 2002;**53**:185-191

[90] Bender ML, Komiyama M. Cyclodextrin Chemistry. Berlin: Springer; 1978

[91] Saenger W. Cyclodextrin inclusion compounds in research and industry. Angewandte Chemie. 1980;**19**:344-362

[92] Marques HMC. A review on cyclodextrin encapsulation of essential oils and volatiles. Flavour and Fragrance Journal. 2010;**25**:313-326

[93] Waleczek KJ, Marques HMC, Hempel B, Schmidt PC. Phase solubility studies of pure (−)-alpha-bisabolol and camomile essential oil with beta-cyclodextrin. European Journal of Pharmaceutics and Biopharmaceutics. 2003;**55**:247-251

[94] Garnero C, Zoppi A, Genovese D, Longhi M. Studies on trimethoprim: Hydroxypropyl-β-cyclodextrin: Aggregate and complex formation. Carbohydrate Research. 2010;**345**: 2550-2556

[95] Chandler RF, Hooper SN, Harvey NJ. Ethnobotany and phytochemistry of yarrow *Achillea millefolium* compositae. Economic Botany. 1982;**36**:203-215. DOI: 10.1007/BF02858720

[96] Benedek B, Geisz N, Jager W, Thalhammer T, Kopp B. Choleretic effects of yarrow (*Achillea millefolium* L.) in the isolated perfused rat liver. Phytomedicine. 2006;**13**:702-706. DOI: 10.1016/j.phymed.2005.10.005

[97] Keser S, Celik S, Turkoglu S, Yilmaz O, Turkoglu I. Determination of antioxidant properties of ethanol and water extracts of *Achillea millefolium* L. (Yarrow). Asian Journal of Chemistry. 2011;**23**:3172-3176

[98] Keser S, Celik S, Turkoglu S, Yilmaz O, Turkoglu I. Antioxidant activity, total phenolic and flavonoid content of water and ethanol extracts from *Achillea millefolium* L. Turkish Journal of Pharmaceutical Sciences. 2013;**10**:385-392

[99] Hill LE, Gomes C, Taylor TM. Characterization of beta-cyclodextrin inclusion complexes containing essential oils (trans-cinnamaldehyde, eugenol, cinnamon bark, and clove bud extracts) for antimicrobial delivery applications. LWT – Food Science and Technology. 2013;**51**:86-93

[100] Rakmai J, Cheirsilp B, Torrado-Agrasar A, Mejuto JC, Simal-Gandara J. Physicochemical characterization and evaluation of bio-efficacies of black pepper essential oil encapsulated in hydroxypropyl-beta-cyclodextrin. Food Hydrocolloids. 2017;**65**:157-164

[101] Li X, Jin Z, Wang J. Complexation of allyl isothiocyanate by α- and β-cyclodextrin and its controlled release characteristics. Food Chemistry. 2007;**103**:461-466

[102] Ayala-Zavala JF, Soto-Valdez H, Gonzalez-Leon A, Alvarez-Parrilla E, Martin-Belloso O, Gonzalez-Agular GA. Microencapsulation of cinnamon leaf (*Cinnamonum zeylanicum*) and garlic (*Allium sativum*) oils in β-cyclodextrin. Journal of Inclusion Phenomena and Macrocyclic Chemistry. 2008;**60**:359-368

[103] Ponce-Cevallos PA, Buera MP, Elizalde BE. Encapsulation of cinnamon and thyme essential oils components (cinnamaldehyde and thymol) in β-cyclodextrin: Effect of interactions with water on complex stability. Journal of Food Engineering. 2010;**99**:70-75

[104] Piercey MJ, Mazzanti G, Budge SM, Delaquis PJ, Paulson AT, Hansen LT. Antimicrobial activity of cyclodextrin entrapped allyl isothiocyanate in a model system and packaged fresh-cut onions. Food Microbiology. 2012;**30**:213-218

[105] Songkro S, Hayook N, Jaisawang J, Maneenuan D, Chuchome T, Kaewnoppart N. Investigation of inclusion complexes of citronella oil, citronellal and citronellol with β-cyclodextrin for mosquito repellent. Journal of Inclusion Phenomena and Macrocyclic Chemistry. 2012;**72**:339-355

[106] Qiu N, Cheng X, Wang G, Wang W, Wen J, Zhang Y, Chen L. Inclusion complex of barbigerone with hydroxypropyl-cyclodextrin: Preparation and in vitro evaluation. Carbohydrate Polymers. 2014;**101**:623-630

[107] Anaya-Castro MA, Ayala-Zavala JF, Muñoz-Castellanos L, Hernández-Ochoa L, Peydecastaing J, Durrieu V. β-Cyclodextrin inclusion complexes containing clove (*Eugenia caryophyllata*) and Mexican oregano (*Lippia berlandieri*) essential oils: Preparation, physicochemical and antimicrobial characterization. Food Packaging and Shelf Life. 2017;**14**:96-101. DOI: 10.1016/j.fpsl.2017.09.002

[108] Santos EH, Kamimura JA, Hill LE, Gomes CL. Characterization of carvacrol beta-cyclodextrin inclusion complexes as delivery systems for antibacterial and antioxidant applications. LWT Food Science and Technology. 2015;**60**:583-592

[109] Miguel MG, Dandlen SA, Figueiredo AC, Pedro LG, Barroso JG, Marques MH. Comparative evaluation of the antioxidant activities of thymol and carvacrol and the corresponding beta cylodextrin complexes. Acta Horticulturae. 2009;**853**:363-368

[110] Kamimura JA, Santos EH, Hill LE, Gomes CL. Antimicrobial and antioxidant activities of carvacrol microencapsulated in hydroxypropyl-beta-cyclodextrin. LWT – Food Science and Technology. 2014;**57**:701-709

[111] Wang J, Cao Y, Sun B, Wang C. Physicochemical and release characterisation of garlic oil-β-cyclodextrin inclusion complex. Food Chemistry. 2011;**127**:1680-1685

[112] Rakmai J, Cheirsilp B, Mejuto JC, Simal-Gandara J, Torrado-Agrasar A. Antioxidant and antimicrobial properties of encapsulated guava leaf oil in hydroxypropyl-beta-cyclodextrin. Industrial Crops and Products. 2018;**111**:219-225

[113] Bhandari BR, D'Arcy BR, Bich LLT. Lemon oil to β-cyclodextrin ratio effect on the inclusion efficiency of β-cyclodextrin and the retention of oil volatiles in the complex. Journal of Agricultural and Food Chemistry. 1998;**46**:1494-1499

[114] Mourtzinos I, Salta F, Yannakopoulou K, Chiou A, Karathaos VT. Encapsulation of olive leaf extract in beta cyclodextrin. Journal of Agricultural and Food Chemistry. 2007;**55**:8088-8094

[115] Gong L, Li T, Chen F, Duan X, Yuan Y, Zhang D, Jiang Y. An inclusion complex of eugenol into β-cyclodextrin: Preparation, and physicochemical and antifungal characterization. Food Chemistry. 2016;**196**:324-330

[116] Chun JY, You SK, Lee MY, Choi MJ, Min SG. Characterization of β-cyclodextrin self-aggregates for eugenol encapsulation. International Journal of Food Engineering. 2012;**8**:1-19

[117] Wang T, Li B, Si H, Chen L. Release characteristics and antibacterial activity of solid-state eugenol/β-cyclodextrin inclusion complex. Journal of Inclusion Phenomena and Macrocyclic Chemistry. 2011;**71**:207-213

[118] Seo JE, Min SG, Choi MJ. Release characteristics of freeze-dried eugenol encapsulated with β-cyclodextrin by molecular inclusion method. Journal of Microencapsulation. 2010;**27**:496-505

[119] Kotronia M, Kavetsou E, Loupassaki S, Kikionis S, Vouyiouka S, Detsi A. Encapsulation of oregano (*Origanum onites* L.) essential oil in β-cyclodextrin (β-CD): Synthesis and characterization of the inclusion complexes. Bioengineering. 2017;**4**:74-89

[120] Fang Z, Comino P, Bandari B. Effect of encapsulation of d-limonene on the moisture adsorption property of β-cyclodextrin. LWT – Food Science and Technology. 2013;**51**:164-169

[121] Tao F, Hill LE, Peng Y, Gomes CL. Synthesis and characterization of β-cyclodextrin inclusion complexes of thymol and thyme oil for antimicrobial delivery applications. LWT – Food Science and Technology. 2014;**59**:247-255. DOI: 10.1016/j.lwt.2014.05.037

[122] Del Toro-Sanchez CL, Ayala-Zavala JF, Machi L, Santacruz H, Villegas-Ochoa MA, Alvarez-Parrilla E, Aguilar G. Controlled release of antifungal volatiles of thyme essential oil from β-cyclodextrin capsules. Journal of Inclusion Phenomena and Macrocyclic Chemistry. 2010;**67**:431-441

[123] Albarca R, Rodríguez FJ, Guarda A, Galotto MJ, Burna JE. Characterization of beta-cyclodextrin inclusion complexes containing and essential oil component. Food Chemistry. 2016;**196**:968-975

[124] Lawtrakul L, Inthajak K, Toochinda P. Molecular calculations on β-cyclodextrin inclusion complexes with five essential oil compounds from *Ocimun basilicum* (sweet basil). Science Asia. 2014;**40**:145-151

[125] Rakmai J, Cheirsilp B, Torrado-Agrasar A, Simal-Gandara J, Mejuto JC. Encapsulation of yarrow essential oil in hydroxypropyl-beta-cyclodextrin: Physiochemical characterization and evaluation of bio-efficacies. CyTA Journal of Food. 2004;**15**:409-417

[126] Zeng Z, Fang Y, Ji H. Side chain influencing the interaction between β-cyclodextrin and vanillin. Flavour and Fragrance Journal. 2012;**27**:378-385

[127] Karathanos VT, Mourtzinos I, Yannakopoulou K, Andrikopoulos NK. Study of the solubility, antioxidant activity, and structure of inclusion complex of vanillin with beta-cyclodextrin. Food Chemistry. 2007;**101**:652-658

[128] Numanoğlu U, Şen T, Tarimci N, Kartal M, Koo OMY, Önyüksel H. Use of cyclodextrins as a cosmetic delivery system for fragance materials: Linalool and benzyl acetate. AAPS PharmSciTech. 2007;**8**:34-42

[129] Haiyee ZA, Saim N, Said M, Illias RM, Mustapha WAW, Hassan O. Characterization of cyclodextrin complexes with turmeric oleoresin. Food Chemistry. 2009;**114**:459-465. DOI: 10.1016/j.foodchem.2008.09.072

[130] Kfoury M, Landy D, Ruellan S, Auezova L, Greige-Gerges H, Fourmentin S. Determination of formation constants and structural characterization of cyclodextrin inclusion complexes with two phenolic isomers: Carvacrol and thymol. Beilstein Journal of Organic Chemistry. 2016;**12**:29-42. DOI: 10.3762/bjoc.12.5

[131] Kfoury M, Auezova L, Fourmentin S, Greige-Gerges H. Investigation of monoterpenes complexation with hydrosypropyl-β-cyclodextrin. Journal of Inclusion Phenomena and Macrocyclic Chemistry. 2014;**80**:51-60. DOI: 10.1007/s10847-014-0385-7

[132] Jiang ZT, Tan J, Tan J, Li R. Chemical components and molecular microcapsules of folium Artemisia essential oil with β-cyclodextrin derivatives. Journal of Essential Oil-Bearing Plants. 2016;**19**:1155-1169. DOI: 10.1080/0972060X.2016.1185973

[133] Belscak-Cvitanovic A, Stojanovic R, Manojlovic V, Komes D, Cindric IJ, Nedovic V, Bugarski B. Encapsulation of polyphenolic antioxidants from medicinal plant extracts in alginate-chitosan system enhanced with ascorbic acid by electrostatic extrusion. Food Research International. 2011;**44**:1094-1101

[134] Dordevic V, Balanc B, Belscak-Cvitanovic A, Levic S, Trifkovic K, Kalusevic A, Kostic I, Komes D, Bugarski B, Nedovic V. Trends in encapsulation technologies for delivery of food bioactive compounds. Food Engineering Reviews. 2015;**7**:452-490. DOI: 10.1007/s12393-014-9106-7

[135] Guimaraes AG, Oliveira MA, Alves RS, Menezes PP, Serafini MR, Araujo AAS, Bezerra DP, Junior LJQ. Encapsulation of carvacrol, a monoterpene present in the essential oil of oregano, with β-cyclodextrin, improves the pharmacological response on cancer pain experimental. Chemico-Biological Interactions. 2015;**227**:69-76

[136] Le Bas D, Rysanek N. Structural aspect of cyclodextrins. In: Duchene D, editor. Cyclodextrins and their Industrial Uses. Vol. 105-211. Paris: Editions de Sante; 1987. pp. 351-393

[137] Batzdorf T, Muller-Goymann CC. Release of ketoprofen from aqueous systems in the presence of hydrophilic β-cyclodextrin derivatives. Indian Journal of Pharmaceutical Sciences. 1993;**55**:857-860

[138] Szente L. Analytical methods for cyclodextrins, cyclodextrin derivatives and cyclodextrin complexes. Supramolecular Chemistry. 1996;**3**:78-253

[139] Wang X, Luo Z, Xiao Z. Preparation, characterization, and thermal stability of β-cyclodextrin/soybean lecithin inclusion complex. Carbohydrate Polymer. 2014;**101**: 1027-1032

[140] Liu H, Yang G, Tang Y, Cao D, Qi T, Qi Y, Fan G. Physicochemical characterization and pharmacokinetics evaluation of β-caryophyllene/β-cyclodextrin inclusion complex. International Journal of Pharmaceutics. 2013;**450**:304-310. DOI: 10.1016/j.ijpharm.2013.04.013

[141] Gomes LMM, Petito N, Costa VG, Falcao DQ, Araujo KGL. Inclusion complexes of red bell pepper pigments with β-cyclodextrin: Preparation, characterisation and application as natural colorant in yogurt. Food Chemistry. 2014;**148**:428-436

[142] Szejtli J. Cyclodextrins and their Inclusion Complexes. Budapest: Akademiai Kiado; 1982. pp. 115-122

[143] Yuan C, Jin Z, Li X. Evaluation of complex forming ability of hydroxypropyl-beta-cyclodextrins. Food Chemistry. 2008;**106**:50-55

[144] Higuchi T, Connors KA. Phase solubility techniques. Advances in Analytical Chemistry and Instrumentation. 1965;**4**:117-122

[145] Connors KA. The stability of cyclodextrin complexes in solution. Chemical Reviews. 1997;**5**:1325-1357

[146] Zengin H, Baysal AH. Antibacterial and antioxidant activity of essential oil terpenes against pathogenic and spoilage-forming bacteria and cell structure-activity relationships evaluated by SEM microscopy. Molecules. 2014;**19**:17773-17798

[147] Dai J, Zhu L, Yang L, Qiu J. Chemical composition, antioxidant and antimicrobial activities of essential oil from *Wedelia prostrata*. EXCLI Journal. 2013;**12**:479-490

[148] Nawar WF. Lipids. In: Fennema O, editors. Food Chemistry. New York: Marcel Dekker, Inc.; 1996. pp. 225-320

[149] Misharina TA, Polshkov AN, Ruchkina EL, Medvedeva IB. Changes in the composition of the essential oil of marjoram during storage. Applied Biochemistry and Microbiology. 2003;**39**:311-316

[150] Trombetta D, Castelli F, Sarpietro MG, Venuti V, Cristani M, Daniele C, Saija A, Mazzanti G, Bisignano G. Mechanisms of antibacterial action of three monoterpenes. Antimicrobial Agents and Chemotherapy. 2005;**49**:2474-2478

[151] Soković M, Glamočlija J, Marin PD, Brkić D, Van Griensven LJLD. Antibacterial effects of the essential oils of commonly consumed medicinal herbs using an in vitro model. Molecules. 2010;**15**:7532-7546

[152] Lambert RJW, Skandamis PN, Coote PJ, Nychas GJE. A study of the minimum inhibitory concentration and mode of action of oregano essential oil, thymol and carvacrol. Journal of Applied Microbiology. 2001;**91**:453-462. DOI: 10.1046/j.1365-2672.2001.01428.x

[153] Soković M, Tzakou O, Pitarokili D, Couladis M. Antifungal activities of selected aromatic plants growing wild in Greece. Nahrung. 2002;**46**:317-320

[154] Couladis M, Tzakou O, Kujundzić S, Soković M, Mimica-Dukić N. Chemical analysis and antifungal activity of *Thymus striatus*. Phytotherapy Research. 2004;**18**:40-42. DOI: 10.1002/ptr.1353

[155] Soković M, Grubišić D, Ristić M. Chemical composition and antifungal activity of the essential oils from leaves, calyx and corolla of *Salvia brachyodon Vandas*. Journal of Essential Oil Research. 2005;**17**:227-229. DOI: 10.1080/10412905.2005.9698884

[156] Šiler B, Živković S, Banjanac T, Cvetković J, Živković JN, Ćirić A, Soković M, Mišić D. Centauries as underestimated food additives: Antioxidant and antimicrobial potential. Food Chemistry. 2014;**147**:367-376. DOI: 10.1016/j.foodchem.2013.10.007

[157] Soković M, van Griensven LJLD. Antimicrobial activity of essential oils and their components against the three major pathogens of the cultivated button mushroom, Agaricus bisporus. European Journal of Plant Pathology. 2006;**116**:211-224. DOI: 10. 1007/s10658-006-9053-0

Chapter 12

β-Cyclodextrin as an Ionophore for Membrane Electrode

Mohsen M. Zareh

Additional information is available at the end of the chapter

http://dx.doi.org/10.5772/intechopen.73597

Abstract

This chapter included an explanation of the role of cyclodextrin as an ionophore for the preparation of voltammetric and potentiometric sensors for different compounds. There was a surveying of the different recorded cyclodextrin-based sensors for drug, organic, or inorganic cations. It also explained the performance characteristics for the first reported sensor for nickel based on β-cyclodextrin as an example.

Keywords: cyclodextrin, sensors, membrane electrodes, potentiometric, voltammetric

1. Introduction

The ion-selective membrane electrodes are based on selecting an appropriate matter which works like a sensing part. The sensing materials are the main component of the potentiometric ion sensors. They are responsible for forming a type of discrimination in the electrode behavior toward one ion rather than others. There are several types of the sensing material. They might have host-guest structure, which allows the inclusion of the analyzed ionic material. Another type is able to form a complex with the analyte ion. The third type is based on ion exchange reaction with the ion under investigation.

A potential difference will be aroused when the analyte ion can penetrate across the phase boundary between the two phases (analyte solution and internal reference solution) (**Figure 1**). Accordingly, an electrochemical equilibrium will be formed due to different potentials at both sides of the membrane. The potential difference (E) across the membrane is described by the Nernst equation:

$$E = E^{\circ} + (RT/ZF) \ln a \quad (1)$$

where E° is the standard cell potential, R is the general gas constant, Z is the valency of the analyte ion, F is the Faraday constant, and a is the activity of the analyte ion.

Cyclodextrin is produced from starch by enzymatic conversion. They are formed of a cycle of oligosaccharides. There are three types of cyclodextrin according to the size of the ring. They may be composed of 5-, 6-, or 7-membered ring. They are named like α, β, and γ, respectively (**Figure 2**).

Typical cyclodextrins contain a number of glucose monomers ranging between 6 and 8. They create a cone shape, which is suitable for the inclusion of different cations (**Figure 3**).

Many researches were recorded where cyclodextrin was used for preparing sensor membrane electrode for the determination of organic compounds. The studies about the application of CD as an ionophore for sensors of inorganic cations were rarely found. The recorded sensors were either voltammetric sensors or potentiometric sensors.

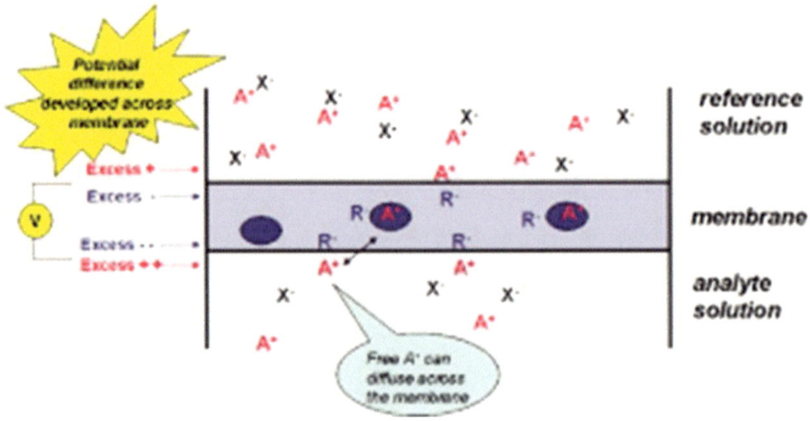

Figure 1. The response of membrane toward different ions.

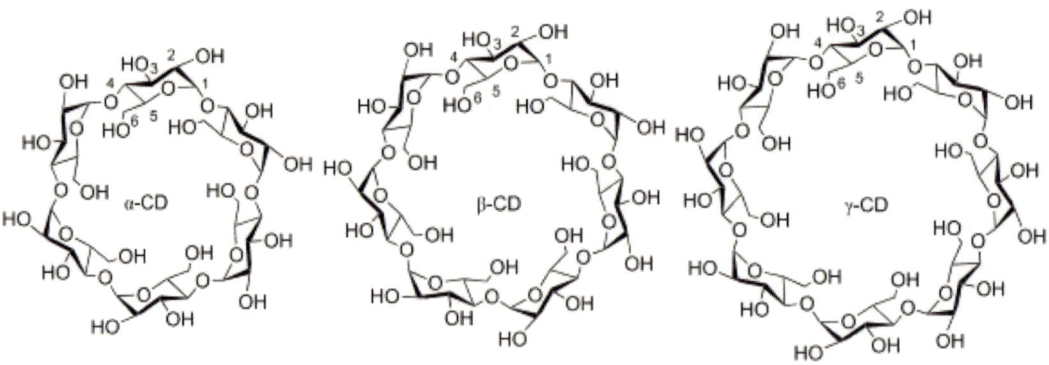

Figure 2. Structure formula of α, β, and γ-cyclodextrin.

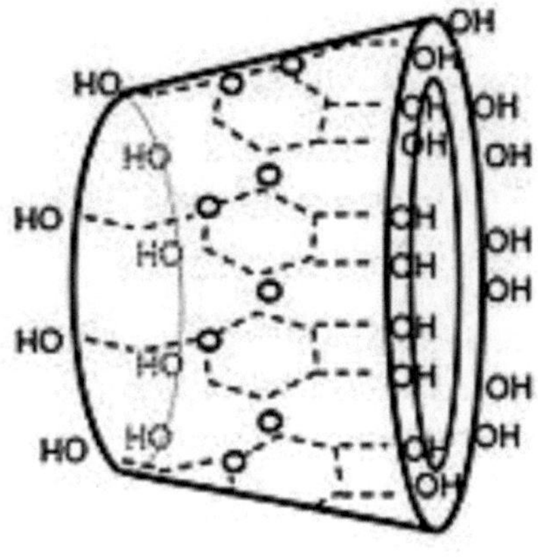

Figure 3. γ-CD toroid structure showing spatial arrangement.

1.1. Voltammetric sensors

Voltammetric sensors are an electrochemical sensor based on measurements of the diffusion current at polarization potential. Several examples of voltammetric sensors were recorded in the literature. They were used for the determination of amine-containing organic/pharmaceutical compounds.

Komiyama [1] prepared cyclodextrin electrode for the determination of p-nitrophenolate. The electrode was prepared by dipping Pt-wire into cyclodextrin polymer. Stefan et al. [2] used a multidextrin as a sensor for preparing selective membrane for *S*-captopril. Lenik and Nieszporek [3] constructed a modified glassy carbon ibuprofen electrode.. The electrode was based on multi-walled carbon nanotubes and β-cyclodextrins. It showed wider linear range (3.2×10^{-5} –1×10^{-2} mol L^{-1}) and lower limit of detection (1.25×10^{-5} mol L^{-1}). It was applied like a sensor for GC. The determination of some neurotransmitters such as dopamine (DA), epinephrine (EP), norepinephrine (NEP), levodopa (L-DOPA), 3,4-dihydroxy-phenyl acetic acid (DOPAC) and serotonin (ST) was applied by reduced graphene oxide (RGO) composite glassy carbon electrode modified with ionic liquid crystal (ILC), 1-butyl-1-methylpiperidinium hexa-fluoro-phosphate, and cyclodextrin [4].

CD-based sensor was applied for the determination of an antibiotic chloramphenicol by Sun et al. [5]. They used a voltammetric sensor on mesoporous carbon @polydopamine and β-CD. It showed a response for chloramphenicol in the range of 5×10^{-7} – 5×10^{-5} M. Li et al. [6] determined o- and p-nitrophenol by using reduced graphene oxide-CD-Chitosan electrochemical sensor. It was based on the host-guest recognition sites on CD and abundant functional groups of Chitosan. It showed linear range for 0.12–0.2 and 4–50 μM for NPH. But for pNPh it was from 0.06 to 0.16 μM and 5 to 40 μM. Yi et al. [7] applied electropolymerization for βCD and L-arginin on the surface of carbon nanotubes @graphene nanoribbons (CNTs@GNRs) for preparing a modified electrode for 2- and 4-aminophenols (AP). A linear response ranging from 25.0 to 1300.0 nM for both 2- and 4-AP and the detection limits (DLs) of 2- and 4-AP obtained in this work were 6.2 and 3.5 nM, respectively.

1.2. Potentiometric sensors

They are a type of sensors based on measuring the potential changes with concentration in accordance with the Nernst equation.

Odashima et al. [8] used a long derivative chain of β-cyclodextrin for preparing a new sensory element for the determination of organic amines. Stefan et al. [9] used either α or γ-cyclodextrin-based electrodes for the determination of R-baclofen. The detection limits were 7×10^{-9} mol^{-1} L^{-1} for α-cyclodextrin-based electrode and 1.44×10^{-10} mol^{-1} L^{-1} for γ-cyclodextrin-based electrode. (2-hydroxypropyl)-β-cyclodextrin, heptakis(2,3,6-tri-O-methyl)-β-cyclodextrins and heptakis(2,3,6-tri-O-benzoyl)-β-cyclodextrin were applied for preparing diclofenac electrode by Lenik [10]. Staden and Nejem [11] determined L-vesamicol by CD derivative-based potentiometric sensor. The detection limit was of the order of 10^{-10} M. They showed a linear range between 10^{-9} and 10^{-7} M. Lenik and Łyszczek [12] determined naproxen by functionalized β-CD. It is based on inclusion complex of naproxen with β-CD-derivative. The linear range was 5.0×10^{-5}–1.0×10^{-2} mol L^{-1}. Staden and Holo [13] determined L-histidine by an entioselective potentiometric membrane electrode based on CD. They applied carbon paste impregnated with α-, β-, 2-hydroxyl-3-trimethylammoniopropyl-β-(as chloride salt) and γ-cyclodextrins (γ-CDs). The recovery of L-his in the presence of D-his was higher than 99.10%. L-proline electrode was constructed based on carbon paste impregnated with α, β, or γ cyclodextrin [14]. The detection limit of the membrane was in the region of 10^{-10}–10^{-9} M.

All the aforementioned cyclodextrin applications were examples of sensors for drug analysis. The application of cyclodextrin for sensors for inorganic cations is rarely found. We choose one of these sensors that was sensing for nickel which was constructed by Zareh et al. [15], to study as an example of CD sensors. Two membrane compositions (**Table 1**) were prepared for the optimization to get the best. Electrode type I comprised DEP, while II comprised NPOE. The behavior of DEP-containing electrodes showed better Nernstian slope value than NOPE-containing membrane electrode. Both electrode types showed the same linear range of 5.01×10^{-5} –10^{-2} M. **Figure 4** shows the calibration graphs for each electrode type.

The effect of inner filling (IF) solution was studied by the application of the electrode type I containing three types of IF-solutions (A, B, and C). They were corresponding to compositions (0.1 M KCl + 0.1 M $NiCl_2$), (0.01 M KCl + 0.01 M $NiCl_2$), and (0.001 M KCl + 0.001 M $NiCl_2$), respectively. The slopes of Ni electrode were 30.9, 28.17, and 29.7 mV/decade, for electrodes with IF A, B, and C, respectively. When A and B IF was applied, the lower linear range reached 5.0×10^{-5} M. In the case of IF solution type C, the linear range was not less than 5×10^{-4} M. **Table 2** summarizes the obtained results.

The response time was estimated according to the IUPAC definition [16]. It is defined as the time between the instant when the Ni electrode was brought into contact with sample solution and the first instant when emf/time slope becomes equal to the limiting value selected on the basis of experimental conditions. Dynamic response was studied for both electrode types containing DEP or NPOE. From **Figures 5** and **6**, the response times for electrode types I and II were between 2 and 3 s for both types of electrodes for the tested concentrations 10^{-2}–10^{-4} M. This fast response for both types will help in the application of both electrode types for real measurements.

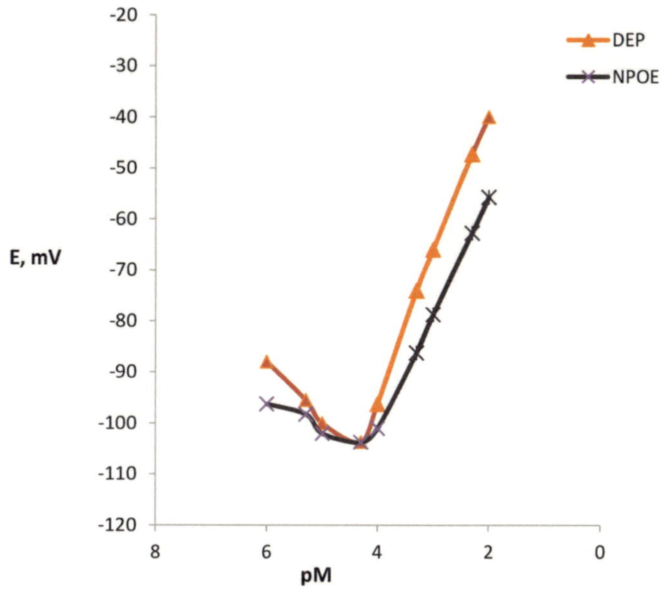

Figure 4. Calibration graphs for Ni-selective electrodes with membrane contains either (a) DEP or (b) NPOE plasticizers.

Composition, w/w %	PVC, mg	βCDX, mg	DEP, mg	NPOE, mg	% Ionophore	Slope, mV/ Decade	Detection limit, M
I-Membrane	63.6	12.8	126.4	0	6.31	28.76	3.98×10^{-5}
II-Membrane	60	14	0	121	7.18	23.33	6.3×10^{-5}

Table 1. Composition of membrane types I and II used for preparing Ni^{2+}-selective electrode.

Inner filling solution	Slope, mV/Decade	Linear range, M	R^2
A-(10^{-1} M $NiCl_2$ + 10^{-1} M KCl)	30.89	10^{-2}–5×10^{-5}	0.9987
B-(10^{-2} M $NiCl_2$ + 10^{-2} M KCl)	28.17	10^{-2}–5×10^{-5}	0.9986
C-(10^{-3} M $NiCl_2$ + 10^{-3} M KCl)	29.68	10^{-2}–5×10^{-4}	0.9931

Table 2. Effect of inner filling on the performance of Ni-selective electrode based on β-CDX with DEP plasticizer.

The detection limit (DL) of an ISE can be defined as the cross point of line segments fitted to linear segments of *emf* versus $log\ a_{Ni2+}$ [16]. For membrane types I and II, the values of the detection limits were 3.98×10^{-5} and 6.3×10^{-5} M.

The process of the electrode response is based on a nonpolarized electrochemical equilibrium. There are two equilibrium steps to explain the response mechanism of the proposed Ni-selective electrode. The first step is the equilibrium between Ni^{2+} in membrane and solution sites. The second step is the equilibrium of formation of Ni^{2+}-β-CDX into the membrane site. The steps of this mechanism can be represented below:

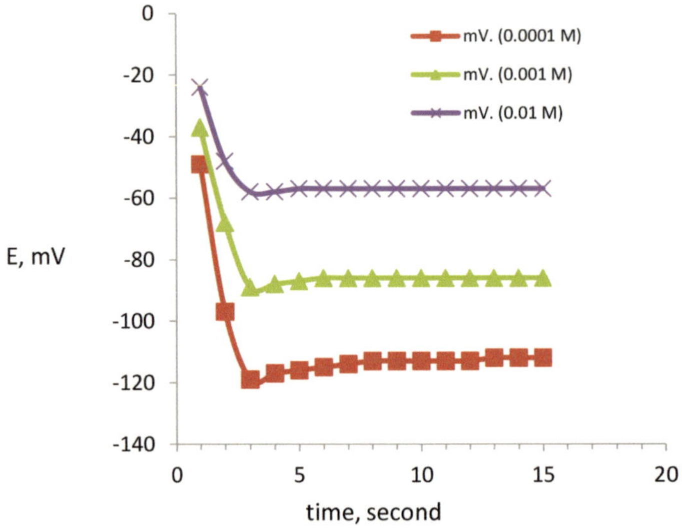

Figure 5. Dynamic response for Ni-selective electrodes with membrane type I containing DEP, for different concentrations (a) 0.01 M, (b) 0.001 M, and (c) 0.0001 M Ni^{2+} solutions.

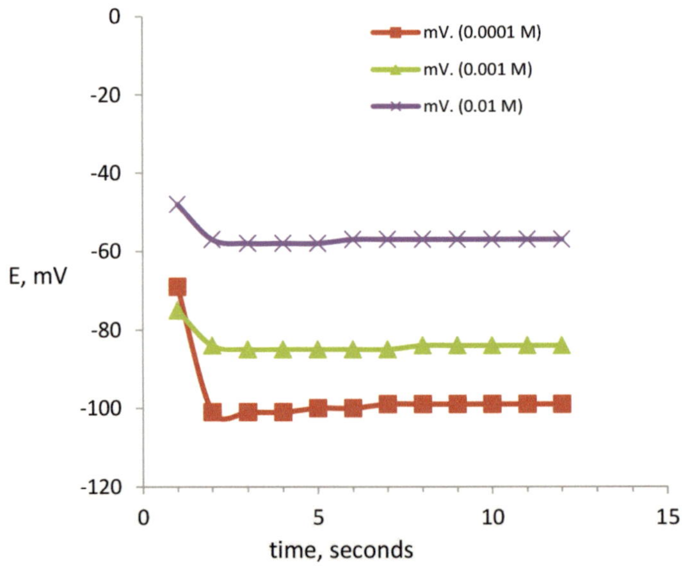

Figure 6. Dynamic response for Ni-selective electrodes with membrane type II containing NPOE, for different concentrations (a) 0.01 M, (b) 0.001 M, and (c) 0.0001 M Ni^{2+} solutions.

$$[\mathrm{Ni}^{2+}]_s \rightleftharpoons [\mathrm{Ni}^{2+}]_m$$

$$[\mathrm{Ni}^{2+}]_m + \beta - \mathrm{CDX}_m \rightleftharpoons [\mathrm{Ni}^{2+} - \beta\text{-CDX}]^{2+}{}_m$$

$$\Delta E = \mathrm{E_m} - \mathrm{E_s} = (\mathrm{RT/F})\ \ln\ \{\mathrm{k_i}\ \mathrm{a_{Ni}}{}^{2+}\ [1 + \mathrm{K_f}\ (\mathrm{carrier})]/(\mathrm{sites})\}$$

2. pH-effect

The potential changes versus different pH values for the DEP-membrane electrode type I and type II were tested. In acidic part, the break was observed at pH values 3.8 and 3.2 for type I and II electrodes, respectively. The sensitivity of β-CDX toward the H^+ is the reason for the potential break in acidic medium. The break in acidic medium was due to the inclusion of H^+ into ionophore in the membrane site. Above pH 3.2 and 3.8, the potential was not changed whatever the pH value for types I and II, respectively. This was due to the absence of the H^+ ion concentration, which disturbs the equilibrium. Representative curves for nickel with membrane types I and II are shown in **Figure 7**.

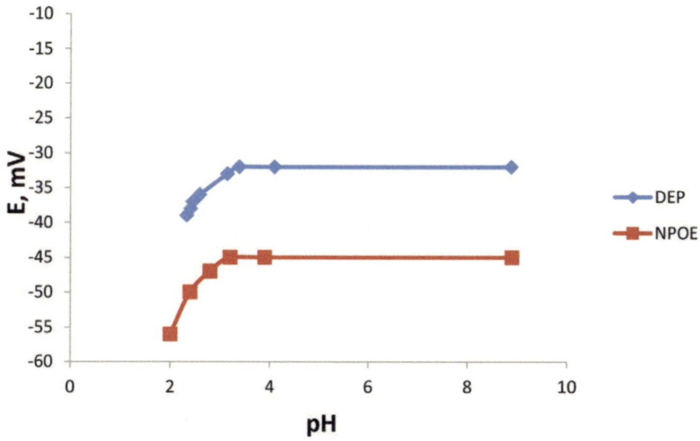

Figure 7. pH-effect for Ni electrode for 0.001 M Ni^{2+} solutions for electrodes comprising (a) DEP and (b) NPOE.

3. Selectivity properties

The selectivity of Ni electrode with different membrane types I and II was calculated according to the SSM [16]. **Table 3** shows the obtained values of the selectivity coefficient ($K^{Pot}_{Ni^{2+}, j^{z+}}$). From the results, it can be reported that most of the tested interferents for Ni-ISE type I showed perfect selectivity. When the electrode type I was used, the values of the selectivity coefficient toward divalent cations were so perfect to consider this electrode selective for Ni^{++} cation. The recorded values for most of the tested divalent cations were of the order of 10^{-3}. In case of trivalent cations, this type of electrode showed better selectivity coefficient values (of order 10^{-5}). When electrode type II was used, the selectivity coefficient values were higher than that for electrode type I. It showed values of order 10^{-2} for trivalent cations and 10^{-1} for most of the divalent cations. This can be attributed to that in type II electrode the NPOE has active sites to interact with cationic species which lowers the selectivity toward Ni^{++} [17]. The tested monovalent cations (Na^+, K^+, and NH_4^+) showed interference with the measurements with either type I or type II. Accordingly, it is recommended that measurements with the proposed electrodes should be conducted in the absence of these cations.

Interferent	$K^{Pot}_{Ni^{2+}, j^{z+}}$	
	I-DEP	II-NPOE
Ir^{3+}	7.67×10^{-5}	1.4×10^{-2}
Fe^{3+}	4.38×10^{-5}	1.3×10^{-2}
Zn^{2+}	1.1×10^{-3}	6.1×10^{-2}
Co^{2+}	1.8×10^{-3}	3.5×10^{-2}
Cu^{2+}	2.1×10^{-3}	3.0×10^{-1}
Ba^{2+}	2.2×10^{-3}	3.0×10^{-1}
Mg^{2+}	1.6×10^{-3}	1.5×10^{-1}
Ca^{2+}	2.2×10^{-3}	2.0×10^{-1}
Cd^{2+}	2.2×10^{-3}	1.6×10^{-1}
Hg^{2+}	1.7×10^{-3}	2.0×10^{-1}
Mn^{2+}	1.5×10^{-3}	1.7×10^{-1}
Pb^{2+}	2.9×10^{-3}	1.2×10^{-1}

Table 3. Selectivity coefficient values for nickel electrodes based on β-CDX incorporating (DEP) [I], and (NPOE) [II] as plasticizers.

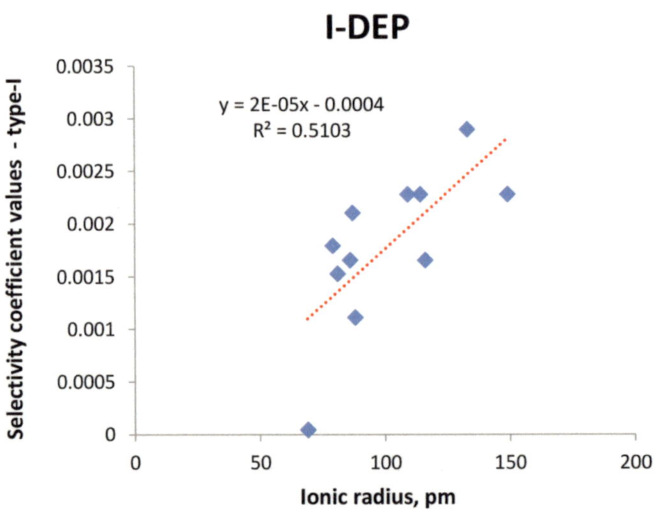

Figure 8. Correlation between ionic radius (pm) of the tested cations and selectivity coefficient (K^{Pot} Ni^{2+}, j^{z+}) for Ni-ISE with membrane type I for 0.0001 M solutions.

The relation between ionic radius [18] of interferent cations and the values of selectivity coefficient for both electrode types is shown in **Figures 8** and **9**. It was found that there was an increase in the selectivity coefficient values with increasing the ionic radius of the tested cations. This was true for both electrode types I and II. The increment values in case of type I were less than those in case of type II. This was attributed to that the increase in ionic volume was suitable for the β-CDX cavity.

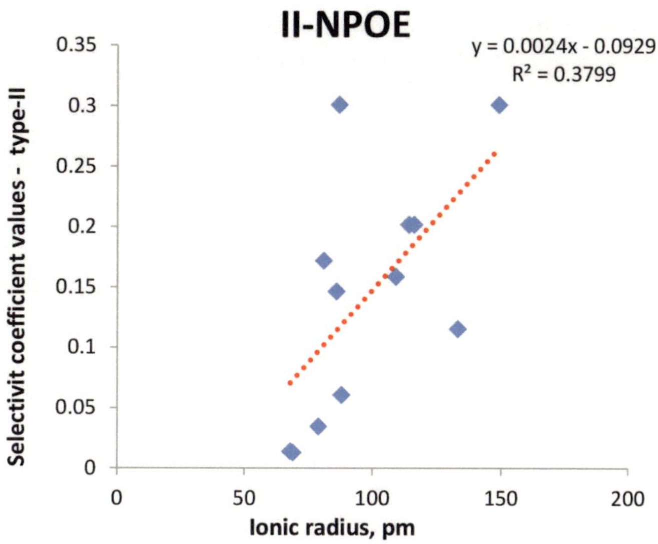

Figure 9. Correlation between ionic radius (pm) of the tested cations and selectivity coefficient ($K^{Pot}\ Ni^{2+}{}_{,\ j}{}^{z+}$) for Ni-ISE with membrane type II for 0.0001 M solutions.

4. Determination of nickel in its samples

In this chapter, two types of samples containing nickel were used. They were representative for food samples and stainless steel samples. Five steel samples (A–D) and one food (E) sample were chosen. On the one hand, 0.1 g of each stainless steel sample was dissolved into aqua-Regia, heated at 105°C, and diluted to 250 ml using bi-distilled water. This solution was measured directly. On the other hand, 0.5 g chocolate sample (E) was dissolved in 100 ml after digestion with HNO_3, $HClO_4$, and H_2O_2. In this case, 1 ml was diluted to 50 ml, and the result solution was subjected to potential measurements using the proposed Ni-selective electrode. The obtained Ni values into the stainless steel samples (A–D) were between 1.467 and 7.354 ppm. The chocolate sample E showed Ni content 14.707 ppm. All the obtained values agreed with the values given by AAS analysis of the same samples [19]. The obtained values of

No.	Sample	Ni^{2+}, ppm		RSD*, %
		AAS method	Ni-ISE method	
A	Test tube holder	1.437	1.467	1.95
B	Shaving blade	6.005	7.354	2.51
C	Screwdriver	4.181	4.640	2.13
D	Coin (1/4 pound)	4.61	3.686	1.81
E	Chocolate (Cadbury Dairy Milk)	12.158	14.707	2.34

*(4-determinations).

Table 4. Determination of nickel in its samples using the proposed Ni-ISE.

Ni in chocolate agreed with previously recorded values [20]. **Table 4** shows the obtained results for analysis by using both the proposed electrode and an AAS method for the same samples.

5. Conclusions

Cyclodextrin (CD) is one of the important reagents that is able to form inclusion complex with a variety of compounds. This property helps its use as an ionophore for preparing electrochemical sensors for many organic pharmaceutical cations. The recorded electrochemical sensors were either potentiometric or voltammetric. Up to date, the application of CD for preparing inorganic sensors was rarely found.

Author details

Mohsen M. Zareh

Address all correspondence to: mohsenzareh@hotmail.com

Department of Chemistry, Faculty of Science, Zagazig University, Zagazig, Egypt

References

[1] Komiyama M. Preparation of cyclodextrin membrane-modified electrodes as sensor materials. Macromolecular Materials and Engineering. 1988;**163**:205-207

[2] Stefan R-I, van Staden JF, Aboul-Enein HY, van Staden JF. Maltodextrins as new chiral selectors in the design of potentiometric, enantioselective membrane electrodes. Fresenius' Journal of Analytical Chemistry. 2001;**370**:33-37

[3] Lenik J, Nieszporek J. Construction of a glassy carbon ibuprofen electrode modified with multi-walled carbon nanotubes and cyclodextrins. Sensors and Actuators B: Chemical. 2017;**255**:2282-2289

[4] AttaEkram NF, El-AdsYousef H, Galal MA. Determination of some neurotransmitters at cyclodextrin/ionic liquid crystal/graphene composite electrode. Electrochimica Acta. 2016; **199**:319-331

[5] Sun Y, Wei T, Jiang M, Xu L, Xu Z. Voltammetric sensor for chloramphenicol determination based on a dual signal enhancement strategy with ordered mesoporous carbon@polydopamine and β-cyclodextrin. Sensors & Actuators, B: Chemical. 2018;**255**:2155-2162

[6] Li C, Wu Z, Yang H, Deng L, Chen X. Reduced graphene oxide-cyclodextrin-chitosan electrochemical sensor: Effective and simultaneous determination of o- and p-nitrophenols. Sensors & Actuators, B: Chemical. 2017;**251**:446-454

[7] Yi Y, Zhu G, Wu X, Wang K. Highly sensitive and simultaneous electrochemical determination of 2-aminophenol and 4-aminophenol based on poly(l-arginine)-β-cyclodextrin/carbon nanotubes@graphene nanoribbons modified electrode. Biosensors and Bioelectronics. 2016;**77**:353-358

[8] Odashima K, Hashimoto H, Umezawa Y. Potentiometric discrimination of organic amines by liquid membrane electrodes based on a long alkyl chain derivative of β-cyclodextrin. Microchimica Acta. 1994;**113**:223-238

[9] Stefan R-I, van, Alexander S, Rat'ko A. Enantioselective, potentiometric membrane electrodes based on cyclodextrins: Application for the determination of R-baclofen in its pharmaceutical formulation. Talanta. 2006;**69**:1049-1053

[10] Lenik J. A new potentiometric electrode incorporating functionalized β-cyclodextrins for diclofenac determination. Materials Science and Engineering: C. 2014;**45**:109-116

[11] van Staden R-IS, Nejem RM. Cyclodextrins-based enantioselective, potentiometric membrane electrodes for l-vesamicol assay in serum samples. Sensors and Actuators B: Chemical. 2006;**117**:123-127

[12] Lenik J, Łyszczek R. Functionalized β-cyclodextrin based potentiometric sensor for naproxen determination. Materials Science and Engineering: C. 2016;**61**:149-157

[13] Stefan-van Staden R-I, Holo L. Enantioselective, potentiometric membrane electrodes based on cyclodextrins for the determination of L-histidine. Sensors and Actuators. 2007; **120**:399-402

[14] Kenneth I, Raluca O, Stefan I. Enantioselective potentiometric membrane electrodes based on α-, β- and γ-cyclodextrins as chiral selectors for the assay of L-proline. Talanta. 2005;**66**:501-504

[15] Zareh MM, Zordek W, Elgendy K, Zaky M, Keshk A, Abd-Alhady A. β-cyclodextrin as a selective Green-ionophore for determination of nickel by membrane electrode. Sensor Letters. 2017;**15**:1-7

[16] Buck RP, Lindner E. Recommendations for nomenclature of ionselective electrodes (IUPAC Recommendations 1994). Pure and Applied Chemistry. 1994;**66**:2527

[17] Zareh MM. Blank Membranes versus Ionophore-based Membranes for the Selective Determination of H^+. Analytical Sciences. 2009;**25**:1135-1131

[18] Shannon RD. Revised effective ionic radii and systematic studies of interatomic distances in halides and chalcogenides. Acta Cryst. 1976;**A32**:751-767

[19] Manual Book, Buck Scientific 210/211 VGP Atomic Absorption Spectroscopy

[20] Vivian Fern Andes and Daksesh Parikh. Toxicologist D.M. Saxena Finds Nickel Content Above Safety Limit in 11 Brands. December 31, 2012

Chapter 13

Removal of an Azo Textile Dye from Wastewater by Cyclodextrin-Epichlorohydrin Polymers

Paola Semeraro, José Antonio Gabaldón, Paola Fini,
Estrella Núñez, José Antonio Pellicer, Vito Rizzi and
Pinalysa Cosma

Additional information is available at the end of the chapter

http://dx.doi.org/10.5772/intechopen.72502

Abstract

Native cyclodextrins (CDs), α-, β- and γ-CDs, were employed to synthetise three different cyclodextrin-based polymers using epichlorohydrin (EPI) as a cross-linker. These polymers were applied as adsorbent material to remove an azo textile dye, Direct Blue 78 (DB78), from water. The formation of inclusion complexes between the alone CDs and DB78 molecules were first studied in aqueous solutions. Then, adsorption experiments of the dye were performed by means of cyclodextrin/epichlorohydrin (CD/EPI) polymers. The effects of various parameters, such as contact time, adsorbent dosage, initial dye concentration, pH and temperature, were examined to determine the better adsorption conditions. The equilibrium isotherms and the adsorption kinetics were also analysed using opportune mathematic models. The chemical-physical characteristics and the morphology of the adsorbent polymers were, respectively, observed by differential scanning calorimetry and field emission scanning electron microscope. The CD/EPI polymers showed a very good ability in the removal of DB78 from aqueous solution; indeed, the maximum efficiencies in the dye removal were found to be about 99% for β-CD/EPI polymer and about 97% for γ-CD/EPI polymer, at pH 6 and 25°C conditions. It is possible to assume that the good adsorbent aptitude of CD/EPI polymers is due to their double peculiarity to include the dye in the inner cavity of CDs and to adsorb the dye on their porous surfaces by physical interaction.

Keywords: inclusion complexes, electrochemical measurements, adsorption process, textile dye removal, cyclodextrin/epichlorohydrin polymers, thermal analysis

1. Introduction

1.1. Textile dyes

Textile and clothing industries generate a remarkable pollution in natural water due to the discharge of large amounts of dye chemicals in the effluents [1]. These dyes give an undesirable colour to the water body, reducing the sunlight penetration and influencing the photochemical and biological activities of aquatic life [2]. Dye molecules present several chemical structures and, depending on functional groups of their chromophore, are classified as azo, anthraquinone, styryl, acridine, nitro, nitroso, benzodifuranone, diphenylmethane, triphenylmethane, azine, xanthene, cyanine, phthalocyanine, hemicyanine, diazahemicyanine, triarylmethane, stilbene, or oxazine dyes [2]. However, the azo compound class accounts for about 65–70% of all classes of dyes [3], and the azo dyes are the most common synthetic molecules released into the environment. It is now recognised that some azo dyes, under certain conditions, produce aromatic amines which are toxic, allergenic, carcinogenic, and mutagenic [4, 5]. Due to the hazard of reduction products arising from the use of azo dyes, the European Union (EU) AZO Colourants Directive 2002/61/EC already came into force in September 2003, and replaced by REACH regulation, regulated the restrictions on the marketing and use of certain dangerous azo dyes. In addition, these contaminants are highly soluble in water and are very difficult to degrade being stable to light irradiation, heat, and oxidation agents. Therefore, the conventional wastewater treatment systems are not able to remove them [6], and it is necessary to treat the industrial effluents before releasing it into the environment. Hence, in recent years, numerous treatments on the removal of azo dyes from the effluents have been studied. There are different methods for the treatment of wastewaters including chemical, physical, and biological technologies [7]. All these methods present both advantages and disadvantages, which are shown in detail in **Table 1** [1, 8–10].

Among all techniques described in **Table 1**, the adsorption is one of the most efficient and popular methods for the removal of textile dyes from industrial effluents [11] and activated carbon is the most common material used for dye removal by adsorption. This material due to its ability to adsorb cationic, acid dyes, and mordant, and, to a slightly lesser extent, dispersed, direct, and reactive dyes [9]. However, commercially available activated carbons are very expensive, and so it is opportune to use low-cost carbons that are able to absorb pollutants from wastewater. In the last years, the research is pointing towards the use of more efficient and inexpensive adsorbent materials for the treatment of coloured effluents. A wide variety of low-cost materials, such as biosorbents and by-products of industry and agriculture [12–15], are being evaluated as viable substitutes for activated carbon to remove dyes. Industrial and agricultural wastes are indeed very interesting adsorbent materials with good adsorption capacity, high selectivity, low cost, easy regeneration, and free availability. A recent paper [15] reported that oil mill solid waste, previously treated, is able to reduce significantly the amount of an azo direct dye in industrial textile wastewater. In particular experimental conditions, this material can adsorb the 100% of the dye in solution with the possibility to recycle both the dye and the adsorbent [15]. Also, natural and biodegradable polymers showed good biocompatibility and high efficiency in dyes adsorption. Indeed, it was demonstrated that

Methods	Advantages	Disadvantages
Chemical treatments		
Oxidative process	Simplicity of application	Oxidising agent, usually hydrogen peroxide (H_2O_2), needs to be activated by some means such as ultra violet light
H_2O_2 + Fe(II) salts (Fenton's reagent)	Fenton's reagent is a suitable chemical means for treatment of wastewaters	Generation of sludge containing concentrated impurities
Ozonation	Ozone, a very good oxidising agent, can be applied in its gaseous state and does not increase the volume of wastewater	Continuous ozonation is required due to its short half-line (20min) with cost increase
Photochemical	No production of sludge and great reduction of foul odours	Formation of by-products
Sodium hypochlorite (NaOCl)	Attack at the amino group of dye molecules with azo-bond cleavage	Release of aromatic amines
Electrochemical destruction	No consumption of chemicals, non-hazardous production of breakdown compounds and no accumulation of sludge	High cost of electricity
Biological treatments		
Decolourisation by white-rot fungi	Degradation of dyes by white-rot fungi using enzymes	Unreliable enzyme production due to the unfamiliar environment of liquid fermentations
Other microbial cultures (mixed bacterial)	Decolourisation of dye mixtures in 24–30 h by anaerobic bacteria and decolourisation of diazo dyes in 15 days by mixed bacterial cultures	Under aerobic conditions, azo dyes are not readily metabolised
Adsorption by living/dead microbial biomass	Great affinity for binding between microbial species and several molecules such as anthraquinone, phthalocyanine, and azo dyes	Not effective for all dyes
Anaerobic textile-dye bioremediation systems	Decolourisation of solutions containing azo and other water-soluble dyes	Production of methane and hydrogen sulphide by anaerobic breakdown
Physical treatments		
Adsorption	Removal of wide variety of dyes	Same adsorbent materials are very expensive
Membrane filtration	Removes all dye types	Production of concentrated sludge
Ion exchange	No loss of adsorbent on regeneration and recycling of solvent after use	Not effective for all dyes and high cost
Irradiation	Effective break down of some dyes and phenolic molecules by radiation	Request of great quantities of dissolved O_2 which affect the cost. Applicability only at a laboratory scale
Electrokinetic coagulation	Economically feasible method for excellent removal of direct dyes	Not effective for acid dyes removal and production of large amounts of sludge

Table 1. Advantages and disadvantages of dye removal methods.

chitosan films [16], chitosan/polyamide nanofibres [17], and alginate-chitosan beads [18] are used as efficient and economic adsorbents for the removal of direct and anionic textile dyes. Numerous experiments are, moreover, conducted to evaluate the possibility to use some polysaccharides, in particular, starch and starch derivatives, as adsorbents for wastewater treatment [19, 20]. Since it was established that the good adsorption properties of polymers derived from starch towards dyes, in this study, cyclodextrin-based polymers were used to remove an azo textile dye, Direct Blue 78 (DB78), from wastewater. In **Figure 1** is shown the chemical structure of DB78, a tri-azo compound characterised by the presence of three azo bonds (—N=N—) with four sulphonate groups.

1.2. Cyclodextrins

Cyclodextrins (CDs) are natural cyclic oligosaccharides, derived from starch, that present a truncated cone structure with an inner relatively apolar cavity and an external hydrophilic face [21]. Due to this characteristic conformation, CDs are host molecules able to include in their cavity, a high range of guest molecules, with appropriate dimensions, through the formation of host-guest inclusion complexes [22]. The native CDs, named, α-, β-, and γ-CDs, are respectively constituted by 6, 7 and 8 glucopyranose, connected by α(1,4)-linkages. CDs can be employed both in their native form and in functionalised form, after opportune chemical modifications. Attributable to their numerous and specific properties, CDs are widely employed in several areas such as pharmaceutical, biomedical, biotechnological, and industrial sectors [22, 23]. Several studies also reported that CDs and CD-based materials are used in removal of dyes [18, 19], organic pollutants, and heavy metals from water, soil, and atmosphere [23, 24]. Moreover, in a previous study [25], the interaction between some azo textile dyes and some commercial cyclodextrins was already demonstrated. Therefore, in this chapter, the study on the removal of DB78 dye from wastewater, by using cyclodextrins, is described in detail. However, since the most of CDs are highly soluble in water, insoluble CD-based materials were employed as dye adsorbent. Indeed, after the adsorption process, these materials can be easily removed from treated solutions obtaining clean water.

1.3. Cyclodextrin-based polymers

Among the numerous preparation methods of water insoluble CD-based materials, cross-linked polymers, obtained by copolymerisation of CDs and coupling agents, have received great attention. The most employed cross-linking agent is epichlorohydrin (1-chloro-2,3-epoxypropane)

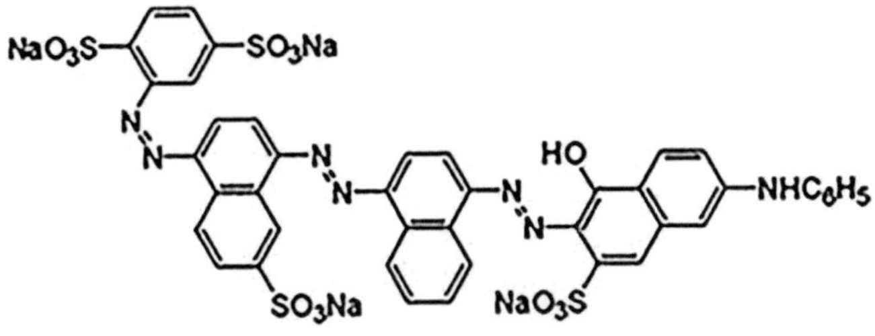

Figure 1. Chemical structures of Direct Blue 78.

[26]. This cross-linker, shortly named EPI, is a bi-functional coupling agent, which contains two reactive functional groups, an epoxide group and a chloroalkyl moiety. EPI form bonds with polysaccharide molecules in cross-linking step and/or with itself in polymerisation step and the hydroxyl groups of native CDs, at the 2-, 3- and 6-positions of glucose units that are available and reactive to form linkages. The -OH groups in 6-positions are more reactive than those in 3-positions; however, their reactivity depends on the reaction conditions, such as temperature and alkalinity, to allow complete alkoxide formation [26]. Indeed, the secondary hydroxyl groups, which have pKa values of around 12.2 (at 298 K), can be deprotonated with hydroxide or hydride to form alcoholate sites. Consequently, typical methods used to synthesise CDs-based polymers require the addition of NaOH, NaH or $NaBH_4$ [27]. Despite its toxicity for humans, animals, algae, and bacteria and its potential pollutant characteristics for the environment, EPI is widely used to synthetise CD/EPI polymers [28] due to simplicity and low cost of the synthesis. On the other hand, a careful purification of these polymers allows to eliminate free EPI and other residual solvents making them good and non-toxic drug delivery systems for pharmacological formulations [29]. Furthermore, the CD/EPI polymers present high adsorption properties, high efficiency in pollutant removal and are recyclable and easily recoverable [26–30]. Despite the β-CDs are the most common cyclodextrins used to produce CD-based polymers, in this study, α-, β- and γ-CDs were employed, and their respective polymers were synthetised.

2. Experimental section

2.1. Preparation of DB78/CD solutions

To verify the formation of inclusion complexes between dye and CDs, aqueous solutions of α-, β- and γ-CDs were respectively added to DB78 solution, at different molar ratio. Stock solutions of DB78 and α-, β- and γ-CD were already prepared in distilled water and the desired volumes of these solutions were mixed and diluted to the chosen final volume to obtain the DB78/CD solutions. They were maintained under stirring for 10 min, at room temperature, to ensure the inclusion complex formation, and then studied by electrochemical measurements.

2.2. Preparation of CD/EPI polymers

The α-CD/EPI, β-CD/EPI, and γ-CD/EPI polymers were prepared dissolving opportune amounts of the respective CDs in water, in presence of sodium borohydride. The mixtures were vigorously stirred at 50°C until the reactants were dissolved. Then, NaOH (40% w/w) solution was added and an excess of epichlorohydrin was slowly added dropwise. The mixtures were vigorously stirred and heated gently at 50°C. About after 5 hours, the solutions started to be viscous, and gelatinous solids were obtained. Then acetone was added, and the systems were maintained under stirring and heating for 10 min. After cooling, the insoluble polymers obtained were poured into water, filtered and the resulting solid was purified by several Soxhlet extractions. Next, the CD/EPI polymers were dried in oven, at 50°C for 12 h, crushed and utilised as adsorbent materials to remove DB78 from aqueous solution. **Figure 2** shows the scheme of CD/EPI polymer synthesis.

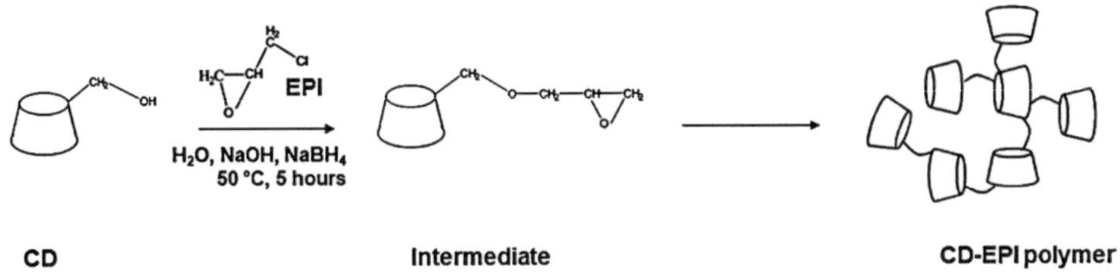

Figure 2. Scheme of CD/EPI polymer synthesis.

2.3. Instruments

Electrochemical measurements were performed in a standard three-electrode cell using hanging mercury drop electrode (HDME) as working electrode. An Ag/AgCl, KCl sat electrode and a Pt rod were used as reference and counter electrodes, respectively. A $LiClO_4$ 0.1 M solution was used as supporting electrolyte. Voltammograms were recorded by means of the AUTOLAB PGSTAT10 potentiostat interfaced with a personal computer. Absorption spectra were recorded from 200 to 600 nm using a Shimadzu UV-1601 spectrophotometer. Calorimetric measurements were performed using an LKB 2277 Thermal Activity Monitor Isothermal Microcalorimeter equipped with an LKB 2277–204 flow mixing cell. The photographs of samples were collected by using a field emission scanning electron microscope (Merlin Compact/VP, Carl Zeiss Microscopy, Germany) with a secondary electron detector using an acceleration voltage of 2 KV and an aperture size of 30 μm.

2.4. Batch adsorption experiments

Batch mode experiments were carried out to study the dye adsorption processes by CD/EPI polymers. The required amounts of adsorbent were added to fixed volume of dye solution, at opportune concentration, under constant condition of agitation rate (170 rpm), pH and temperature. At predetermined time intervals, the dye concentration in solution was evaluated by UV–Vis absorption measurements. Different variables, such as contact time, adsorbent dosage, initial dye concentration, pH and temperature, were analysed to recognise the optimum adsorption states. These experiments were performed by varying the parameter under evaluation and maintaining the other parameters constant. Values of dye removal (%) and amount of dye adsorbed onto adsorbent q_t (mg/g) at time t were respectively calculated using the following expressions:

$$\% = \frac{(C_i - C_t)}{C_i} 100 \tag{1}$$

$$q_t = \frac{(C_i - C_t)\, V}{m} \tag{2}$$

where C_i and C_t (mg/L) are the dye concentration in solution at initial and at t adsorption time, respectively. V (L) is the initial volume of dye solution and m (g) is the mass of adsorbent. All tests were achieved in triplicate and the mean values were reported.

2.5. Adsorption equilibrium isotherms

Adsorption isotherms, by means of accurate mathematical models, allow to evaluate the adsorption behaviour and to describe how the adsorbate interacts with the adsorbent [31]. Among all isotherm models developed, the more common models, Langmuir and Freundlich models were used in this study. The Langmuir adsorption isotherm model presumes that the adsorption occurs on homogeneous sites of adsorbent surface forming a saturated monolayer of adsorbate on the outer surface of adsorbent and that the adsorption of each molecule onto the surface has equal adsorption activation energy [32, 33]. The Freundlich adsorption isotherm is an empirical equation which describes heterogeneous systems having unequal available sites on adsorbent surface with different adsorption energies [31, 32]. The adsorption isotherms were evaluated adding different amounts of CD/EPI polymers to dye solutions and maintaining the systems at constant temperature of 25°C under continuous stirring until the equilibrium was achieved. Values of dye concentration were measured before and after the adsorption processes and the obtained experimental data were fitted with Langmuir and Freundlich models. The values of the linear regression correlation coefficient R^2 give information about the best-fit model.

The linearised form of Langmuir is represented by Eq. (3):

$$\frac{1}{q_e} = \frac{1}{q_m} + \frac{1}{b\,q_m}\frac{1}{C_e} \quad (3)$$

where q_e (mg/g) is the amount of the dye adsorbed on polymer at equilibrium, q_m (mg/g) is the maximum monolayer amount of DB78 adsorbed per unit mass of adsorbent, C_e (mg/L) is the concentration of dye in solution at equilibrium and b is the constant related to the affinity of the binding sites (L/mg). From the intercept and slope of the plot $1/q_e$ versus 1/Ce, it is possible to obtain the values of q_m and b, respectively. Moreover, the Langmuir isotherm can be expressed in terms of a dimensionless constant separation factor, R_L that is defined by the following equation:

$$R_L = \frac{1}{(1 + b\,C_0)} \quad (4)$$

where C_0 is the initial concentration of adsorbate (mg/L) and b (L/mg) is Langmuir constant. The value of R_L indicates the trend of the adsorption process, indeed, the isotherm can be either favourable ($0 < R_L < 1$), unfavourable ($R_L > 1$), linear ($R_L = 1$) or irreversible ($R_L = 0$). The Langmuir values, q_m, b, and R_L are presented in **Table 2**.

The linear form of Freundlich equation is:

$$\ln q_e = \ln K_F + \frac{1}{n}\ln C_e \quad (5)$$

where q_e (mg/g) is the amount of DB78 adsorbed at equilibrium, C_e (mg/L) is the concentration of the dye in solution at equilibrium, K_F (L/g) is the Freundlich constant related to the maximum adsorption capacity of adsorbent and n (dimensionless) is the heterogeneity factor. The values of K_F and n, reported in **Table 2**, were calculated respectively by the intercept and slope of the linear plot ln q_e versus ln C_e. The magnitude of n gives an indication of the favourability of adsorption process: when n = 1, the adsorption is linear, when n > 1, the adsorption is a favourable adsorption condition [31, 33].

Polymers	T (K)	Langmuir				Freundlich		
		b (L/mg)	q_m (mg/g)	R_L	R^2	K_F (L/g)	n	R^2
β-CD/EPI	298	0.425	4.988	0.028	0.988	1.649	3.520	0.760
	323	0.205	11.775	0.057	0.991	2.264	2.237	0.828
	353	0.237	12.183	0.050	0.999	2.354	2.273	0.908
γ-CD/EPI	298	0.026	14.156	0.377	0.999	0.501	1.463	0.985
	323	0.075	15.954	0.143	0.981	1.326	1.728	0.931
	353	0.062	23.207	0.168	0.993	1.566	1.586	0.965

Table 2. Adsorption isotherm values.

2.6. Thermodynamic analysis

Thermodynamic parameters, such as Gibb's free energy change (ΔG°) (J mol^{-1}), enthalpy change (ΔH°) (J mol^{-1}) and entropy change (ΔS°) (J mol^{-1} K^{-1}), allow to comprehend the nature of adsorption process and the effect of temperature on adsorption. These parameters can be calculated using the following relations [34]:

$$\Delta G^{\circ} = -RT \ln K_C \tag{6}$$

where R is the universal gas constant (8.314 J mol^{-1} K^{-1}), T is the solution temperature (K) and K_c is defined as:

$$K_C = \frac{C_i}{C_e} \tag{7}$$

$$\Delta G^{\circ} = \Delta H^{\circ} - T\Delta S^{\circ} \tag{8}$$

Therefore, Eqs. (6) and (8) can be rewritten as:

$$\ln K_C = \frac{\Delta S^{\circ}}{R} - \frac{\Delta H^{\circ}}{RT} \tag{9}$$

ΔH° and ΔS° were obtained by plot of Eq. (9), while the ΔG° values were determined from Eq. (8).

3. Results and discussion

3.1. Electrochemical measurements

Before testing the ability of azo dye removal by CD/EPI polymers, the interactions between DB78 and α-, β- and γ-CDs were investigated in solution by electrochemical measurements. Generally, electrochemical studies of different azo dyes show that the electroreduction of the

N=N double bond occurs at the hydrazo stage (HN-NH), via the consumption of $2e^-/2H^+$, or at the amine stage ($-NH_2$), via the consumption of $4e^-/4H^+$, in one or two steps depending on the chemical structure of the investigated azo compound, the nature of adjacent substituents and the pH of the medium [35]. Furthermore, the electrochemical reduction of azo compounds is an irreversible process complicated by preceding and following chemical reactions leading to the cleavage of the azo bond and resulting in various degradation products [36]. In **Figure 3**, the cyclic voltammetry measurements of DB78 are reported. It presented three cathodic peaks, located in the range from −0.2 to −1.0 V. The first two weak waves (I and II) were positioned at −0.15 and −0.70 V respectively, while the more intense wave (III), were located at about −0.80 V. Ep, I and Ep, II are both attributable to the azo moieties electroreduction [37, 38]. The different potential for the azo group reduction is due to the different substituents present in ortho position respect to it. The first peak can be attributed to the electroreduction of the azo group with the ortho -OH group that facilitate the electroreduction due to its electron-donating effect.

The electrochemical behaviour of DB78 in presence of increasing CDs concentration was then analysed. Although the addition of α-CD did not greatly influence the cyclic voltammograms of DB78 (data not showed), it is not possible to affirm that there is no interaction between dye and α-CD, but that this technique did not allow to obtain detailed information. On the contrary, the addition of β-CD and γ-CD, at increasing molar ratio, showed regular changes in the cyclic voltammograms of DB78. Indeed, in **Figure 4a** and **b**, it is possible to observe a strong increment of current intensity values at the increasing of the CD amount, particularly in the case of γ-CD, while no shifts of the potential peaks were detected. These regular variations indicate that the dye was reduced with more difficulty because its involvement in the inclusion complex. The inclusion of the azo groups of dye inside the cavity of the CDs prevents the interaction with the electrode and reduces the diffusion coefficient of the molecule determining the reduction of the peak current intensity. Consequently, the electrochemical measurements confirmed the formation of inclusion complexes between DB78 and β-CD and between DB78 and γ-CD.

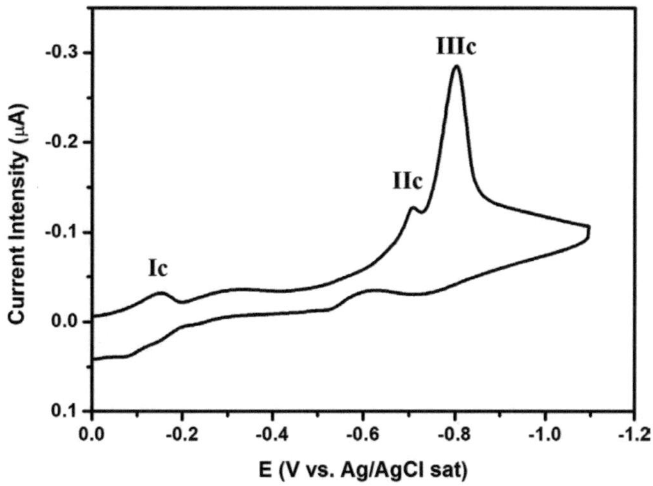

Figure 3. Cyclic voltammetry at HMDE of aqueous solution containing Direct Blue 78.

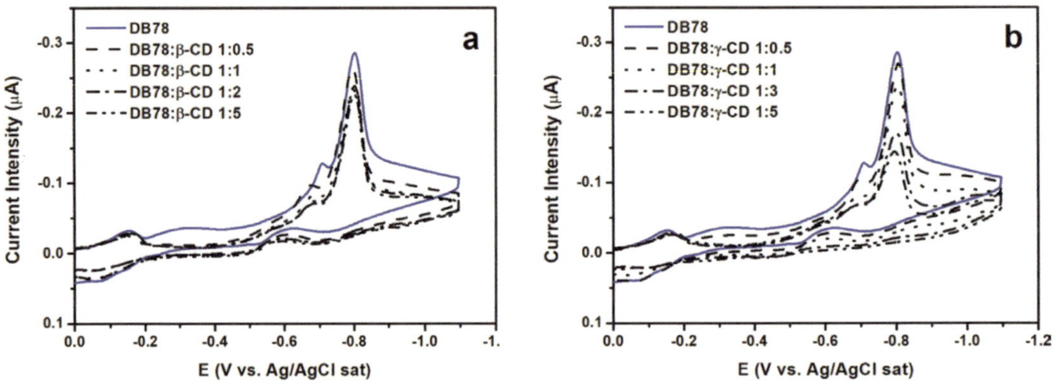

Figure 4. Cyclic voltammetry at HMDE of aqueous solution containing DB78 in presence of increasing amounts of CDs. (a) DB78/β-CD and (b) DB78/γ-CD at different molar ratios.

3.2. Dye adsorption efficiency by different CD/EPI polymers

To evaluate the more appropriate material able to adsorb DB78, three different types of adsorbents, α-CD/EPI, β-CD/EPI, and γ-CD/EPI polymers, were used. Ten milliliters of dye (11.00 mg/L) at pH 6 and 25°C were analysed using 1.00 g of polymers as adsorbent. **Figure 5a** shows that β-CD/EPI polymer presented a better ability to remove DB78 from solution than the other polymers. The dye removal efficiency was 98.90% with β-CD/EPI polymer, in contrast to 97.25% and only 92.70% when γ-CD/EPI and α-CD/EPI polymers were respectively used. Consequently, all adsorption experiments were carried out on β-CD/EPI and γ-CD/EPI polymers. It is possible to suppose that the adsorption is based not only on physical adsorption process in the polymers networks but also on inclusion complex formation [39]. Therefore, β- and γ-CD, which are characterised by a wider cavity, can form more host-guest supramolecular interaction with dye than α-CD. However, β-CD/EPI, despite the intermediate size of β-CD between α- and γ-CD, showed the better efficiency in the removal of dye. This behaviour is due to the highest complexing ability and stability with cross-linking agents of β-CD [40].

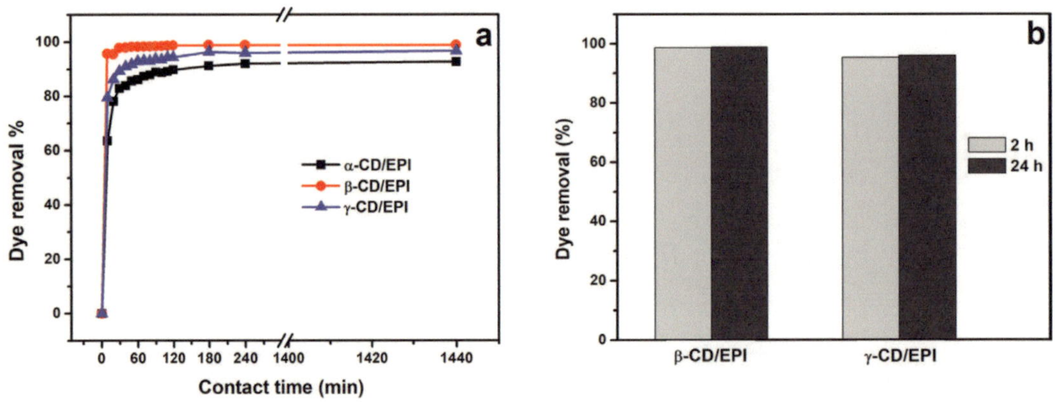

Figure 5. Adsorption measurements using 10 mL of DB 78 (11.00 mg/L) and 1.00 g of polymers at pH 6 and 25°C. (a) Adsorption comparison between α-CD/EPI, β-CD/EPI and γ-CD/EPI polymers and (b) effect of contact time.

3.3. Effect of contact time

To determine the effect of contact time on adsorption processes, 10 mL of DB 78 (11.00 mg/L) was maintained for 24 h under continuous stirring with 1.00 g of β- and γ-CD/EPI polymers at pH 6 and 25°C. The concentrations of dye in solution were measured at several times. **Figure 5b** shows that both polymers presented the maximum dye removal after 2 h of adsorption process and no further changes were observed after 24 h. Therefore, it is possible to affirm that the time required to achieve the equilibrium was about 2 h. During this time, the complete saturation of active sites of polymers was reached.

3.4. Effect of adsorbent dosage

The amount of the adsorbent used in these experiments is another important parameter that affects the uptake of dye. Indeed, a quantitative removal cannot be achieved when the polymer is less than the optimum amount. To optimise the smallest quantity of polymer able to adsorb the greater amount of DB78, increasing dosage of adsorbents, from 0.05 to 1.25 g, was added into 10 mL of dye solution (11.00 mg/L). The systems, maintained at pH 6 and 25°C, were stirred, until equilibrium achievement, and the remaining amount of dye in solutions were measured. In **Figure 6a** and **b** are presented the effect of adsorbent dosage on β-CD/EPI and γ-CD/EPI polymers, respectively. It is possible to observe that both polymers have the same behaviour: the percentage of dye removal increased with the increase in dosage of polymers, due to the major availability of adsorbent surface sites [18]. In presence of β-CD/EPI polymer (**Figure 6a**), the removal of dye from the initial solutions increased from 41.20 to 98.90% as the adsorbent dosage increased from 0.05 to 1.00 g. When γ-CD/EPI polymer (**Figure 6b**) was used as an adsorbent, the removal of DB78 increased from 52.01 to 97.25% as the adsorbent dosage increased from 0.05 to 1.00 g. A further increase in dosage of polymers (1.25 g) did not improve the removal of both dye since the systems were achieved the maximum adsorption efficiency. Therefore, 1.00 g of polymers was used for further measurements.

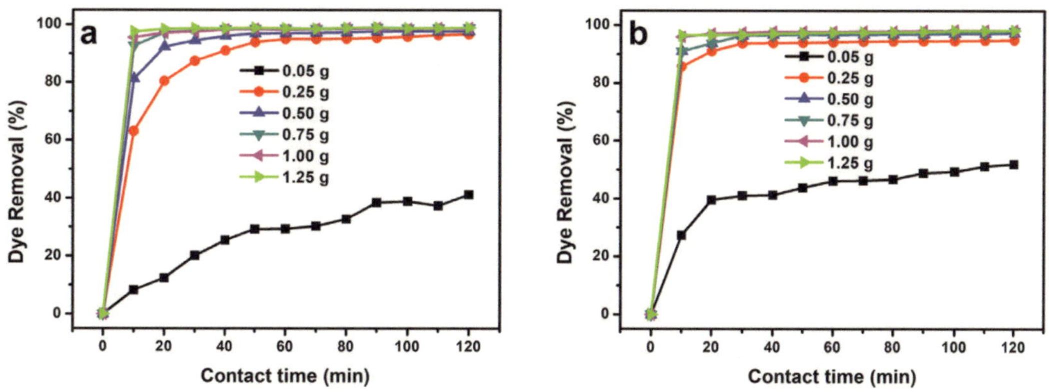

Figure 6. Effect of adsorbent dosage using 10 mL of DB78 (11.00 mg/L) and increasing dosage of polymers, from 0.05 to 1.25 g, at pH 6 and 25°C. (a) Adsorption measurements with β-CD/EPI polymer and (b) adsorption measurements with γ-CD/EPI polymer.

3.5. Effect of initial dye concentration

To study the effect of initial dye concentration on adsorption mechanism onto CD-based polymers, increasing the concentrations of DB78 solutions were used. The experiments were performed at pH 6 and 25°C, using a constant volume of dye solution (10 mL) and a constant dosage (1.00 g) of β-CD/EPI polymer and of γ-CD/EPI polymer. Experimental results show that the amount of dye adsorbed onto adsorbent, q_t (mg/g), increased with the increase in initial concentration of dye. This behaviour was more evident in the case of β-CD/EPI polymer (**Figure 7a**), where the amount of dye adsorbed onto polymer at equilibrium, q_e, improved from 0.32 to 1.99 mg/g as the initial concentration of DB78 increased from 11.00 to 70.00 mg/L. In the case of γ-CD/EPI polymer (**Figure 7b**), q_e increased from 0.24 to 1.28 mg/g when the initial concentration of dye was incremented from 11.00 to 70.00 mg/L. This occurs because the increase in the initial concentrations of dye induces the optimisation of favourable interaction raising the driving force, able to overcome the resistance to the mass transfer of dye between the aqueous and the solid phase [41]. Furthermore, these measurements demonstrate again the better adsorption ability of β-CD/EPI polymer than γ-CD/EPI polymer.

3.6. Effect of initial pH

To study the influence of pH on the adsorption of azo dye onto the two polymers, experiments were carried out at pH 2, 6 and 11 with a contact time of 2 h. In **Figure 8a** and **b** are reported the results respectively obtained with β-CD/EPI and γ-CD/EPI polymers. Generally, the initial pH of solution plays a significant role in the chemistry of adsorbent and dye, however, in this case, no significant changes in the adsorption process were observed at different pH conditions. Indeed, for both polymers, when acid and basic conditions were used, no important variations in the adsorption efficiency were observed. However, the highest percentage of dye removal was obtained at pH 6. It is possible to suppose that at alkaline pH, the presence of excess —OH ions compete with the anionic dye for the adsorption sites. Indeed, as the pH of

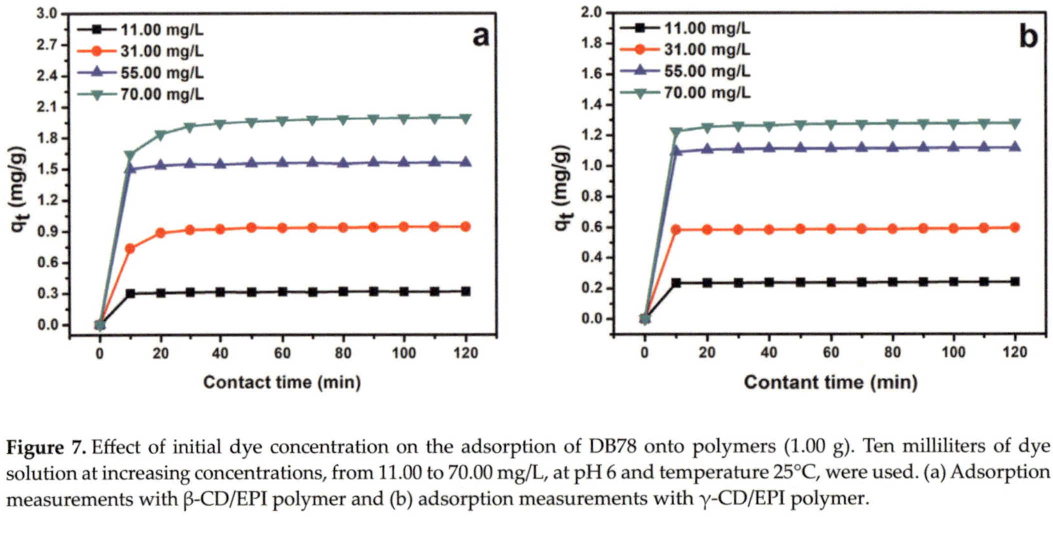

Figure 7. Effect of initial dye concentration on the adsorption of DB78 onto polymers (1.00 g). Ten milliliters of dye solution at increasing concentrations, from 11.00 to 70.00 mg/L, at pH 6 and temperature 25°C, were used. (a) Adsorption measurements with β-CD/EPI polymer and (b) adsorption measurements with γ-CD/EPI polymer.

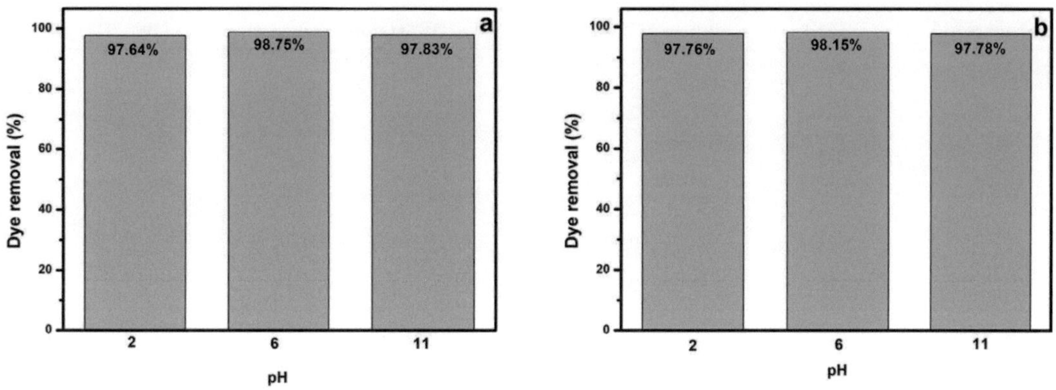

Figure 8. Effect of initial pH on the adsorption of DB78 onto polymers (1.00 g). Ten milliliters of dye solution (55.00 mg/L), at pH 2, 6, 11 and temperature 25°C, were used. (a) Adsorption measurements with β-CD/EPI polymers and (b) adsorption measurements with γ-CD/EPI polymers.

the system increases, the number of negatively charged sites increases as well, and the number of positively charged sites decreases. A negatively charged surface site on the adsorbent does not support the adsorption of anionic dye due to electrostatic repulsion [42]. On the other hand, at low values of pH, the sulphonate groups of dye are protonated and the number of positively charged sites increases, inducing again electrostatic repulsion. Therefore, all experiments were performed at pH 6 that is the natural pH of DB78 aqueous solution.

3.7. Adsorption equilibrium isotherms

The adsorption isotherms of DB78 onto CD/EPI polymers were determined at pH 6 maintaining the systems at constant temperature of 25, 50 and 80°C. Various quantities of adsorbent, from 0.05 to 1.00 g, were added to 10 mL of dye (80.00 mg/L) and the adsorption process was maintained until the reaching of equilibrium state. The Langmuir and Freundlich values are listed in **Table 2**, respectively and the value of the linear regression correlation coefficient R^2 is used to determine the best-fit model. Based on R^2, the results show that the adsorption process with both polymers was better represented by Langmuir isotherm model than the Freundlich equation. The applicability of Langmuir isotherm describes a monolayer and homogeneous adsorption of the dye onto the surface of polymers, where the adsorption of each molecule onto the surface has equal adsorption activation energy [31]. These results agree with a study, reported in the literature [42], where some azo dyes have been removed by β-cyclodextrin-based polymers. Furthermore, these measurements show that increasing the temperature from 25 to 80°C induced a higher maximum adsorption capacity. Since the R_L values were between 0 and 1, it possible to underline that β-CD/EPI and γ-CD/EPI polymers are good and favourable adsorbent for DB78 removal.

3.8. Thermodynamic analysis

The thermodynamic parameters for the adsorption of DB78 dye wastewater on β-CD/EPI and γ-CD/EPI polymers are summarised in **Table 3**. The negative values of $\Delta G°$ indicated that the dye adsorption by these polymers is a spontaneous and a favourable process. Since the

Polymers	T (K)	ΔG° (kJ mol⁻¹)	ΔS° (J mol⁻¹ K⁻¹)	ΔH° (kJ mol⁻¹)
β-CD/EPI	298	−11.263	54.352	4.942
	323	−12.622		
	353	−14.252		
γ-CD/EPI	298	−9.847	39.266	1.860
	323	−10.907		
	353	−12.006		

Table 3. Thermodynamic parameters.

obtained values of free energy change were in the range of −9.85 to −12.01 kJ mol^{-1}, for the β-CD/EPI polymer, and in range of −11.26 to −14.25 kJ mol^{-1}, for the γ-CD/EPI polymer, it is possible to affirm that the adsorption was principally physical. Indeed, some studies reported that the adsorption is classified as physical adsorption when the ΔG° values range between −20 and 0 kJ mol^{-1}, and as chemical adsorption when ΔG° values range from −80 to −400 kJ mol^{-1} [34]. The positive values of ΔS°, for both polymers, showed that the disorder of the systems increased at the solid solution interface during the adsorption of DB78 on polymers. Also, the ΔH° values for β-CD/EPI and γ-CD/EPI polymers were 4.94 and 1.86 kJ mol^{-1}, respectively. These positive values indicate that the adsorption followed an endothermic process as in agreement with results derived from the isotherm measurements.

3.9. Thermal analysis

The thermal analysis of β-CD/EPI polymer, γ-CD/EPI polymer, and their respective polymers loaded with DB78 was performed with differential scanning calorimetry (DSC) under N_2 atmosphere with heating rate of 20°C/min. As shown in **Figure 9a**, the DSC thermograms of β-CD/EPI polymer exhibited an endothermic peak at about 280°C [43]. After the interaction of this polymer with DB78, the thermogram presented a double endothermic peak at about 250 and 280°C. Since the first signal corresponds to the decomposition temperature of the only dye, it is possible to affirm that DB78 exhibits a thermal instability even after adsorption. This result allows to hypothesise that the interaction between DB78 and the polymers did not occur only in the internal cavities of cyclodextrins but also in the pores present on the external surface of polymer. In **Figure 9b**, the DSC thermograms of γ-CD/EPI polymer loaded with DB78 were no longer exhibit the typical thermal decomposition phenomena, shown in the thermograms of DB78 and γ-CD/EPI polymer. It confirms the interaction between DB78 and γ-CD/EPI: the inclusion of the dye into the γ-CD by weak forces stabilises both the adsorbent and the adsorbate.

3.10. Morphologic study

CD-based polymers were observed by field emission scanning electron microscope (FESEM) to examine their morphology. In **Figure 10a** and **c**, FESEM images of unloaded β-CD/EPI and

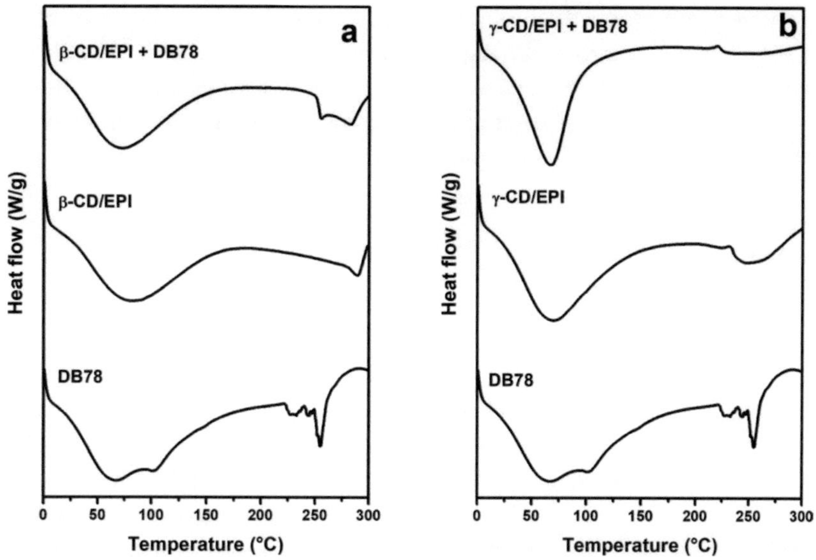

Figure 9. Thermograms obtained by DSC analysis. (a) Thermograms of DB78, β-CD/EPI polymer and DB78 loaded β-CD/EPI polymer and (b) thermograms of DB78, γ-CD/EPI polymer and DB78 loaded γ-CD/EPI polymer.

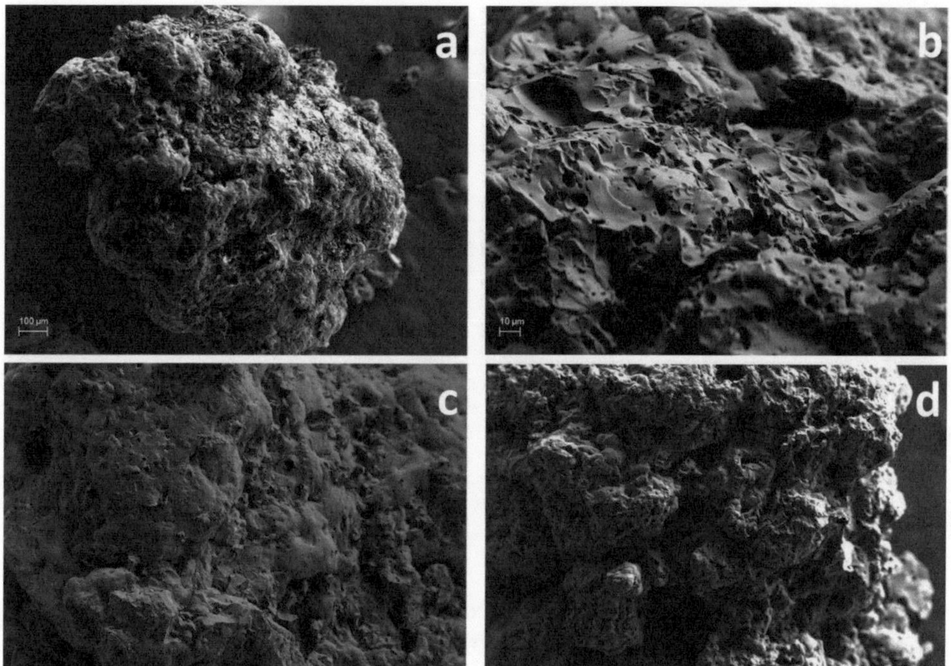

Figure 10. Scanning electron microscopy images of polymers. (a) Unloaded β-CD/EPI polymer, (b) DB78 loaded β-CD/EPI polymer, (c) unloaded γ-CD/EPI polymer and (d) DB78 loaded γ-CD/EPI polymer.

γ-CD/EPI polymers are respectively showed. It is possible to observe that these materials presented a very porous, rough and irregular structure which cavities are able to adsorb the DB78 molecules. Moreover, the presence of loaded dye molecules on polymers did not affect significantly the morphology of the samples, as reported in **Figure 10b** and **d**, confirming the weak and physical interaction of adsorption process.

4. Conclusion

Results of adsorption show that β- and γ-CD/EPI polymers exhibited good adsorption properties towards azo dye Direct Blue 78 (**Figure 11a**) and the maximum efficiencies in dye removal, performed at pH 6, 25°C, with an initial dye concentration equal to 11.00 mg/L and using 1.00 g of adsorbents, were found to be about 99% for β-CD/EPI polymer and about 97% for γ-CD/EPI polymer, respectively. The proposed adsorption mechanism involved several kinds of interactions such as physical adsorption in the polymer network, hydrogen bonding and formation of inclusion complex due to the presence of CD molecules through host-guest interactions. As illustrated in **Figure 11b**, this adsorption method allows, after 2 h of treatment with polymers, to obtain clean water that could be reused in further industrial processes of fabric dyeing. Furthermore, these polymers could be promising adsorbents for industrial application due to their low cost of production and their possible recycling in different adsorption cycles.

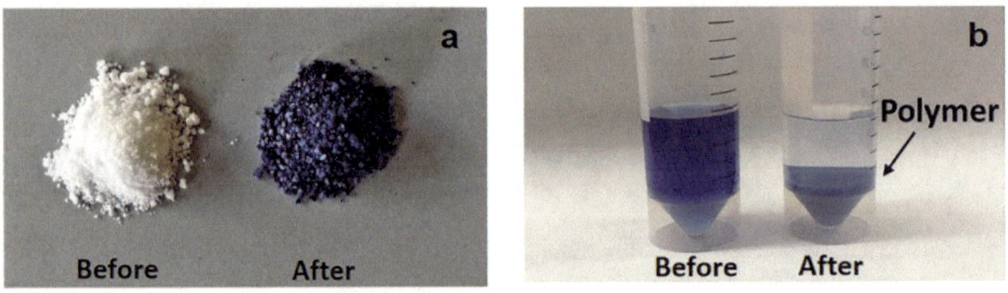

Figure 11. Adsorption of DB78 (11.00 mg/L) onto β-CD/EPI polymer at pH 6 and at constant temperature of 25°C before and after 2 h of treatment. (a) Image of β-CD/EPI polymer before and after the adsorption process and (b) image of DB78 solution in presence of β-CD/EPI polymer before and after the adsorption process.

Acknowledgements

This study was financed by the European project "DYES4EVER" (Use of cyclodextrins for treatment of wastewater in textile industry to recover and reuse textile dyes, LIFE12 ENV/ES/000309) within the LIFE+ 2012 "Environment Policy and Governance project application" program. We gratefully acknowledge the skilful technical assistance of Mr. Sergio Nuzzo and the precious collaboration of Dr. Maria Josefa Yáñez (Catholic University San Antonio of Murcia, Spain) for SEM measurements.

Author details

Paola Semeraro[1], José Antonio Gabaldón[2], Paola Fini[3], Estrella Núñez[2], José Antonio Pellicer[2], Vito Rizzi[1] and Pinalysa Cosma[1,2*]

*Address all correspondence to: pinalysa.cosma@uniba.it

1 Department of Chemistry, University of Bari Aldo Moro, Bari, Italy

2 Department of Food Technology and Nutrition, Catholic University San Antonio of Murcia, Murcia, Spain

3 Department of Chemistry, National Research Council CNR-IPCF, UOS Bari, Bari, Italy

References

[1] Yagub MT, Sen TK, Afroze HM. Dye and its removal from aqueous solution by adsorption: A review. Advances in Colloid and Interface Science. 2014;**209**:172-184. DOI: 10.1016/j.cis.2014.04.002

[2] Ejder-Korucu M, Gürses A, Doğar Ç, Sharma SK, Açıkyıldız M. Removal of organic dyes from industrial effluents: An overview of physical and biotechnological applications. In: Sharma SK, editor. Green Chemistry for Dyes Removal from Wastewater: Research Trends and Applications. 1st ed. Hoboken, New Jersey: John Wiley & Sons, Inc and Salem, Massachusetts: Scrivener Publishing LLC; 2015. pp. 1-22. DOI: 10.1002/9781118721001.ch1

[3] Gupta VK, Suhas. Application of low-cost adsorbents for dye removal—A review. Journal of Environmental Management. 2009;**90**:2313-2342. DOI: 10.1016/j.jenvman.2008.11.017

[4] Peláez-Cid AA, Herrera-González AM, Salazar-Villanueva M, Bautista-Hernández A. Elimination of textile dyes using activated carbons prepared from vegetable residues and their characterization. Journal of Environmental Management. 2016;**181**:269-278. DOI: 10.1016/j.jenvman.2016.06.026

[5] Edison TNJI, Atchudan R, Sethuraman MG, Lee YR. Reductive-degradation of carcinogenic azo dyes using *Anacardium occidentale* testa derived silver nanoparticles. Journal of Photochemistry & Photobiology, B: Biology. 2016;**162**:604-610. DOI: 10.1016/j.jphotobiol.2016.07.040

[6] Tripathi P, Srivastava VC, Kuma A. Optimization of an azo dye batch adsorption parameters using Box–Behnken design. Desalination. 2009;**249**:1273-1279. DOI: 10.1016/j.desal.2009.03.0102009

[7] Ghoreishi S, Haghighi R. Chemical catalytic reaction and biological oxidation for treatment of non-biodegradable textile effluent. Chemical Engineering Journal. 2003;**95**:163-169. DOI: 10.1016/S1385-8947(03)00100-1

[8] Salleh MAM, Mahmoud DK, Karim WAWA, Idris A. Cationic and anionic dye adsorption by agricultural solid wastes: A comprehensive review. Desalination. 2011;**280**:1-13. DOI: 10.1016/j.desal.2011.07.019

[9] Robinson T, McMullan G, Marchant R, Nigam P. Remediation of dyes in textile effluent: A critical review on current treatment technologies with a proposed alternative. Bioresource Technology. 2001;**77**:247-255. DOI: 10.1016/S0960-8524(00)00080-8

[10] Pearce CI, Lloyd JR, Guthrie JT. The removal of colour from textile wastewater using whole bacterial cells: a review. Dyes and Pigments. 2003;**58**:179-196. DOI: 10.1016/S0143-7208(03)00064-0

[11] Vakili M, Rafatullah M, Salamatinia B, Abdullah AZ, Ibrahim MH, Tan KB, Gholami Z, Amouzgar P. Application of chitosan and its derivatives as adsorbents for dye removal from water and wastewater: A review. Carbohydrate Polymers. 2014;**113**:115-130. DOI: 10.1016/j.carbpol.2014.07.007

[12] Crini G. Non-conventional low-cost adsorbents for dye removal: A review. Bioresource Technology. 2006;**97**:1061-1085. DOI: 10.1016/j.biortech.2005.05.001

[13] Rizzi V, D'Agostino F, Fini P, Semeraro P, Cosma P. An interesting environmental friendly cleanup: The excellent potential of olive pomace for disperse blue adsorption/desorption from wastewater. Dyes and Pigments. 2017;**140**:480-490. DOI: 10.1016/j.dyepig.2017.01.069

[14] Rizzi V, D'Agostino F, Gubitosa J, Fini P, Petrella A, Agostiano A, Semeraro P, Cosma P. An alternative use of olive pomace as a wide-ranging bioremediation strategy to adsorb and recover disperse orange and disperse red industrial dyes from wastewater. Separations. 2017;**4**:29-41. DOI: 10.3390/separations4040029

[15] Rizzi V, Mongiovì C, Fini P, Petrella A, Semeraro P, Cosma P. Operational parameters affecting the removal and recycling of Direct Blue industrial dye from wastewater proposing bleached oil mill waste as alternative adsorbent material. International Journal of Environmental, Agriculture and Biotechnology. 2017;**2**:1560-1572. DOI: 10.22161/ijeab/2.4.15

[16] Rizzi V, Longo A, Fini P, Semeraro P, Cosma P, Franco E, García R, Ferrándiz M, Núñez E, Gabaldón JA, Fortea I, Pérez E, Ferrándiz M. Applicative study (Part I): The excellent conditions to remove in batch direct textile dyes (Direct Red, Direct Blue and Direct Yellow) from aqueous solutions by adsorption processes on low-cost chitosan films under different conditions. Advances in Chemical Engineering and Science. 2014;**4**:454-469. DOI: 10.4236/aces.2014.44048

[17] Dotto GL, Santos JMN, Tanabe EH, Bertuol DA, Foletto EL, Lima EC, Pavan FA. Chitosan/polyamide nanofibers prepared by Forcespinning® technology: A new adsorbent to remove anionic dyes from aqueous solutions. Journal of Cleaner Production. 2017;**144**:120-129. DOI: 10.1016/j.jclepro.2017.01.004

[18] Semeraro P, Fini P, D'Addabbo M, Rizzi V, Cosma P. Removal from wastewater and recycling of azo textile dyes by alginate-chitosan beads. International Journal of Environmental, Agriculture and Biotechnology. 2017;**2**:1835-1850. DOI: 10.22161/ijeab/2.4.48

[19] Ozmen EY, Sezgin M, Yilmaz A, Yilmaz M. Synthesis of β-cyclodextrin and starch based polymers for sorption of azo dyes from aqueous solutions. Bioresource Technology. 2008;**99**:526-531. DOI: 10.1016/j.biortech.2007.01.023

[20] Tan P, Hu Y. Improved synthesis of graphene/β-cyclodextrin composite for highly efficient dye adsorption and removal. Journal of Molecular Liquids. 2017;**242**:181-189. DOI: 10.1016/j.molliq.2017.07.010

[21] Duchêne D, Bochot A. Thirty years with cyclodextrins. International Journal of Pharmaceutics. 2016;**514**:58-72. DOI: 10.1016/j.ijpharm.2016.07.030

[22] Voncina B. Application of cyclodextrins in textile dyeing. In: Hauser P, editor. Textile Dyeing. Europe: InTech; 2011. pp. 373-392. DOI: 10.5772/19147

[23] Singh M, Sharma R, Banerjee UC. Biotechnological applications of cyclodextrins. Biotechnology Advances. 2002;**20**:341-359. DOI: 10.1016/S0734-9750(02)00020-4

[24] Crini G, Morcellet M. Synthesis and applications of adsorbents containing cyclodextrins. Journal of Separation Science. 2002;**25**:789-813. DOI: 10.1002/1615-9314(20020901)25:13<789::AID-JSSC789>3.0.CO;2-J

[25] Semeraro P, Rizzi V, Fini P, Matera S, Cosma P, Franco E, García R, Ferrándiz M, Núñez E, Gabaldón JA, Fortea I, Pérez E, Ferrándiz M. Interaction between industrial textile dyes and cyclodextrins. Dyes and Pigments. 2015;**119**:84-94. DOI: 10.1016/j.dyepig.2015.03.012

[26] Morin-Crini N, Crini G. Environmental applications of water-insoluble β-cyclodextrin–epichlorohydrin polymers. Progress in Polymer Science. 2013;**38**:344-368. DOI: 10.1016/j.progpolymsci.2012.06.005

[27] Krause RWM, Mamba BB, Bambo FM, Malefetse TJ. Cyclodextrin polymers: Synthesis and application in water treatment. In: Jie H, editor. Cyclodextrins: Chemistry and Physics. Transworld Research Network; 2010. pp. 185-208. ISBN: 978-81-7895-430-1

[28] Vélaz I, Isasi JR, Sánchez M, Uzqueda M, Ponchel G. Structural characteristics of some soluble and insoluble β-cyclodextrin polymers. Journal of Inclusion Phenomena and Macrocyclic Chemistry. 2007;**57**:65-68. DOI: 10.1007/s10847-006-9221-z

[29] Poornima KN, Deveswaran R, Bharath S, Basavaraj BV, Madhavan V. Synthesis and evaluation of β-cyclodextrin-epichlorohydrin inclusion complex as a pharmaceutical excipient. Journal of Fundamental and Applied Sciences. 2015;**7**:203-221. DOI: 10.4314/jfas.v7i2.5

[30] Hongguo J, Zujin Y, Xiantai Z, Yanxiong F, Hongbing JI. Immobilization of β-cyclodextrin as insoluble β-cyclodextrin polymer and its catalytic performance. Chinese Journal of Chemical Engineering. 2012;**20**:784-792. DOI: 10.1016/S1004-9541(11)60249-8

[31] Crini G. Kinetic and equilibrium studies on the removal of cationic dyes from aqueous solution by adsorption onto a cyclodextrin polymer. Dyes and Pigments. 2008;**77**:415-426. DOI: 10.1016/j.dyepig.2007.07.001

[32] Moussavi G, Mahmoudi M. Removal of azo and anthraquinone reactive dyes from industrial wastewaters using MgO nanoparticles. Journal of Hazardous Materials. 2009;**168**: 806-812. DOI: 10.1016/j.jhazmat.2009.02.097

[33] Hameed BH, Mahmoud DK, Ahmad AL. Equilibrium modeling and kinetic studies on the adsorption of basic dye by a low-cost adsorbent: Coconut (*Cocos nucifera*) bunch waste. Journal of Hazardous Materials. 2008;**158**:65-72. DOI: 10.1016/j.jhazmat.2008.01.034

[34] Ali IH, Alrafai HA. Kinetic, isotherm and thermodynamic studies on biosorption of chromium(VI) by using activated carbon from leaves of *Ficus nitida*. Chemistry Central Journal. 2016;**10**:36. DOI: 10.1186/s13065-016-0180-1

[35] Ghoneim MM, El-Desoky HS, Amer SA, Rizk HF, Habazy AD. Electroreduction and spectrophotometric studies of some pyrazolyl-azo dyes derived from 3-acetylamino-1-phenyl-5-pyrazolone in buffered solutions. Dyes and Pigments. 2008;**77**:493-501. DOI: 10.1016/j.dyepig.2007.07.018

[36] Mandić Z, Nigović B, Simunić B. The mechanism and kinetics of the electrochemical cleavage of azo bond of 2-hydroxy-5-sulfophenyl-azo-benzoic acids. Electrochimica Acta. 2004;**49**:607-615. DOI: 10.1016/j.electacta.2003.09.015

[37] Matsue T, Osa T, Evans DH. Determination of some physical constants of cyclodextrin complexes by electrochemical methods. Journal of Inclusion Phenomena. 1984;**2**:547-554. DOI: 10.1007/bf00662220

[38] Retna RC, Ramaraj R. Electrochemistry and photoelectrochemistry of phenothiazine dye-β-cyclodextrin inclusion complexes. Journal of Electroanalytical Chemistry. 1996; **405**:141-147

[39] Crini G. Studies on adsorption of dyes on beta-cyclodextrin polymer. Bioresource Technology. 2003;**90**:193-198. DOI: 10.1016/S0960-8524(03)00111-1

[40] Tejashri G, Amrita B, Darshana J. Cyclodextrin based nanosponges for pharmaceutical use: A review. Acta Pharmaceutica. 2013;**63**:335-358. DOI: 10.2478/acph-2013-0021

[41] Yagub MT, Sen TK, Ang HM. Equilibrium, kinetics, and thermodynamics of methylene blue adsorption by pine tree leaves. Water, Air, & Soil Pollution. 2012;**223**:5267-5282. DOI: 10.1007/s11270-012-1277-3

[42] Yilmaz E, Memon S, Yilmaz M. Removal of direct azo dyes and aromatic amines from aqueous solutions using two β-cyclodextrin-based polymers. Journal of Hazardous Materials. 2010;**174**:592-597. DOI: 10.1016/j.jhazmat.2009.09.093

[43] Liu H, Cai X, Wang Y, Chen J. Adsorption mechanism-based screening of cyclodextrin polymers for adsorption and separation of pesticides from water. Water Research. 2011;**45**: 3499-3511. DOI: 10.1016/j.watres.2011.04.004